Tillage

FMO

**FUNDAMENTALS OF
MACHINE OPERATION**

PUBLISHER

Fundamentals of Operation (FMO) is a series of manuals created by Deere & Company. Each book in the series is conceived, researched, outlined, edited, and published by Deere & Company. Authors are selected to provide a basic technical manuscript which is edited and rewritten by staff editors.

PUBLISHER: DEERE & COMPANY SERVICE TRAINING, Dept. F, John Deere Road, Moline, Illinois 61265; DIRECTOR OF SERVICE: Robert A. Sohl.

FUNDAMENTAL WRITING SERVICES EDITORIAL STAFF
Managing Editor: Louis R. Hathaway
Editor: Laurence T. Hammond
Editor: Alton E. O'Banion
Promotions: Annette M. LaCour

AUTHOR: *Frank Buckingham* is an agricultural engineer and freelance writer of numerous articles on agricultural machinery.

CONTRIBUTING WRITER: *Harold Thorngren,* Factory Marketing (retired), John Deere Plow & Planter Works.

CONTRIBUTING WRITER: *Bruno Johannsen,* Chief Engineer Moldboard Plows (Retired), John Deere Plow & Planter Works.

COPY EDITOR: *Vernon D. Hagelin,* copywriter (retired), Deere & Company Advertising Department.

CONSULTING EDITORS: *Roland F. Espenschied,* Ed.D., professor of Agricultural Engineering and Vocational Agriculture Service, University of Illinois, has written many publications and prepared other types of audio-visual materials in agricultural mechanization during his 34 years as a teacher and teacher-educator. Professor Espenschied has spent 24 years specializing in agricultural mechanics in the Agricultural Engineering Department of the University of Illinois following ten years of teaching vocational agriculture.

Thomas A. Hoerner, Ph.D. is a professor and teacher-educator in agricultural mechanics at Iowa State University. Dr. Hoerner has 26 years of high school and university teaching experience. He has authored numerous manuals and instructional materials in the Agricultural Mechanics area.

Keith R. Carlson has 20 years of experience as a high school vocational agriculture instructor. Mr. Carlson is the author of numerous instructor's guides. All instructor's kits for the FMO texts are being prepared by Mr. Carlson who is president of Agri Education, Inc., an agricultural consulting firm in Stratford, Iowa.

SPECIAL ACKNOWLEDGEMENTS: The author and editors wish to thank the following Deere & Company reviewers for their assistance: R.A. Sohl, Director of Service; Technical Center personnel; Marketing and Engineering personnel at the following John Deere Factories: Plow and Planter Works, Des Moines Works, and Welland Works; and a host of other John Deere people who gave extra assistance and advice on this project.

CONTRIBUTORS: The publisher is grateful to the following individuals and companies who helped by providing some of the photographs used in this text.

Sunflower Manufacturing Co., Beloit, Kansas—pg. 141, Intro.; pg. 143, Fig. 4; pg. 144, Fig. 6

Krause Plow Corp., Inc., Hutchinson, Kansas—pg. 142, Fig. 1

Richardson Mfg. Co., Cawker City, Kansas—pg. 143, Figs. 2, 3, 5

Glencoe, Portable Elevator Division Dynamics Corporation of America, Bloomington, Illinois—pg. 176, Fig. 6.

KMN Modern Farm Equipment, Westwood, New Jersey—pg. 184, Figs. 1, 2; pg. 187, Figs. 6, 8; pg. 188, Fig. 9; pg. 191, Fig. 15; pg. 192, Fig. 17; pg. 193, Fig. 19; pg. 195, Fig. 21

Howard Rotavator, Harvard, Illinois—pg. 183, Intro.; pg. 185, Fig. 4; pg. 186, Fig. 5; pg. 187, Fig. 7; pg. 189, Figs. 10, 11, 12; pg. 191, Fig. 14; pg. 192, Fig. 18; pg. 194, Fig. 20

Fuerst Brothers, Rhinebeck, New York—pg. 271, Fig. 5

Vicon Farm Mach., Inc., Chesapeake, Virginia—pg. 281, Fig. 33

PUBLISHER: Fundamentals of Machine Operation (FMO) texts and visuals are published by John Deere Service Publications, Dept. F., John Deere Road, Moline, Illinois 61265.

FOR MORE INFORMATION: This text is part of a complete series of texts and visuals on agricultural machinery called Fundamentals of Machine Operation (FMO). For more information, request a free FMO Catalog of Manuals and Visuals. Send your request to John Deere Service Publications, Dept. F., John Deere Road, Moline, Illinois 61265.

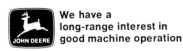

We have a long-range interest in good machine operation

Contents

1
TILLAGE

Tillage in Perspective 2
Evaluating Tillage Operations 3
Primary Tillage 4
Secondary Tillage 5
Tillage Practices 5
Eliminating Operations 6
Effect on Yields 7
Tillage History 8

2
FIELD EFFICIENCY

Machine Capacity and Performance 14
Operating Speed 16
Time Spent in Operation 18

3
TRACTION, FLOTATION
AND SOIL COMPACTION

The Problem 26
Traction 27
Cast-Iron and Liquid Ballast 29
Flotation 30
Compaction 32

4
TOOLBARS

Toolbar Hitches 36
Sled Carriers 40
Standards and Shanks 45

5

MOLDBOARD PLOWS

What the Plow Does 52
Moldboard Plow Bottoms 53
Moldboard Plow Frames 65
Choosing a Moldboard Plow 66
Hitching-In Furrow or On Land? 73
Principles of Moldboard Plow Operation 73
Basic Moldboard-Plow Adjustments 76
Power Required for Plowing 81
Tractor Load-and-Depth and Draft-Control Systems 82
Adjustments for Two-Way Plows 84
Tractor Preparation for Plowing 85
Moldboard Plow Preparation and Maintenance 87
Field Operation for One-Way Plows 89
Planning Field Layout 89
Operating Sequence 91
Plowing in Heavy Trash 91
Field Operation for Two-Way Plows 95
Troubleshooting 95
Transport and Safety 102

6

DISK PLOWS

Types and Sizes 106
Principles of Operation 107
Integral One-Way Disk Plows 109
Reversible Disk Plows 111
Hitching and Adjusting 112
Transport and Safety 114
Field Operation 114
Disk Plow Attachments 114
Troubleshooting 116

7

CHISEL PLOWS

Chisel Plows Versus Field Cultivators 122
Chisel Plows Versus Moldboard Plows 123
The Versatile Chisel Plow 123
Chisel-Plow Types and Sizes 126
Transport and Gauge Wheels 130
Principles of Chisel Plow Operation 132
Tractor Preparation 135
Machine Preparation and Maintenance 135
Field Operation 136
Transport and Safety 136
Troubleshooting 137

8
STUBBLE-MULCH PLOWS

Types and Sizes 142
Principles of Wide-Sweep Plow Operation 144
Tractor Preparation 144
Machine Preparation and Maintenance 144
Field Operation 145
Transport and Safety 145
Troubleshooting 146

9
DISK TILLERS

Tiller Types and Sizes 151
Principles of Disk-Tiller Operation 154
Tractor Preparation 157
Tiller Preparation and Maintenance 158
Field Operation 160
Tandem Hitching 164
Transport and Safety 165
Safety 166
Troubleshooting 167

10
STUBBLE-MULCH TILLERS

Less Draft 174
Stable Operation 176
Principles of Stubble-Mulch Tiller Operation 177
Tractor Preparation 178
Stubble-Mulch Tiller Preparation and Maintenance 178
Field Operation 179
Transport and Safety 180
Troubleshooting 181

11
ROTARY TILLERS

Power Requirements 185
Common Rotary Tillage Methods 185
Poor Weather 186
Moisture Conservation 186
Special Applicator or Conditions 186
Rotary Tiller Types and Sizes 187
Blade Types 188
Location of the Drive 188
Principles of Rotary Tiller Operation 188
Tractor Preparation 191
Machine Preparation and Maintenance 192
Field Operation and Adjustments 193
Transport and Safety 195
Troubleshooting 195

12
LISTERS AND BEDDERS

Profile Planting 201
Several Planting Methods 202
Lister and Bedder Types and Sizes 204
Protecting Lister Bottoms 206
Principles of Operation 208
Machine Preparation and Maintenance 208
Field Operation 209
Hitching 209
Depth Control 210
Troubleshooting 211

13
SUBSOILERS

Chisel Plows 214
Mole-Drain Requests 215
Subsoiler Types and Sizes 215
Principles of Subsoiler Operation 217
Subsoiler Preparation and Maintenance 218
Field Operation 218
Tractor Preparation 219
Troubleshooting 220
Transport and Safety 221

14
DISK HARROWS

Wide Versatility 224
Secondary Tillage 224
Larger Tractors, Larger Harrows 227
Folding the Wings 228
Cone-Shaped Blades 232
Comparing Weight 234
Principles of Operation 237
Tractor Preparation 242
Disk-Harrow Preparation and Maintenance 243
Field Operation 246
Transport and Safety 248
Troubleshooting 249

15
ROLLER HARROWS AND PACKERS

Roller-Harrow and Packer Types and Sizes 258
Principles of Roller-Harrow and Packer Operation 263
Tractor Preparation 263
Machine Preparation and Maintenance 264
Field Operation 264
Transport and Safety 265
Troubleshooting 265

16
TOOTH-TYPE HARROWS

Types, Sizes, and Principles of Operation 269
Main-Frame Styles 272
Attachments for Other Implements 274
Tractor and Implement Preparation 282
Field Operation 284
Transport and Safety 286
Troubleshooting 286

17
FIELD CULTIVATORS

Types and Sizes 292
Principles of Field Cultivator Operation 299
Tractor Preparation 300
Field Cultivator Preparation and Maintenance 300
Field Operation 301
Transport and Safety 301
Troubleshooting 302

18
ROD WEEDERS

Types and Sizes 306
Principles of Rod-Weeder Operation 310
Tractor Preparation 312
Rod-Weeder Preparation and Maintenance 312
Field Operation 312
Transport and Safety 313
Troubleshooting 314

19
ROW-CROP WEED-CONTROL EQUIPMENT

Row-Crop Cultivation 318
Weed-Control Equipment, Types and Sizes 320
Principles of Operation 335
Tractor Preparation 338
Preparation and Maintenance 339
Field Operation 339
Transport and Safety 342
Troubleshooting 343

APPENDIX

Suggested Readings 349
Useful Information 349
Glossary 355
Index 361

1
Tillage

EARLY MODEL MOLDBOARD PLOW

Fig. 1—A Plow Used When America's Heartland Was Settled

TILLAGE IN PERSPECTIVE

Tillage has been defined as *those mechanical, soil-stirring actions carried on for the purpose of nurturing crops.* The goal of proper tillage is to provide a suitable environment for seed germination, root growth, weed control, soil-erosion control, and moisture control—avoiding moisture excesses and reducing stress of moisture shortages.

Tillage requires well over half of the engine power expended on American farms, and it has been estimated that more than 250 billion tons of soil are tilled each year in this country. Many of the implements used, and much of the need for all this soil movement, have

long been taken for granted. However, there have been more changes in tillage implements and methods in the last 100 years than in previous recorded history, and more growth in mechanization in the last 25 years than in the preceding century (Fig. 1).

The primary objective of any cropping program is continued profitable production, so most farmers prefer to follow proven practices with readily available equipment. This offers reasonable assurance of predictable results with least risk.

But no tillage operation can be justified merely on basis of tradition or habit. Any tillage practice which doesn't return more than its cost by increasing yield

Fig. 2—Residue Management Is an Important Part of Tillage

Fig. 3—Bare Soil Dries and Warms Faster in Spring

and improving soil conditions should be eliminated or changed. Contrary to previous beliefs, soil needs to be worked only enough to assure optimum crop production and weed control. Any tillage activity beyond that is of questionable value.

EVALUATING TILLAGE OPERATIONS

From a realistic management standpoint, each tillage operation must be evaluated on the basis of its contribution to one or more of the following goals:

Management Of Crop Residue

Reduce interference with subsequent operations. Bury residue, mix it into the soil (Fig. 2), leave it on the surface, or combine two or three of these effects.

Soil Aeration

Provide optimum air availability for plant growth without creating large voids among clods or soil particles.

Weed Control

Kill growing weeds. Bury weed seeds found on the surface. Discourage growth of more weeds by not bringing more seeds to the surface for germination. Permit effective use of chemicals. Smother weeds under mulch.

Incorporation Of Fertilizer

Provide maximum mixdown of fertilizer with the soil. However studies show that some uncultivated, minimum-tilled crops with surface mulch utilize surface-applied fertilizers just as effectively. Use accurate soil tests.

Moisture Management

Avoid excessive moisture at planting, and promote optimum moisture level during the growing season.

Provide runoff control and loose surface for good moisture infiltration. Reduce surface evaporation during growing season. Provide drainage in heavy-rainfall areas and uniform surface to avoid ponding. Avoid or reduce soil compaction (such as plow sole), which restricts water movement. Prepare the soil for irrigation.

Insect Control

Bury residue to help control the European cornborer and some other pests. However, pest control alone cannot justify burial of all trash if other control measures—such as shredding residue, crop rotation, resistant varieties, and pesticides—are economically available.

Temperature Control For Seed Germination

Dry, bare soil warms faster in the spring than mulch-covered soil (Fig. 3). More erosion and water loss through evaporation occurs. Colder soil temperature under mulch may be critical only in abnormally cold or wet seasons. Most locally adapted crops can tolerate more temperature variations than need to be controlled by altering tillage practices.

Improvement Of Soil Tilth

Incorporate organic matter into the soil. Work the soil only at proper moisture content. Maintain adequate fertility level. Encourage growth of soil organisms by adding extra nitrogen. (Soil organisms basically need the same conditions of warmth and moisture as germinating seeds and small plants.) Reduce soil density to improve root growth and increase moisture capacity.

Provide Good Seed-Soil Contact

Provide firm seed contact with moist soil for five to ten days for germination and root growth. (Corn must absorb 30 percent of its weight in water to germinate; soybeans need 50 percent.) But unnecessary soil compaction must be avoided.

Fig. 4—Bedding Prepares Land for Planting and Irrigation

Prepare Surface For Other Operations

Build or level beds for the next crop (Fig. 4). Prepare soil surface for efficient harvest operation. Eliminate tracks or ruts from previous harvest, but don't make the surface too smooth, for a smooth surface may reduce water infiltration and increase runoff and erosion.

Erosion Control

Provide loose, mulch-covered surface (best for erosion control). Use contour tillage, contour planting, ridging, and no-till planting on erosion-prone soils. Protect soil from wind and water erosion. Remember that the finest, most-fertile soil particles are usually lost first to wind and water erosion.

Energy Conservation

To reduce energy input, many tillage operations are being eliminated or combined with other operations. On many farms moldboard plows are being replaced by chisel plows or heavy disk harrows which require less energy per acre (hectare). This is often called reduced tillage or conservation tillage if more residue is left on the surface.

Tillage provides temporary (one season or less) modification of such soil conditions as density, aeration, and moisture-holding capacity. Minerals and organic matter in the soil and climatic conditions—rainfall, freezing, and thawing—have longer-range effects.

Tillage is normally classified as *primary* or *secondary*.

PRIMARY TILLAGE

PRIMARY TILLAGE cuts and shatters soil and may bury trash by inversion, mix it into the tilled layer, or leave it basically undisturbed. Primary tillage is a more-aggressive, relatively deeper operation, and usually leaves the surface rough. Implements commonly used for primary tillage include:

- *Moldboard, chisel, and wide-sweep plows*
- *Bedders and listers*
- *Subsoilers, chisels, and rippers*

Fig. 5—Secondary Tillage Works Soil to Shallower Depth

- Disk tillers
- Offset and heavy tandem disk harrows and plowing disks
- Stubble-mulch tillers
- Rotary tillers

SECONDARY TILLAGE

SECONDARY TILLAGE works the soil to shallower depth (Fig. 5), provides additional pulverization, levels and firms the soil, closes air pockets, kills weeds, and helps conserve moisture. Secondary-tillage tools include:

- Disks and spring-, spike-, and tine-tooth harrows
- Field conditioners and cultivators
- Disk/field cultivator combinations
- Rod weeders
- Roller packers and roller harrows

- Disk tillers (some summer-fallow operations)
- Rotary hoes
- Row-crop cultivators
- Other similar weed-control implements

TILLAGE PRACTICES

Current tillage operations are frequently given a variety of confusing labels. Many of the terms are only loosely defined, and often mean different things in different areas. Some of the titles include:

- Conventional tillage
- Minimum tillage
- Optimum tillage
- Reduced tillage
- Economy seedbed
- Conservation tillage
- Mulch tillage
- Till-plant and zero-till planting

Except for till-plant and zero-till planting, some equipment may be used for many of the other systems, perhaps in different combinations or sequences.

Let's look at conventional and minimum tillage methods.

CONVENTIONAL TILLAGE

The number of operations performed in conventional tillage varies by crop and area. Conventional tillage may typically involve many of these practices:

- Shredding or disking crop residue
- Plowing to pulverize soil
- Listing or bedding to shape soil
- On spring plowing, immediate use of tooth-type harrows to close air pockets and level and firm the surface
- Disk harrowing or field cultivating
- Harrowing with tooth-type harrow
- After planting, harrowing with spike- or tine-tooth harrow
- Rotary hoeing to break soil crust or uproot weeds
- One or more row-crop cultivations, depending on crop, area, and weed problems

MINIMUM/OPTIMUM/ REDUCED/ECONOMY TILLAGE

These systems involve principles, rather than well-defined practices. System objectives include:

- Reducing energy input and labor requirement for crop production.
- Conserving soil moisture and reducing erosion.

- *Providing optimum seedbed and rootbed growth areas, rather than homogenizing the entire soil surface.*
- *Keeping trips over field to minimum.*

CONSERVATION AND MULCH TILLAGE

Major objectives of conservation and mulch-tillage systems include soil and moisture conservation. Energy conservation is another goal, achieved through reduction of tillage operations. Both systems usually leave crop residue on the surface, and each operation is planned and conducted to maintain continuous soil coverage by residue or growing plants.

Major tools are the chisel plow, wide-sweep plow, field cultivator (Fig. 6), and perhaps the disk tiller or disk harrow. Rod weeders are commonly used in small-grain areas to control weeds in summer fallow.

Conservation and mulch tillage are sometimes used to describe zero-till planting systems.

TILL-PLANT AND ZERO-TILL PLANTING

In these systems, crop residue usually is shredded and planting is done with no preliminary tillage. The till-planter has a large sweep with trash rods to clear a path through the row area of the previous crop. Special openers, disk coverers, and packer wheels are used for obtaining desired seed placement. Cultivation is usually done with disk tillers or rotary cultivators to ridge soil around the plants for increased support or to provide channels for irrigation. However, in some cases no mechanical cultivation is provided and weed control is entirely dependent on herbicides.

Zero- or no-till planters usually have fluted coulters or similar openers to cut surface trash and perhaps till a narrow strip ahead of each row unit. Coulters are adjustable to provide the desired degree of tillage. Planters usually have disk openers and packing wheels for firming the limited amount of loose soil over the seed.

Zero-till planting can permit row-cropping of land too steep for conventional tillage. If weather interferes with work on soils which must normally be fall-plowed, no-till planting could be better than delaying work until soil dries in the spring. Zero-till also leaves surface residue which may help prevent crusting (Fig. 7).

Adequate weed control is usually the biggest problem with zero-till planting and often requires heavy use of herbicides and cultivation with disk hillers. This may place the cultivator in the role of the primary or major tillage implement in the entire cropping system.

Use of zero-till planters is increasing in planting double-crop soybeans or sorghum after small-grain harvest in mid-south states. This reduces the time between harvesting and planting to provide the maximum growing season for the second crop. It reduces loss of soil moisture through evaporation which would result from conventional tillage of stubble.

If planting is done in untilled soil, extra weight usually must be added to the planter for adequate penetration by coulters and furrow openers. Special, heavy-duty zero-till planters should be used, instead of adding fluted coulters to conventional planters, which are not designed for this type of operation.

ELIMINATING OPERATIONS

These objectives may be met by carefully selecting operations in conventional tillage which may be elimi-

Fig. 6—Field Cultivators Are Used in Many Conservation Practices

FIELD CULTIVATOR

Fig. 7—Zero-Till Planters Work in Untilled or Previously Worked Land

ZERO-TILL PLANTER

Fig. 8—Chisel Plows Often Are Used for Primary Tillage

nated or combined. For instance, a farmer may elect to shred and disk residue in one pass, or disk and harrow in one operation before planting.

Residue may be shredded without disking, or a heavy offset or tandem disk may be used to cut trash and provide primary tillage in one operation, usually in the fall. Another disking in the spring, or perhaps a once-over with a field cultivator, prepares the soil for planting. Where moldboard plowing is eliminated, fluted coulters or disk openers are usually used on the planter to aid in cutting through trash. Disk hillers are often used for cultivation.

Chisel plows replace moldboard plows in some reduced-tillage operations (Fig. 8). Chiseling soybean ground instead of disking or moldboard plowing may in some cases reduce soil erosion. More residue remains exposed, and soil is left rough to resist blowing and runoff.

Reducing the number of field operations saves fuel, cuts labor needs, usually permits earlier planting, and can lower equipment investment if some machines are eliminated. Reduced wheel traffic also reduces soil compaction.

Rotary tillers are sometimes called minimum-tillage implements because they can prepare a seedbed in a single operation. However, "minimum" refers to reduced field operations rather than reduced soil tilth. Rotary tillers can provide maximum tilth. Forward speed, rotor speed, and tractor power must be matched properly. A fast once-over in the fall with a rotary tiller can shred and partially incorporate crop residue and leave the soil relatively rough. A second pass in the spring leaves the soil ready to plant. Strip rotary tillers can till a narrow strip, incorporating fertilizer and pesticides and planting in one operation. In many minimum-tillage operations, planters are mounted on or pulled behind field cultivators, chisel plows, or subsoilers to reduce operations.

EFFECT ON YIELDS

Yields resulting from these tillage systems vary from equal to or better than conventional tillage to total failure. Good management involves selection of the system for particular soil and climatic conditions, selection and operation of equipment, and to some extent having understanding of and faith in the system. If seeing trash on the surface before planting and cultivating worries the operator, he will likely reject such tillage systems. Some farmers have said that the biggest obstacle is changing traditional attitudes about what has to be done and how it should look.

Weed control can be a problem with these systems, and the greater dependence on chemical weed control may offset savings in equipment and operating costs. If reduced-tillage cost savings exceed the value of possible yield reductions, net profit may be increased by accepting slightly lower yields.

SUMMARY: TILLAGE PRACTICES

Regardless of the system used, more soil in the United States is over-tilled each year than is under-tilled. Over-tilling may result in excessive breakdown of soil-particle size, avoidable erosion, unnecessary compaction from wheel traffic, and wasted time and fuel.

Overworked seedbeds are a relatively modern phenomenon. When soil was tilled and crops were planted and harvested by hand, there was no time or energy to overwork the soil. Even with horses, there was little inclination to perform unnecessary tillage. With the introduction of tractor power, tillage became easier than ever before, and many farmers decided that if a little tillage was good, more would be better. They generally failed to recognize that overworking the soil could damage its structure and perhaps cause compaction.

Farming today is far different than it was just after World War II. Improved communications and increased application of sound business principles have accelerated the pace at which new ideas and practices are accepted.

TILLAGE HISTORY

Agricultural development has paralleled the civilization of man from the time he learned to domesticate plants and animals. As man turned to farming, he was no longer forced to depend on his skill and strength as a hunter to survive. He learned to produce and store food for lean seasons. He began to settle in one spot, instead of being a nomad. He developed villages, cities, and nations as the food produced by one farmer fed more and more people.

As he became civilized, man had more time to create tools and improve his standard of living—to make his work easier. So he gradually improved on the forked sticks and sharp stones he found naturally, and turned them into tools for breaking and stirring the soil and improving his crop yields. Crude tillage implements date back thousands of years, and most have evolved into the modern machines used today.

Some historic events in tillage-tool improvement include:

6000 B.C.—Egyptian drawings show a forked stick, apparently with a stone point, being used as a hoe or mattock. Later one fork was left longer for pulling by slaves or animals.

900 B.C.—Elisha was found "plowing with twelve yoke of oxen before him." (Kings 19:19)

The plow changed little in the next 26 centuries until the 18th century A.D. In **1721** the Norfolk wheel-plow had a cast-iron share and rounded iron moldboard.

1730—The Roman plow was brought to northern Europe.

1750's—The Essex plow had an iron moldboard.

1760—The curved moldboard appeared on Suffolk swing plow.

1797—Charles Newbold obtained the first United States patent for a plow. The idea of a one-piece cast-iron share, moldboard, and landside was rejected because of high replacement cost. Farmers thought iron plows poisoned the soil and caused weeds to grow.

1798—Thomas Jefferson designed a moldboard plow from mathematical computations. He hoped to design an ideal shape for all soils.

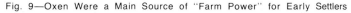

Fig. 9—Oxen Were a Main Source of "Farm Power" for Early Settlers

Fig. 10—Huge Steam Tractors Pulled Large Gang Plows During the Late 1800's

1813—R. B. Chenoworth of Baltimore patented a cast-iron plow with separate share, moldboard, and landside.

1813, 1819—Jethro Wood received patents on cast-iron plows. He developed a curved moldboard to turn soil in even furrows.

1820—The horse hoe with one large shovel was developed for row-crop cultivation. A second shovel was added to the design about **1850.**

1833—John Lane made the first steel plow from three sections of an old handsaw.

1837—John Deere used an old sawmill blade to shape a one-piece steel plowshare and moldboard over a log pattern (Fig. 1). His highly polished steel plow turned the sticky prairie soil where cast-iron plows failed.

1846—The first one-horse wheel cultivator was used.

1847—The disk plow was patented.

1856—M. Farley patented a single-bottom, wheeled riding sulky plow.

1856—A 2-horse, straddle-row walking cultivator was patented; it cultivated both sides of the row simultaneously.

1860's—The Civil War, with shortage of labor and booming demand for food and fiber, spurred improvements in all types of farm equipment.

1860's—Chisel cultivators came into use.

1863—Riding cultivators became a commercial success.

1864—F. S. Davenport patented a 2-bottom horse-drawn gang plow.

1868—John Lane, who made a steel plow from sections of hand saw, patented a soft-center steel moldboard (currently used in making most plow moldboards).

1869—The springtooth harrow was patented.

1877—The disk harrow with concave blades was patented.

1880—Keystone Manufacturing Co. started factory production of disk harrows in Sterling, Ill.

1880—Commercial lister production was underway.

1884—The first 3-wheel riding plow was used.

1890—The disk plow was developed for practical use.

1900's—The 2-row horsedrawn cultivator came into use.

1910—The rod weeder was developed.

1912—The rotary hoe was produced commercially.

1914-1918—World War I labor shortage and demand for agricultural products accelerated farm-mechanization.

1918—B. F. Avery Co. built a tractor-mounted row-crop cultivator.

1920's—Mechanical power lifts were developed for moldboard plows.

1924—The offset disk harrow was developed.

1927—The disk tiller, or one-way wheatland disk plow, began to sell in large numbers.

1930—Swiss-made rotary tillers were introduced in the United States.

1930's—Power lifts for cultivators were developed. Harry Ferguson developed the integral plow and tractor 3-point hitch (in England). These were brought to the United States in **1939.**

1935—The National Tillage Machinery Laboratory was established by the United States Department of Agriculture at Auburn, Ala.

1941—Hydraulic remote control of drawn implements was introduced.

Fig. 11—Motorized Row-Crop Cultivator During the Early 1900's

1941-1945—World War II intensified demand for mechanized agricultural production of more crops with less labor.

1949—The wheel disk harrow was introduced.

1953—Safety-trip beam plows were in use.

1960's—Rotary cultivators and selective crop thinners were introduced.

1970's—Combination disk/chisels (stubble-mulch tiller) and disk/field cultivators were introduced and perfected.

1980's—On board computers and electronic sensors increase efficiency.

Parallel development of other equipment and practices or processes have increased the pressure for development and acceptance of many new tillage methods and machines.

To name a few (not in chronological order):

1. Invention of the steam engine which helped spark the industrial revolution.

2. Development of steam, gasoline and diesel tractors.

3. Invention of cotton gin.

4. Invention of reaper, cornpicker, and grain combine.

5. Adaptation of rubber tires to farm tractors.

6. Invention of mechanical planters and drills.

7. Development of balers, forage harvesters, and other hay tools.

8. Increasing fuel prices after 1973.

There actually have been few revolutionary developments in tillage equipment. Most changes have been the result of modification, improvement, and evolution of earlier equipment designs and ideas. Some changes have been quickly accepted but others died in infancy. Nevertheless, the continued change and growth in equipment and practices have helped the American farmer become the most-efficient food producer in history.

CHAPTER QUIZ

1. Define tillage.

2. (Choose one) Tillage absorbs (a) one-fourth, (b) more than one-half, or (c) most of the power expended annually on American farms.

3. List six contributions of tillage to crop production.

4. (Fill blanks) _____ tillage cuts and shatters soil. _____ tillage pulverizes and levels the surface prior to planting.

5. (True or false) Conventional tillage involves the same implements and operations for all crops and soil conditions.

6. What damage can result from overworking soil?

7. (True or false) The steel plow was developed and first used in England.

2
Field Efficiency

INTRODUCTION

Today's farmer is constantly striving for more field-operation capacity and efficiency. This is especially true with tillage operations. Weather usually poses as a threat to farmers and can prevent tillage from being completed in time for optimum planting dates. And costs of these field operations are increasing rapidly. Several factors contribute to poor field efficiency, and later, we will discuss how changing these factors can help to achieve optimum field efficiency.

Good field efficiency does not necessarily mean increased capacity; usually, a slight trade-off of efficiency is given to increase capacity. This is over-simplification, so let's look at what we really mean.

Strictly defined, field efficiency of any implement is the ratio of effective field capacity to theoretical field capacity. Here's how *field efficiency* is expressed as an equation:

$$\text{Field efficiency (FE)} = \frac{\text{Effective field capacity (EFC)}}{\text{Theoretical field capacity (TFC)}} \times 100$$

Effective field capacity is the actual amount of work done, while *theoretical field capacity* is the amount of work which would be done if no time were lost.

With W meaning width of implement in feet (or meters), and S meaning miles (or kilometers) per hour, here's the equation for *theoretical field capacity:*

$$\text{TFC} = \frac{W \times S}{8.25} \text{ acres per hour, or: TFC} = \frac{W \times S}{10} \text{ hectares per hour}$$

For example, to calculate theoretical field capacity and field efficiency, take a 20-foot chisel plow operating at 6 miles an hour (10 km/h):

$$\text{TFC} = \frac{W \times S}{8.25} = \frac{20 \times 6}{8.25} = \frac{120}{8.25} = 14.5 \text{ acres per hour, or:}$$

$$\text{TFC} = \frac{W \times S}{10} = \frac{6 \times 10}{10} = \frac{60}{10} = 6 \text{ hectares per hour}$$

However, under actual working conditions we might expect to average 12 acres per hour (5 ha/h) (effective field capacity). Therefore, field efficiency is:

$$\text{FE} = \frac{\text{EFC}}{\text{TFC}} \times 100 = \frac{12}{14.5} \times 100 = .8276 \times 100 = 82.76 \text{ percent, or:}$$

$$\text{FE} = \frac{5}{6} \times 100 = 0.833 \times 100 = 83.3 \text{ percent}$$

For this discussion, the definition of field efficiency will be broadened to consider general machine performance, the use of time, and some of the factors affecting machine capacity (Fig. 1).

Fig. 1—Field Efficiency Considers General Machine Performance

Fig. 2—Big Tractors Provide More Capacity, but Need Lots of Use to Hold Down Costs

Fig. 3—Smaller Tractors Have Less Capacity, but May Have Lower Cost Per Hour

Many machines are used simply because of tradition, local custom, or because the farmer already has them. But modern farmers no longer plow all corn ground just because their fathers and grandfathers did. Farmers today realize that the only economic justification for any given practice is in terms of the value that the practice adds above the operating cost to the total output. If a practice boosts yield four bushels per acre (250 Kg/ha), but operating costs equal the value of five bushels (315 Kg/ha), that practice should be discontinued.

However, machines and practices cannot be judged in complete isolation. What effect will a change, or lack of change, have on the total farming operation? Savings in seedbed-preparation costs from switching to mulch tillage without plowing must be weighed against changes required in planting equipment, weed-control measures, etc.

We will take a brief look at machinery selection on the basis of capacity and performance. For a more complete study of machine capacities, power requirements, and costs, see John Deere's FMO text, *Machinery Management*.

For some operators, selecting a machine may be a matter of buying the biggest one available or the biggest one they can afford (Fig. 2). It may be a "feeling" that a certain size will be adequate for the area to be covered. Or, it may be a decision made after a careful analysis of the entire cropping program, expected value of the crop produced, the time available for each specific task, and the anticipated cost of production.

But, as stated before, to be economically justifiable, any machine must add more income to the farm than it costs to operate the machine (Fig. 3). However, the economic penalty for oversized equipment has often been shown to be less than the penalty for equipment too small to finish work on time.

TRACTOR SIZE

Tractor power should be on the high side of basic requirements to overcome adverse soil conditions and changes in topography, to permit some increase in speed during periods of adverse weather, for future expansion, and for other reasons.

However, selecting a tractor with considerably more power than is necessary can:

1. Waste fuel.

2. Increase operating and depreciation costs.

3. Cause poor implement performance if excess speed is used.

4. Cause excessive wear and damage under adverse soil conditions — stones, very hard soil, etc.

Using a tractor with less power than necessary can:

1. Limit field capacity, causing untimely operation.

2. Cause excessive wheel slippage and tire wear.

3. Provide poor soil pulverization due to too-slow speed.

4. Increase labor costs in relation to other expenses.

5. Cause tractor failure because of continuous overload.

Using a large tractor with a multiple hitch and two or three implements can reduce tractor and labor costs per acre or hectare, increase productivity per man-hour, and increase the probability of timely completion of each operation for maximum yield and optimum return (Fig. 4).

MACHINE CAPACITY AND PERFORMANCE

Basically, three factors control machine capacity and performance.

• *Machine width or size.*

• *Operating speed.*

• *Time spent in operation.*

Now let's look at these factors.

MACHINE WIDTH OR SIZE

We will consider eight basic questions concerning how machine width and size affect machine capacity and performance.

1. How many acres will the machine be expected to cover annually?

A small machine requires more labor per acre (hectare), thus offsetting lower machine costs. Large machines operated an equal number of hours may have similar total costs per acre (hectare) because of reduced labor but higher investment costs.

2. How much time is available to complete the work?

Weather information is available to show the number of days expected in a given period which may be used for field work. Many experts now recommend using equipment that is large enough to permit completion of field work within the most desirable period at least 80 or even 90 percent of the time.

If the operator only counts on average weather, he's in trouble about half the time. Different crops also compete simultaneously for field time and good weather — for instance, hay-making and row-crop cultivation. Some practices, such as plow-plant, require peak periods of labor and machine time which may limit their application in many operations.

Livestock operations, such as dairy and hogs, limit the available field time on some farms and could force the

Fig. 4—Big Fields Call for Large Implements

LARGE IMPLEMENTS FOR BIG FIELDS

selection of larger machines than would otherwise be necessary.

Studies show that each common field implement used in the United States, except for tractors, trucks, and wagons, is seldom actually at work more than 15 days annually.

3. What field size and terrain will be encountered?

Large equipment in small, irregular, or hilly plots is very inefficient no matter how much land must be covered. Use smaller equipment with more flexibility, or consolidate fields if possible (Fig. 5).

4. How about maneuverability and turning radius?

Integral equipment is usually most maneuverable — semi-integral for larger units. But drawn models may be better for really big fields. However, avoid cumbersome machines requiring excessive headland width for turning.

5. What is effective width?

That's the width of work completed on each pass. To assure working all of the soil, some overlapping is inevitable with most machines, with the exception of bedders, listers, planters, and row-crop cultivators which cover a certain number of fixed row spaces.

6. Boost output by increasing size?

Yes, but not in direct proportion to the implement dimension (Fig. 6). Studies show that a 6-row cultivator covers only 35 to 40 percent more ground at the same speed than a 4-row model. Also, a big machine misses

Fig. 6—Larger Implements Boost Output

more work each minute it's idle than does a small one, regardless of the cause of the delay.

7. What about costs?

Fixed costs are related to machine ownership and are an expense whether the machine is used or not. Fixed costs include investment, interest, taxes, insurance, depreciation, and housing. But the machine is left outside, so there's no housing cost — right? Right, but the machine depreciates faster and repair costs are higher that way, so housing is still a valid charge.

Fig. 5—Estimating Capacity Is Important in Order to Match Machine Size to Available Working Time

MATCH IMPLEMENT WITH FIELD CONDITIONS

Variable costs are related to operating the machine and include fuel, lubricants, repairs, maintenance, and labor.

Buying equipment that is too small boosts labor charges to an uneconomic level. But buying equipment that is too large increases fixed costs. However, many economists now recommend buying the larger size if there is doubt between two models.

Also consider the possibility of using another machine or method that could do the same job faster or better at less cost.

8. How much power is required?

Or how much power is available? Certainly, more power is required to operate the same implement in clay and other heavy soils than in sand or loam. But, if a particular implement requires far more power than any other operation on the farm and will have limited usage, better think again before buying that machine and a tractor to match. Consider hiring that particular operation done and save labor for other work. Look at a smaller machine that can be used with available power. Or, rent a tractor with enough power to handle the big job, and maybe even rent the implement to avoid tying up extra capital.

OPERATING SPEED

Seven basic questions concerning how operating speed affects machine capacity and performance will be discussed.

1. What machine is being used?

A disk tiller can be operated faster than a subsoiler, and a disk harrow faster than a rotary tiller, or a rotary hoe faster than most other tillage machines (Fig. 7). See Table 1 at end of this chapter for typical range of field-operating speeds for different tillage machines.

2. How does speed affect capacity?

Doubling the speed cannot double the output because of turning time, adjustments, and repairs. For some operations, stops per acre may be increased as speed

Fig. 7—Field Speeds Range from 1 to 12 Miles Per Hour (1.6 to 19 Km/h) for Different Implements

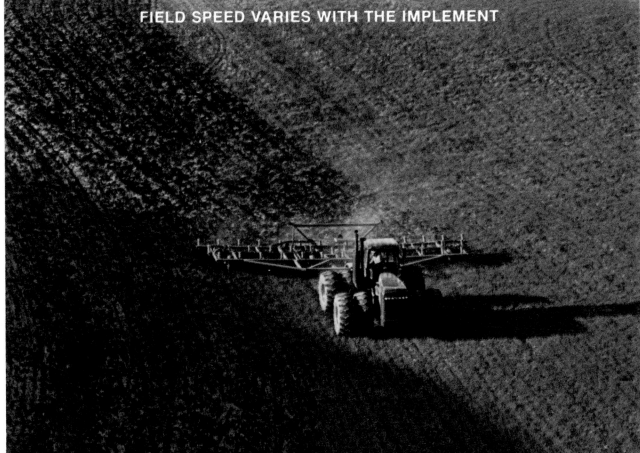

goes up because of more plugging, greater danger of breakdowns, etc.

To estimate the area covered in a 10-hour day, multiply the implement width times the speed. This calculation includes a field efficiency factor of 82.5 which is typical for most tillage tools.

Example: Chisel plow, 20 feet wide, operating at 6 miles per hour. Acres covered in 10-hour day = width x speed = 20 x 6 = 120 Acres.

To estimate hectares worked in 10 hours, multiply implement width in meters times speed in kilometers per hour, times 0.825 (typical field efficiency of tillage tools).

Metric Example: Chisel plow, 6 meters wide, operating at 10 kilometers per hour. Hectares covered in 10-hour day = width times speed times efficiency = 6 × 10 × 0.825 = 49.5 Hectares.

To determine approximate speed in miles per hour (K/h):

1. Mark off a distance of 176 feet (53.7 m) in the field.

2. Drive the measured distance at the speed you would like to operate (with a "running start" before the measured distance).

3. Check the number of seconds required to drive between the markers with a stop watch or watch with a sweep-second hand.

4. Drive the time in seconds into 120 for speed in miles per hour (mph) (Km/h).

5. Adjust the speed, if necessary, to the recommended rate.

The chart below lists the time in seconds for speeds up to either 8 miles or kilometers per hour

FIELD SPEEDS	
Time to Drive 176 Feet or 33.3 Meters	Speed, mph or (Km/h)
120 seconds	1 (1.6)
60 seconds	2 (3.2)
40 seconds	3 (4.8)
30 seconds	4 (6.4)
24 seconds	5 (8)
20 seconds	6 (9.6)
17 seconds	7 (11.2)
15 seconds	8 (12.8)

3. How much power is available?

Whatever the draft, enough power must be available to move the machine fast enough to perform properly. Too much power wastes fuel and may damage implements

Fig. 8—Operator Skill Is Important for Optimum Machine Performance

if not properly matched. But too little power wastes time, may damage the tractor by overloading gears and engine, and causes excessive tire wear.

4. What are soil conditions?

Speed probably will drop in hard, dry soil, and increase when the moisture level is just right. If the soil is too wet or too loose, poor traction may reduce speed. Too much trash requires slower operating speeds to prevent plugging or permit better cutting. Heavy trash also may require a chisel plow with greater clearance or a disk harrow with greater weight. weight.

5. How skilled is the operator?

Operating most modern farm equipment is not really strenuous work, but it is fatiguing because it requires constant operator alertness. Proper operator training is necessary for optimum machine performance and safe operation (Fig. 8).

Some research has shown that the operator's attitude and motivation are more important factors in relation to the amount of work accomplished than are age and total experience. Some operators simply prefer to operate at higher speeds and are able to perform well under such conditions. Others cannot successfully cope with more than minimum operating speeds. A good operator adjusts his field speed to the maximum possible rate for crop, soil, and equipment conditions without sacrificing the quality of work being performed.

6. What does higher speed cost?

Draft of most equipment goes up much faster than speed. For instance, doubling speed may increase draft four times. So speed must be balanced with draft and total output. Excessive speed of some implements may increase fuel consumption, wear, damage, downtime, and time spent in unplugging equipment. If working at 6 miles per hour (10 Km/h) requires stopping every 10 minutes to unplug, it might be possible to travel 4 miles per hour (6.5 Km/h), and unplug once or twice all day, and get more work done. On the other hand, faster plowing pulverizes soil more and may reduce time and cost of later tillage operations and therefore justify the high speed.

7. How does speed affect field efficiency?

Increasing speed increases capacity, but turning time and other stops remain relatively unchanged. Therefore, calculated field efficiency may be reduced even though total capacity is increased. Similarly, reducing speed improves the numerical value of field efficiency but total capacity is reduced. This simply shows it is not always good management to attempt to maximize field efficiency.

TIME SPENT IN OPERATION

Here we will discuss seven basic questions that deal with how time spent in operation affects machine capacity and performance.

1. What working pattern is followed?

Moldboard and disk plows, disk tillers, and offset disk harrows may work either in lands or around the field. Reversible plows (Fig. 9) usually operate back and forth in adjacent passes. But they must all work within certain limits to leave soil properly leveled. Most other tillage machines can operate without difficulty in a wide variety of patterns (Fig. 10).

Adjust working patterns to fit the field size and shape. Minimize total field travel by working the length of fields as much as possible. Consider efficient machine operation when planning soil conservation practices by using parallel terraces. Try to arrange contour strips in multiples of machine width to reduce point rows and backtracking.

2. How does turning affect capacity?

The total number of turns is directly affected by field pattern and machine width. Working across fields or with narrow machines increases the number and total time of turns. Select headland or turnstrip width to permit easy turning, but narrow enough for quick finishing. With wide implements make the turning area about double the implement width, even more for narrow machines. If turning area is narrow and requires backing the tractor to turn, turning time may be increased by 50 percent (Fig. 11). Rough turning areas slow turning time by 10 to 30 percent compared to smooth ground.

Working narrow lands requires less travel time on ends, but it may leave an excessive number of dead furrows when plowing, and requires extra time to mark, open, and finish lands. Wide lands leave fewer dead furrows, but require more travel time on ends and could leave the headland soil heavily compacted.

If an increased number of turns is required, irregularly shaped fields require more working time per acre than rectangular fields whether borders are straight or curved. Consolidate small fields whenever possible. Consider using appropriate soil conservation practices to permit adding previously untilled areas to small fields. Save time by using reversible plows on contoured and oddly-shaped fields.

Fig. 9—Work Pattern Depends on Implement Used and Field Size and Shape

REVERSIBLE PLOW

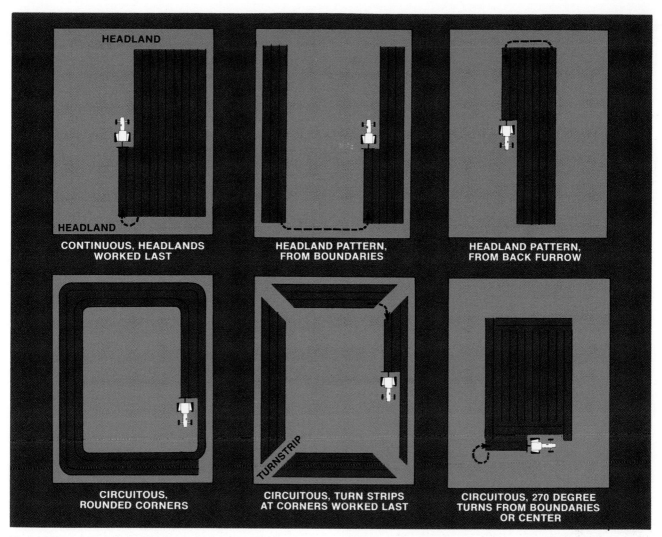

Fig. 10—Typical Field Operating Patterns

Fig. 11—Turning Time Must Be Kept To A Minimum

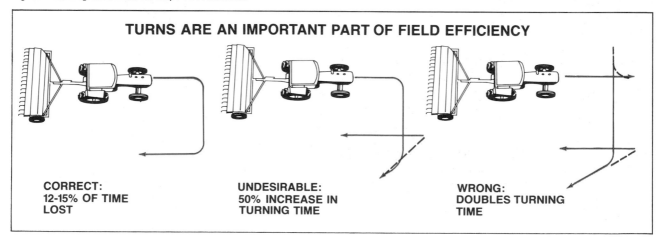

Fig. 12—Proper Maintenance Prevents Many Breakdowns

Fig. 13—Field Fuel Supply Systems Are Necessary

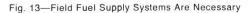

FIELD FUEL SUPPLY

3. How can adjustments and repairs be reduced?

Careful and complete preseason preparation and maintenance and daily servicing, as stressed throughout this manual, can help reduce lost time in the field. A good preventive-maintenance program and making preliminary adjustments before going to the field mean more field time can be spent in actual operation (Fig. 12).

However, major breakdowns can happen at any time and are not closely related to total hours of machine use or age. Some of the time lost in making adjustments and repairs and correcting major breakdowns can be related to the operator's attitude. Well-motivated operators, whether equipment owners or hired labor, can usually accomplish more work with less downtime than less-careful operators.

Time can be saved by selecting equipment with maximum reliability, which will require minimum service. Buying from established dealers with adequate parts-supply and service facilities can minimize lost time for equipment failures.

Field delays caused by adjustments, repairs, or major breakdowns can become critical from the standpoint of timely completion of operations for maximum yield and crop quality (which will be discussed later). Reliability and reduced lost time become even more important in terms of an equipment system in which certain operations are dependent upon completion of previous work. For instance, plowing may be delayed by stalk shredding or disking, planting delayed by secondary tillage, or cultivation held up by planting date.

4. How much time for maintenance?

Large fuel tanks on current tractors permit many hours of operation without refueling. But returning the tractor from remote fields is time consuming, costly, and may be dangerous if it means driving over busy highways. Hauling fuel in cans is slow and may result in dirt contaminating the fuel system, causing additional delays and lost power.

Thus, to avoid delays, speed and efficiency of the field fuel-supply system are imperative (Fig. 13).

Sealed bearings and increased intervals between lubrication of some implements reduce field maintenance time. But lubrication can be overlooked only at the risk of serious equipment failure and costly delays. Lubricant cost is minimal compared to the consequences of not lubricating at the recommended intervals.

5. Is travel time part of field efficiency?

Yes and no. Some experts feel it is unfair to penalize a machine for the geographic location of work areas, which is reasonable. However, if an implement is not readily transportable, or requires a great deal of time to prepare for transport, some of that time could reasonably be charged to the machine as reduced efficiency. Having remote fields may be unwise.

6. How is the operator's personal time involved?

Field delays for personal reasons vary widely and often involve the attitude, health, and environment of the operator. An operator who enjoys a particular job will stop less frequently than someone who finds the work distasteful or dull. The operator's health may require frequent breaks from some operations, particularly under adverse conditions. Improving the operator's environment with an enclosure or cab can be economically worthwhile as well as physically more comfortable. This is especially true if the enclosure permits continued operation under previously unacceptable conditions.

A roll-over protective structure (ROPS) provides a safer operator's environment and can further improve safety conditions by maintaining greater operator alertness (Fig. 14). Shielding the operator from dust, heat, cold, and rain can prolong intervals between rest stops, can reduce fatigue, and can often help attract and retain better-qualified operators.

However, an enclosure may reduce the operator's vision under some circumstances and isolate him from

Fig. 14—ROPS Enclosure Provides Improved Safety Conditions

ROPS ENCLOSURE IMPROVES SAFETY

COMBINING OPERATIONS

Fig. 15—Combining Operations Saves Time, Fuel, and Labor

the noise and "feel" of equipment operation traditionally used in identification of operating problems. Additional means may be required for monitoring of some implements and controlling others.

7. How much labor is available?

A plentiful supply of low-cost labor may seem to justify operation of several smaller implements instead of fewer, larger units because of lower total labor expense. But labor output must also be considered.

Labor cost per unit of work completed may be much higher for low-cost labor than for fewer, better-qualified, and higher-paid operators with the same equipment. Combined labor and machine costs per acre may be significantly lower for highly productive operators on large equipment (Fig. 15) compared to more smaller machines operated by less-skilled, lower-paid operators.

Livestock operations and different crops competing for the same field time also must be considered in determining how much labor is available or will be required for a given operation.

TABLE I

TYPICAL MACHINE PERFORMANCE

Machine	Typical Power Or Draft Required	Typical Speed, MPH (km/h)	Typical Range, Field Effec., %
MOLDBOARD OR DISK	psi (N/cm²) of furrow slice 3-6 (2-4), light soil 5-9 (3.5-6), medium soil 8-14 (5.5-10), heavy soil	3.5-6 (5.5-10)	74-90
CHISEL PLOW	200-800 lb. per ft. (270-1100 N/m) width; or 40-120 lb. per ft. per in. (22-65 N/m/cm) of depth	3.5-6.5 (5.5-10.5)	75-90
DISK TILLER	180-400 lb. per ft. (245-540 N/m) width	4-7 (6.5-11)	70-90
ROTARY TILLER	5-10 hp. per ft. (12-24 kW/m) width	1-5 (1.5-8)	70-90
DISK HARROW, SINGLE	50-100 lb. per ft. (65-135 N/m) width	3-6 (5-10)	70-90
DISK HARROW, TANDEM	100-280 lb. per ft. (135-380 N/m) width	3-6 (5-10)	70-90
DISK HARROW, OFFSET OR HEAVY TANDEM	250-400 lb. per ft. (340-540 N/m) width	3-6 (5-10)	70-90
SPRING-TOOTH HARROW	75-310 lb. per ft. (100-420 N/m) width	3-7 (5-11)	70-90
SPIKE-TOOTH HARROW	20-60 lb. per ft. (27-81 N/m) width	3-7 (5-11)	70-90
FIELD CULTIVATOR	3-5 in. (7.5-12.5 cm) deep with sweeps; 100-300 lb. per ft. width	3-8 (5-13)	75-90
ROW-CROP CULTIVATOR	shallow depth— 40-80 lb. per ft. width; deep cultivation— 20-40 lb. per ft. (11-22 N/m) width/per in. (cm) of depth	2-5.5 (3-9)	68-90
ROLLER PACKERS AND ROLLER HARROWS	20-150 lb. per ft. (27-205 N/m) width	3-7.5 (5-12)	70-90
ROD WEEDER	60-120 lb. per ft. (80-160 N/m) width	4-6 (6.5-10)	70-90
LISTER (FIRM SOIL)	400-800 lb. (1800-3600 N) per bottom	3.5-5 (5.5-8)	70-90
SUBSOILER	lb. per in. (N/cm) of depth per standard: 70-110 (125-195), sandy loam 100-160 (175-280), medium or clay loam	3-5 (5-8)	70-90
ROTARY HOE	30-100 lb. per ft. (40-135 N/m) width	4.5-12 (7-19.5)	70-85

SUMMARY

For optimum field efficiency, machine capacity and performance must be related to timeliness and a cost penalty assessed for potential yield reductions due to lack of work completion by the optimum date. For instance, in various parts of the cornbelt it has been shown that if corn is planted after May 10 to 15 (depending on area) yield drops an average of one bushel per acre (63 Kg/ha) for each day of delay. If the optimum date is May 15, and 100 acres (41 hectares) of corn are planted May 25, the value of 1,000 bushels (25.8 tonnes) of corn (10 days × one bushel per acre × 100 acres) (10 days × 63 Kg/ha × 41 ha/1000 Kg/ tonne) should be charged against equipment operating costs because of delayed planting.

Obviously, such a charge could soon pay the difference for larger equipment or more reliable machines. Similar, less well-defined penalties also apply to planting dates for some other crops and to harvesting of certain crops after various stages of maturity.

From an economic standpoint, the ultimate goal of any tillage system is not necessarily lowest cost of operation, but to provide the maximum possible contribution to total profit of the farm enterprise.

CHAPTER QUIZ

1. What three things control machine capacity and performance?

2. What must a machine do to be economically justified?

3. (True or false) Large machines do not always cost more per acre to operate than similar units.

4. (True or false) Equipment should be sized to complete work within the most desirable time period on the basis of average weather conditions.

5. (True or false) Doubling implement size doubles capacity.

6. What is the difference between effective and theoretical field capacity?

3
Traction, Flotation, and Soil Compaction

WHEEL WEIGHT ADDED

Fig. 1—Adding Wheel Weights Improves Traction

Because comparatively few track-type tractors are currently used for agricultural purposes, this discussion will concern only wheeled tractors.

For agricultural machines *flotation* is the ability of tires to stay on top of the soil surface or resist sinking into the soil. Flotation is also directly related to vehicle weight, soil conditions, and contact area.

Equipment-induced *soil compaction* is the packing or firming of soil caused by wheel traffic or the soil-working elements of some implements. Compaction is usually undesirable because it can restrict movement of air, water, and crop roots in the soil.

Several means are available to increase traction and flotation while reducing the potential for compacting the soil. These will be discussed in the following pages.

INTRODUCTION

Traction, flotation, and soil compaction are closely related, and changes in soil-surface conditions, soil-contact area, and vehicle weight will directly affect all three.

For our discussion, *traction* will be described as the linear force, pull, or draft resulting from torque applied to tractor tires or tracks. Its efficiency depends on nature of the soil surface, contact area between the tire or track and the soil surface, and vehicle weight.

THE PROBLEM

Tractors are built with ample size and power for large farming operations. But, to get maximum benefit from that power, additional weight must be added (Fig. 1) to gain maximum drawbar pull and sufficient traction for large implements, particularly tillage implements.

Large, heavy tractors and implements may cause severe compaction of some soils. In some farming systems, as much as 90 percent of the soil surface may be covered by wheel tracks between plowing and

Fig. 2—Part of Front-End Weight is Transferred to Rear Wheels; Part Balances Rear Loads

IMPROVE TRACTION

BALANCE REAR LOADS

cultivation. By then, much of the soil may be repacked to as much as 90 percent of its density before plowing.

Compaction reduces the ability of the soil to absorb and hold moisture and may severely inhibit root-growth. Reduced root-growth and limited available moisture can seriously reduce potential crop yields.

TRACTION

Getting improved traction may require various methods, and there is no one answer for all situations. Adding ballast (weight) to drive wheels and tractor front end is the most common method of improving traction and helping increase drawbar pull. Adding weight also reduces wheel-slippage, tire wear, and bouncing, and saves time and fuel. Front-end weight (Fig. 2) is also required for tractor stability, particularly with integral and heavy-draft, semi-integral equipment.

Tests show that installing radial tractor tires can increase drawbar power by 7 to 11 percent, and provide 6 to 13 percent more work from the same volume of fuel compared with using conventional bias ply tires.

EFFECTS OF UNEQUAL TRACTION

With normal differential action for the tractor drive wheels, each wheel can rotate independently when turning, but if traction is unequal one wheel may spin. Thus neither wheel can pull more than the one with the poorest traction. A differential lock can solve minor traction problems when one wheel starts slipping. This simply locks both wheels together and power goes to the wheel with the best traction. However, consider the differential lock only as a temporary or emergency traction device and take other steps to overcome continual slippage of one wheel.

EXCESSIVE WHEEL SLIPPAGE

Slippage of one wheel is often a major problem when plowing with one wheel in the furrow. The furrow wheel gets excellent traction, but the land wheel spins on trash, grass, and loose soil. Also, the tractor tilt puts more weight on the furrow wheel. Extra weight is usually added to the land wheel to reduce slippage, but another solution, where possible with drawn plows, is to readjust the plow hitch and operate all tractor wheels on the land. On-land hitching can result in more draft, however, due to increased landside pressure.

Wheel-slippage can be expensive. It wastes time and fuel, causes unnecessary tire wear, and delays field work. For instance, an extra 10-percent wheel-slippage means that it takes 11 hours to do 10 hours' work, or 11 days to finish what should have been done in 10. Cutting slippage can also reduce fuel consumption by 10 percent and tire wear by 40 percent.

But don't try to eliminate all slippage. Some slippage acts as a cushion for the engine and the drive train to soften the impact of sudden overloads. As slippage increases to about 15 percent, traction increases in most soils. But beyond 15 percent traction doesn't increase enough to offset the reduction in speed.

HOW TO MEASURE WHEEL-SLIPPAGE

For a quick measurement of wheel-slippage in the field, mark a spot on the ground and a chalk mark on one rear tractor tire. Then drive the tractor, under load with the implement in its normal operating mode, and count 10 complete rotations of the rear tire and place another mark on the ground (Fig. 3). Repeat the trip without the implement and again count wheel rotation between the two marks (Fig. 4). Estimate the fraction of the last rotation as nearly as possible.

Check the number of rotations counted on the second trip, using the chart below to determine the percentage of rear-wheel slippage.

If less than 8½ rotations are counted, add weight. If more than 9 rotations are counted, remove weight from the rear wheels.

Fig. 3—With Tractor Loaded, Count Wheel Revolutions Between Marks

Fig. 4—Leave Marks on Ground for 10 Full Tire Revolutions with No Load

Fig. 5—Rear Tires Are Carrying Too Much Weight

Fig. 6—Rear Tires Have Too Little Weight

REAR-WHEEL SLIPPAGE CHART

Rotations	Rear-Wheel Slippage Percent	What to Do
10	0	Remove
9½	5	Ballast
9	10	Proper
8½	15	Ballast
8	20	Add
7½	25	Ballast
7	30	

If unable to measure tire-slippage by counting wheel rotations, the tire-tread pattern produced when pulling under load provides an approximate indication.

When too much weight is used, the tire tracks will be sharp and distinct in the soil. There is no evidence of slippage (Fig. 5). The tires are figuratively "geared" to the ground, and do not allow the flexibility of engine operation obtained when some slippage occurs.

Too much weight increases the engine-power needed to move the tractor through the field and reduces the power available for pulling the implement.

When the tires have too little weight, they lose traction. The tread marks are entirely wiped out and forward progress is slowed (Fig. 6).

When the tires have proper weight, a small amount of slippage occurs. The soil between the cleats in the tire pattern is shifted, but the tread pattern is visible (Fig. 7). Proper weighting allows the engine to perform at its best with maximum flexibility for varying loads.

PROPER SLIPPAGE

Fig. 7—Proper Weight on Tires

If slippage exceeds 15 percent take immediate action to reduce spinning. Slippage under 15 percent is barely visible. So, as a rule of thumb, if slippage can be seen, it's too much.

EFFECTS OF ADDING WEIGHT

Adding 100 pounds (45 Kg) of ballast to rear of tractor increases drawbar pull according to the surface conditions as shown below:

- Concrete   66 lbs. (30 Kg) more pull
- Dry clay   55 lbs. (25 Kg) more pull
- Sandy loam   50 lbs. (23 Kg) more pull
- Dry sand   36 lbs. (16 Kg) more pull
- Green alfalfa   36 lbs. (16 Kg) more pull

Keep in mind that draft for the same operation is usually much less in sand than in dry clay, but that any extra weight greatly increases rolling resistance in sand or other loose soil. So, to be effective, ballast must be adjusted for soil type and conditions as well as the implement load. Regardless of weight added, drawbar pull may drop to zero or close to zero under adverse surface conditions.

CAST-IRON AND LIQUID BALLAST

Ballast may be added in the form of cast-iron frame and wheel weights, or by filling tires with calcium-chloride solution or dry ballast. Dry ballast is heavier than calcium-chloride solution and requires special handling equipment. Dry ballast is not widely used; it is more expensive per pound than cast-iron or liquid.

Here are some of the advantages and disadvantages of cast-iron weights and liquid ballast.

Cast-Iron Weights:

Are relatively easy to install and remove

Require no special equipment or servicing

Are not lost if tire is damaged or changed

Have higher initial cost than liquid ballast

May not provide sufficient weight for heavy loads

May cause undesirable increase in tractor width

Liquid Ballast (Calcium-Chloride Solution):

Is readily available at most tire and implement stores

Has lower initial cost than cast-iron or dry ballast

Provides excellent weight distribution

Is contained within the tires—causes no changes in tractor dimensions

Is 30 percent heavier than water alone and will not freeze above -50 degrees F.

Is lost in case of tire or valve damage

Is not easily removed and reinstalled

Requires specialized handling and installation equipment

Slightly reduces bruise resistance of tire (at 75 percent fill) (Fig. 8)

The amount of ballast which can be added safely to a tractor is limited by tire and tractor design. Exceeding recommended weight limits may cause premature tire failure and damage to the tractor frame, and may severely overload the drive train. Check the tractor operator's manual and the tire manufacturer's recommendations for maximum allowable ballast.

EFFECT OF INCREASED SPEED

Adding weight is not the only way to improve traction nor is it always the best. Pulling a lighter load at higher

Fig. 8—Liquid Ballast Should Fill Rear Tires Only 75 Percent Full

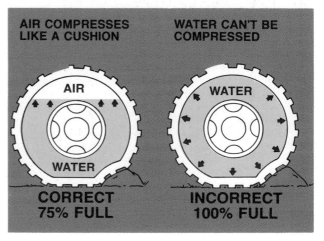

AIR COMPRESSES LIKE A CUSHION

WATER CAN'T BE COMPRESSED

AIR

WATER

WATER

CORRECT 75% FULL

INCORRECT 100% FULL

WEIGHT TRANSFER
IMPROVES TRACTION

Fig. 9—Weight Transfer from Integral and Semi-Integral Implements Improves Traction

speed reduces wheel-slip without increasing soil compaction. Reducing plow size by one bottom may permit shifting up one or two gears. Increasing speed is not recommended if frequent soil obstructions are encountered. High-speed operation may also require switching to a field cultivator or a chisel plow instead of a disk harrow to avoid excessive ridging.

Fig. 10—Power Weight Transfer "Borrows" Weight from Drawn Implements

WEIGHT TRANSFER HITCH

WEIGHT TRANSFER

Integral and semi-integral mounting of implements permits transfer of weight from the implement and tractor front-end to the tractor rear tires to increase traction (Fig. 9). This reduces dependence on cast-iron wheel weights and tire ballast. Power weight-transfer hitches for drawn equipment (Fig. 10) offer additional possibilities for reducing permanently-added tractor weight. These systems add only as much weight as is needed to provide traction at the time. Additional front-end weight is usually required, but those weights are normally quite simple to install and remove.

ADVANTAGES OF 4-WHEEL DRIVE TRACTORS

Four-wheel drive tractors develop more traction per pound of weight (Fig. 11) because:

- *Rolling resistance is reduced on the front wheels*
- *All wheels are driving*
- *All available tractor weight is used to develop traction*

Tractors with all four drive wheels of equal size perform best. But tractors with factory- or field-installed front-wheel drive (Fig. 12) usually perform better than 2-wheel-drive tractors with equal power when the going is tough. Both hydrostatic and mechanical front-wheel drives are available for various tractors.

FLOTATION

In soft, loose soil such as sand, increasing wheel diameter and width is more effective in increasing drawbar pull for a given weight than increasing ballast, partly because of reduced rolling resistance and better flotation.

EFFECTS OF ADDING DUAL WHEELS

Better flotation is also provided by adding dual wheels (Fig. 13), which have become quite common on large tractors in recent years. Adding duals doubles the contact area between tires and soil, but does not increase traction. In fact, equally weighted single and dual wheels of the same size will pull almost exactly the same load in good soil conditions.

Dual wheels have several advantages and disadvantages:

Advantages

More total weight can be added because of the additional load capacity of the extra tires, and the tires and wheels themselves provide some added weight.

Increased soil-contact area increases flotation and with added ballast improves traction in adverse conditions.

FOUR-WHEEL DRIVE TRACTORS DEVELOP MORE TRACTION

Fig. 11—Four-Wheel Drive Tractors Develop Maximum Traction Per Pound of Tractor Weight

Fig. 12—Front-Wheel Drive Increases Traction, Reduces Rolling Resistance

Fig. 13—Dual Wheels Increase Tractor Flotation

FRONT-WHEEL DRIVE

IMPROVED FLOTATION

Tractor stability is improved by increasing width.

Duals improve performance for land preparation—can be removed for row-crop work, haying, etc., if desired.

Often a complete set of duals can be installed for less money than a set of single, larger tire and rim assemblies.

Duals permit use of a tractor under a wider range of conditions, including adverse weather.

Disadvantages

But duals may overload axles, bearings, and gears of some tractors — check manufacturer's recommendations.

Duals may make a tractor more difficult to turn and get through gates, and frequent sharp turns can damage tread on outer tires.

Large duals are difficult to install and remove without proper equipment. To avoid overloading axles, and to permit easier handling when adding or removing wheels, do not fill outer duals with liquid.

COMPACTION

Most soils compact more severely if worked when wet, and the problem increases as moisture gets higher. If soil is tilled when moisture is too high and then permitted to dry, extremely hard clods are formed which are difficult to break up later. Traction is also better at lower soil-moisture levels. So, delaying tillage until soil reaches the proper moisture level will reduce the potential for compaction and make pulverization easier. Also, weeds are easier to kill when soil is dry.

In some cases the plow pan formed by primary tillage tools and the compaction of row middles by wheel traffic during planting, cultivation, and spraying of row crops provides a "window box" (root bed) of soil as deep as the soil was tilled (Fig. 14). In this small area the crop is expected to grow and mature. But the full potential is seldom reached because much of the tilled soil volume may be too dense for normal growth of the plant root by the time roots need more space.

Soil compaction also reduces pore space between soil particles and thus restricts water and air movement into the soil, reduces water holding capacity of the soil, and may restrict root growth into wheel tracks and similar compacted areas thus limiting nutrient uptake and wasting fertilizer.

CONTROLLED-TRAFFIC FARMING

A relatively new practice called controlled-traffic farming can reduce the compaction problem in susceptible soils. With this system, each year all wheel traffic from tillage through harvest is confined to the same tracks across the field. The wheel track area is untilled, which reduces energy requirements, and the soil which is worked requires less energy because it has not been compacted by wheel traffic.

These reduced energy requirements can save fuel and time. Water infiltration can be increased, runoff and soil erosion can be reduced, and in some cases there have been yield increases. However, engineers and soil scientists who have worked with the system don't predict increased yields for all areas and believe that time- and energy-saving may be a greater incentive.

HOW TO CONTROL FIELD TRAFFIC

Here are steps any farmer can take toward controlling traffic in his fields:

1. Match implement wheels to tractor tread and use 4-wheel tractors instead of tricycle models.

2. Reduce the number of trips over the field for seedbed preparation. Pull two or more light-draft implements together rather than making separate trips.

Fig. 14—National Tillage Machinery Lab Studies Show Effects of Compaction

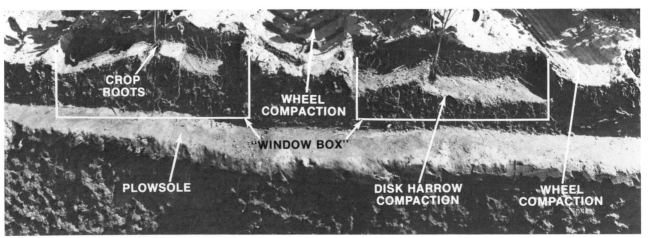

Fig. 15—Keep All Tractor Wheels On-Land to Reduce Compaction

3. Switch to reduced tillage or even no-till planting.

4. If possible keep all tractor wheels on the land when plowing instead of putting one wheel in the furrow (Fig. 15).

5. Work the widest possible swath through the field each time.

6. Remove extra tractor weight when not required for traction. To reduce the need for extra ballast, drive at the maximum recommended speed.

7. Grow crops on ridges or beds to provide more uncompacted soil for root growth. Use the same beds each year and follow established patterns for all wheel traffic.

8. Avoid working soil that is too wet. Use aerial application of chemicals if soil is wet and subject to compaction.

9. Use subsoilers to shatter plow pan.

SUMMARY

Because of their close interrelation, traction, flotation, and compaction must be considered simultaneously to provide the best answers to each problem.

Traction is dependent on weight, speed, and surface conditions. Adding weight helps increase traction, but also may increase soil compaction. Increasing speed with smaller implements makes better use of power without increasing weight. But, higher speeds are limited by:

● *Soil obstructions in some cases*

● *Soil ridging by some implements*

● *Rapidly increasing draft with some implements*

● *Safe operating conditions*

Loose, wet surfaces provide poorest traction. Best traction is on dry, firm surfaces.

Flotation is increased by increasing the tire size, but oversized tires may interfere with some integral equipment. Extra-large tires can cost more than duals, especially if big rims must be added. Duals increase flotation and can carry more weight for better traction, but they don't provide more traction than equally-weighted single tires in good soil conditions.

Compaction can be reduced by removing excess ballast and limiting trips over the field. Confining wheel traffic to the same row middles all season limits the compacted area and may produce better root growth in the remaining soil.

CHAPTER QUIZ

1. What three things determine the amount of traction developed by a particular wheel?

2. What is flotation, as applied to agricultural equipment?

3. Why is soil compaction detrimental to crops?

4. How does a differential lock improve traction?

5. (True or false) Adding 100 pounds of ballast to tractor rear wheels will increase drawbar pull by 100 pounds.

6. List three types of ballast which may be added to a tractor.

7. (True or false) Dual wheels can pull twice as much as the same-size single tire in good soil conditions.

8. (Choose one) Most soils compact more when (wet) (dry).

9. How does speed of operation affect traction and compaction?

4
Toolbars

Fig. 1—Toolbar Equipment Matches Individual Needs

INTRODUCTION

Toolbar tillage and planting began in the Southwest and later moved into the Mississippi Delta area. Only in recent years has there been much interest in toolbar equipment in the Corn Belt. With relatively small tractors and stable 40-inch-row spacing(1016 mm), there formerly was little incentive for Corn-Belt farmers to use toolbars.

But as row spacings began to change, and planting and cultivation became 6-, 8-, 12-, and even 24-row operations, toolbars have taken on new importance in assembling the particular equipment each farmer needs (Fig. 2). Since the late 1960s the variety, size, and use areas of toolbar equipment have grown rapidly.

But as row spacings began to change, and planting and cultivation became 6-, 8-, 12-, and even 16-row operations, toolbars have taken on new importance in assembling the particular equipment each farmer needs (Fig. 2). Since the late 1960s the variety, size, and use areas of toolbar equipment have grown rapidly.

Because of the vast array of possible toolbar-equipment combinations, we will only consider some of the basic components and a few applications. Details will be minimal, and we will not pursue individual principles of operation, preparation, or trouble shooting, but will discuss some general areas of application.

Important: Every tractor has certain lift and draft limitations, and all "ready-made" equipment is likewise designed and constructed to meet definite strength and performance specifications. This permits manufacturers to provide recommendations for matching of implements to specific tractor models.

However, most toolbar-based implements assembled by dealers or farmers have not undergone the extensive testing of factory-built machines. Therefore, it may be difficult to predict the exact size of bar required, the correct number of clamps, or the lift ca-

pacity required to transport the assembled machine. Too much weight can overload the tractor hydraulic system, reduce tire life, and cause poor tractor stability. Excessive draft means unsatisfactory field performance and possible tractor or equipment damage, and could cause tractor instability.

Too much power for the strength of the implement could result in serious equipment damage, especially in adverse operating conditions.

The best method for determining toolbar size and tractor capacity is to compare plans for combinations with similar toolbar implements offered by the manufacturer for certain tractor models. It also is helpful to check with other farmers who may be successfully using an assembly similar to what is being planned.

TOOLBARS

For many years toolbars were solid-steel bars, 2, 2¼, or 2½ inches (50, 57, or 64 mm) square. These bars are usually mounted with the bar diagonals vertical and horizontal (Fig. 3). A typical solid 2¼-inch (5.7 cm) square bar weighs about 17 pounds per foot (25 kg/m) of length and may be up to 30 feet (9 m) long.

Square hollow bars are also available to reduce weight by 55 to 60 percent, cut cost by as much as 20 percent, and retain ample strength for many small jobs (Fig. 4). These also are 2 to 2½ inches (50 to 64 mm) square and often are used interchangeably or in combination with solid bars. The strength of a 2¼-inch (57 mm) hollow bar is rated at about 70 percent of an equal-size solid bar.

A hollow 3½ × 7-inch (89 × 178 mm) flattened-diamond shape is used for some toolbars (Fig. 5). This shape provides considerably more strength than small, square toolbars, and permits use of some of the same attachments by using longer bolts.

Many rear-mounted row-crop cultivators use 4 × 4- or 5 × 7(100 × 100 or 127 × 178 mm) hollow bars for maximum strength per pound. However, these bars are seldom used for other applications, so they will not be discussed further here.

Box-beam toolbars provide outstanding strength-to-weight ratio for large tillage and planting units (Fig. 6). These bars range from 4 × 7 to 7 × 7½ inches (100 × 178 to 178 × 190 mm) for different applications. Wall thickness is from ¼ to ½ inch (12.7 mm) and varies with bar size and length, which may reach 30 feet (9.75 m) or more.

The type of toolbar must be decided on the basis of anticipated use and total weight. Oversized toolbars provide added strength, but increase weight and cost. However, replacement of toolbars which fail due to insufficient strength may increase total costs even more.

TOOLBAR HITCHES

One-piece integral toolbar hitches (Fig. 7) may be used with solid or hollow bars of various lengths, de-

Fig. 2—Unit Planters Are Common Toolbar Attachments

Fig. 3—Solid-Steel Square Bars Provide Excellent Strength for Many Applications

Fig. 4—Square Hollow Bars Reduce Weight and Cost

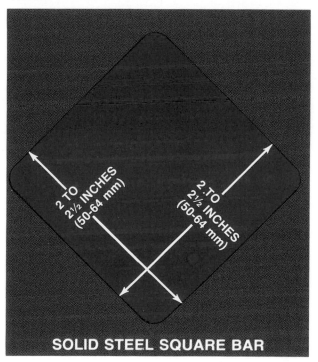

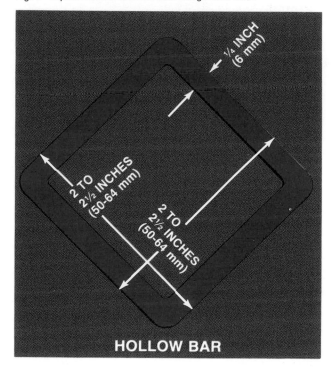

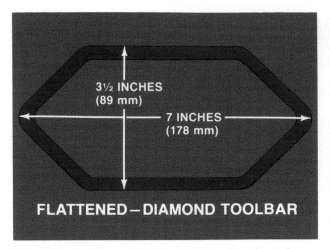

FLATTENED—DIAMOND TOOLBAR

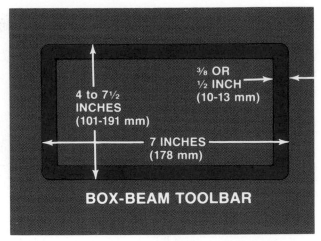

BOX-BEAM TOOLBAR

Fig. 5—Flattened-Diamond Toolbar Is Stronger Than Small, Square Bars

Fig. 6—Box-Beam Toolbars Provide Maximum Strength

Fig. 7—One-Piece Tool Carrier for One or Two Square Bars

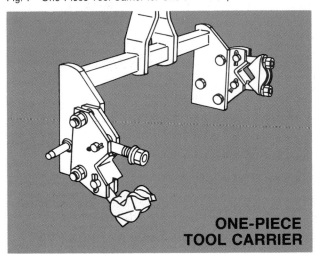

ONE-PIECE TOOL CARRIER

pending on the expected weight and draft load. A second bar may be added, by using spacers, for extra strength or clearance. The hitch may be removed from the bar and reattached to a different bar with other attachments, if desired, to eliminate need for completely changing all equipment.

A three-piece hitch attached directly to the toolbar (Fig. 8) provides conformity with different hitch dimensions. Relocating brackets and switching pins permits operation with tractor hitches of various sizes.

Increased strength is provided by doubling bars and using straddle-mounted hitch pins (Fig. 9). Additional rigidity is provided by heavy mast braces, truss rods, and diagonal braces between bars. Bars can be spaced at different distances for easy placement of a variety of attachments.

Box-beam toolbars require extremely rugged hitches to match the strength of large tractors and the weight

Fig. 8—Three-Piece Hitch Has Versatility

Fig. 9—Doubling Bars Provides More Strength and Rigidity

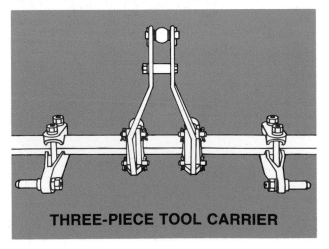

THREE-PIECE TOOL CARRIER

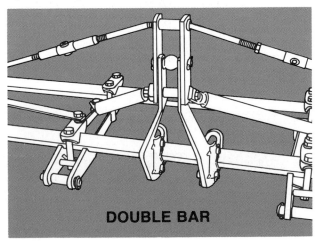

DOUBLE BAR

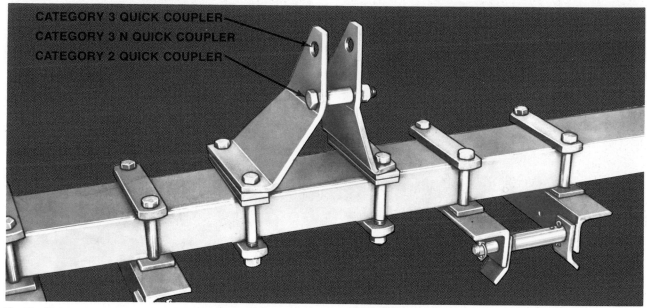

CATEGORY 3 QUICK COUPLER
CATEGORY 3 N QUICK COUPLER
CATEGORY 2 QUICK COUPLER

Fig. 10—Hitch on Box-Beam Bar Can Be Set for Different Tractor Sizes With or Without Quick Coupler

of large implements. Some hitches are attached directly to the box-beam bar (Fig. 10).

These hitches provide:

Straddle-mounted hitch pins.

A choice of hitching directly or with a quick coupler.

Space between clamps for increased freedom in attaching tools to the bar.

Additional versatility is provided for special applications by mounting hitch brackets on a separate bar

Fig. 11—Versatility and Adjustability Are Keys to Toolbar Success

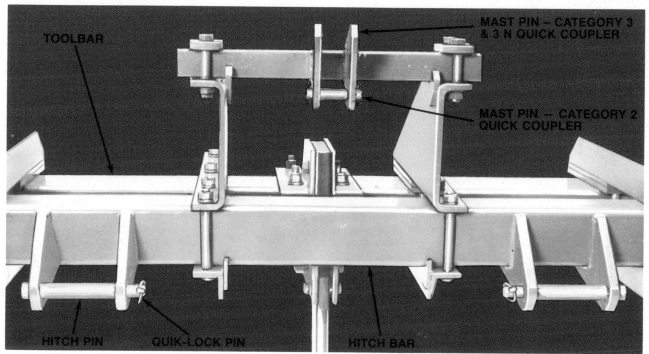

TOOLBAR

MAST PIN – CATEGORY 3 & 3 N QUICK COUPLER

MAST PIN – CATEGORY 2 QUICK COUPLER

HITCH PIN   QUIK-LOCK PIN   HITCH BAR

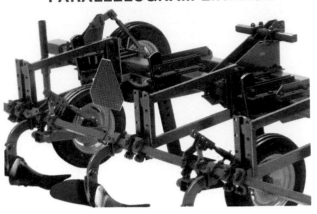

PARALLELOGRAM LINKAGE

RELOCATING CARRIER ON TOOLBAR

Fig. 12—Relocating Carrier On Toolbar Permits Simple Addition of Second Bar

Fig. 13—Flexible, Parallelogram Linkage Permits Independent Control of Bar Height

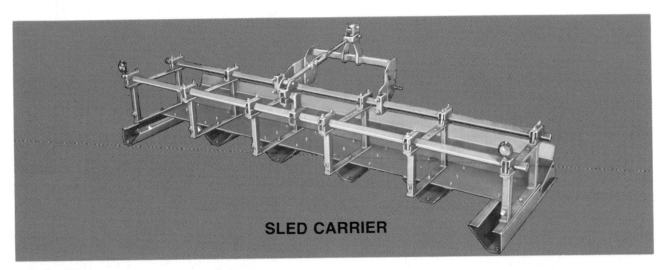

SLED CARRIER

Fig. 14—Sled Carriers Can Make Precision Beds

Fig. 15—Many Different Bed-Shaping Attachments Are Used

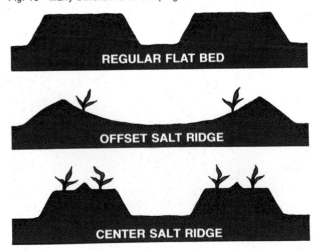

REGULAR FLAT BED

OFFSET SALT RIDGE

CENTER SALT RIDGE

clamped in front of the box-beam toolbar (Fig. 11). Mast side-plates can be moved laterally for maximum flexibility in clamping attachments to the main bar. Shifting the mast and hitch clamps back on the toolbar and using spacer clamps (Fig. 12) permit attachment of an optional square bar.

If heavy attachments (such as subsoilers) are mounted on the box-beam bar, bedder bottoms or other attachments may be clamped to a second bar attached to the box-beam by flexible linkage (Fig. 13). This permits regulation of working depth of the front bar by the tractor load-and-depth system or gauge wheels, while the rear bar is controlled by its gauge wheels, providing independent depth control for front and rear units.

SLED CARRIERS

Sled carriers can be the basic implements for a complete specialized-crop-growing system. With bed-shaping attachments (Fig. 14) they form precision beds in many different row spacing and configurations. Shapers are available for wide and narrow beds, various widths and depths of furrow, and salt-ridge shapers for irrigated areas where salt may damage crops (Fig. 15).

Attaching unit planters (for beets, beans, vegetables, or cotton) permits shaping beds and planting in one simple operation (Fig. 16). Bed shapers sweep dry topsoil from the beds and seeds are placed directly in moist soil for rapid germination. Shapers can also plane off weeds which may have started since the field was bedded, and shape the beds for efficient application of herbicides.

Furrows between beds are uniformly shaped and compacted for optimum flow of irrigation water. They may also be used to guide cultivator attachments for later weed control (Fig. 17). Replacing unit planters and bed shapers with cultivating tools on the same sled carrier permits fast, precision cultivation with minimum plant damage and operator fatigue.

SPACER CLAMPS

Rigid spacer clamps (Fig. 18) permit attachment of a square second bar to box-beam toolbars, or for clamping two, three, or four square bars together (Fig. 19). Many clamp configurations are available to space two bars from 6½ to 20 inches (165-508 mm) apart (Fig. 20). Additional clamps of equal or different size may be used to attach a third or even fourth bar for individualized assembly.

ONCE OVER
BED SHAPING AND PLANTING

Fig. 16—Once-Over Bed-Shaping and Planting Saves Time

Fig. 17—Sled Carrier Can Provide Precision Cultivator Guidance

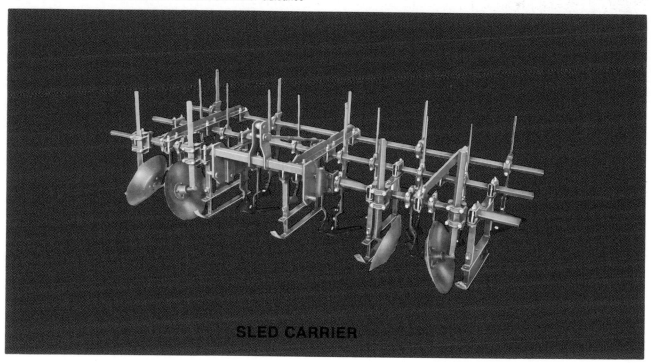

SLED CARRIER

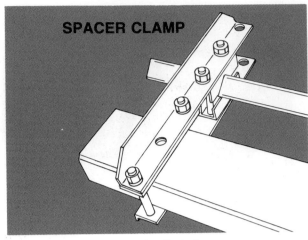

Fig. 18—Clamp Square Bar to Box-Beam Bar

If two square bars are used for added strength, use the shortest spacers available to keep the load close to the tractor for better stability. Multiple toolbars may be used with different tools on each bar (Fig. 21), or one bar may easily be removed for replacement with a different bar and different tools. Multiple toolbars also provide more space for staggering attachments for complete soil coverage or improved trash flow.

A flexible toolbar spacer places the square bar approximately 3 feet (1 m) behind the box-beam bar (Fig. 22) for operations such as simultaneous subsoiling and bedding.

GAUGE WHEELS

Toolbar gauge wheels are used to control implement working depth or toolbar height for specific operations. They are generally available in the following configurations:

Fig. 19—Two, Three, or Four Square Bars Can Be Clamped Together

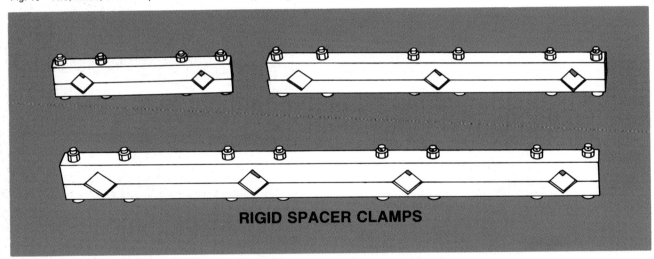

Fig. 20—Bar Spacing Ranges from 6½ to 20 Inches (165-508 mm)

Fig. 21—Multiple Toolbars May Be Used Many Different Ways

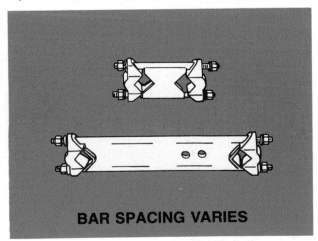

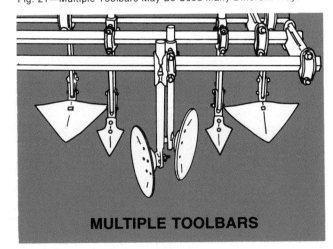

42

1. Dual wheels (Fig. 23) provide increased stability in uneven terrain. Height is adjusted by turnbuckle or hydraulic cylinder. They are used with box-beam toolbars.

2. This single wheel (Fig. 24) has a turnbuckle adjustment (also available with a hydraulic cylinder). It is used with box-beam toolbars.

3. This single wheel (Fig. 25) has a clamp adjustment for economy. It is used with small, square toolbars.

4. This single wheel (Fig. 26) is adjusted by a crank. It is used for square toolbars.

5. These wheels (Fig. 27) have clamp-type brackets and semi-pneumatic tires. They are used for cultivation with square bars.

6. This cast gauge wheel (Fig. 28) is used for bedders. It has a deep rim and broad flanges.

FLEXIBLE TOOLBAR SPACER

Fig. 22—Flexible Toolbar Spacer Clamps to Box-Beam Bar

Fig. 23—Dual Gauge Wheels "Smooth" Operation on Uneven Ground

Fig. 24—Turnbuckle or Hydraulic Control Permits Easy Adjustment

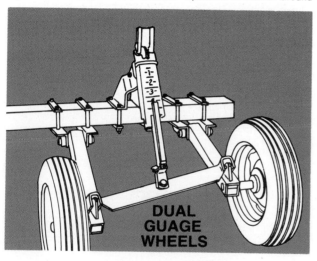

DUAL GUAGE WHEELS

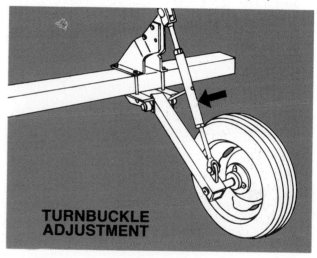

TURNBUCKLE ADJUSTMENT

Fig. 25—Single-Wheel Clamp Adjustment for Economy

Fig. 26—Crank Adjustment Saves Time

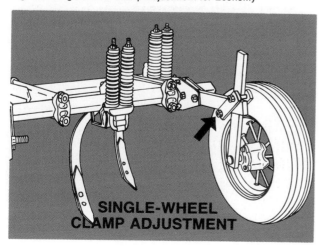

SINGLE-WHEEL CLAMP ADJUSTMENT

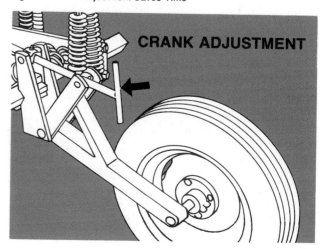

CRANK ADJUSTMENT

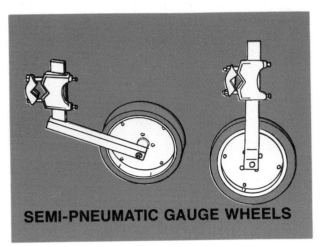

SEMI-PNEUMATIC GAUGE WHEELS

Fig. 27—Small Wheels Control Cultivators

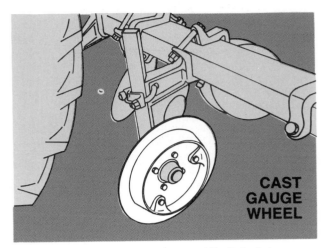

CAST GAUGE WHEEL

Fig. 28—Cast Wheels Provide Stability and Depth Control

Fig. 29—Lift-Assist Wheels Help Carry Heavy Loads in Operation and Transport

Fig. 30—Subsoiler Standards Usually Clamp to Box-Beam Bars

LIFT ASSIST WHEELS

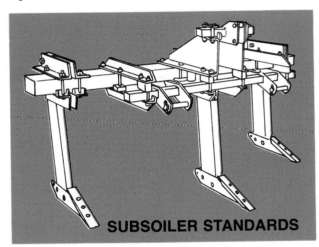

SUBSOILER STANDARDS

Fig. 31—Vegetable-Cultivating Standards Fit Different Tools and Row Arrangements

Fig. 32—Chisel-Plow-Type Shanks Fit Square Bars

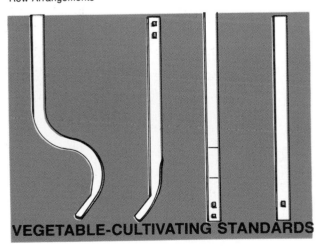

VEGETABLE-CULTIVATING STANDARDS

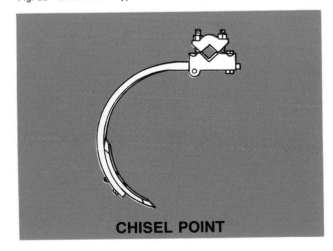

CHISEL POINT

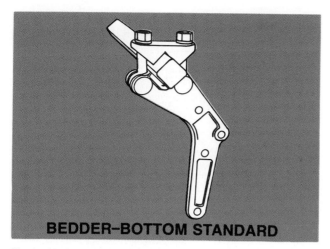

BEDDER-BOTTOM STANDARD

Fig. 33—Bedder-Bottom Shanks Fit Box-Beam or Square Bars

7. A hydraulically-controlled lift-assist wheel is available for extra-heavy loads (Fig. 29), such as fertilizer hoppers.

STANDARDS AND SHANKS

The variety of toolbar shanks and standards is almost endless. They range from subsoiler standards (Fig. 30), that can work 24 inches (610 mm) deep, to vegetable-cultivating standards for weed knives, small shovels, etc., (Fig. 31). In between are shanks for chisel points or sweeps (Fig. 32) and standards for lister or bedder bottoms (Fig. 33). Combination units feature chisel points and bedder bottoms (Fig. 34).

There are also irrigation border disks (Fig. 35), disk-bedder standards with or without subsoilers (Fig. 36), and flexible-frame adjustable 2- or 3-blade disk gangs (Fig. 37), plus many other options.

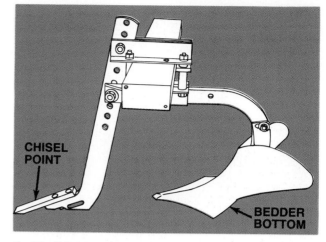

CHISEL POINT

BEDDER BOTTOM

Fig. 34—Chisel Point Breaks Soil Ahead of Lister or Bedder Bottom

Fig. 35—Border Disks Build Irrigation Borders or Reverse to Tear Down Old Borders

BORDER DISK

Fig. 36—Disk-Bedder Standards May Be Used Alone or with a Subsoiler Working Under Each Bed

Fig. 37—Flexible Disk Gangs Adjust to Different Soils

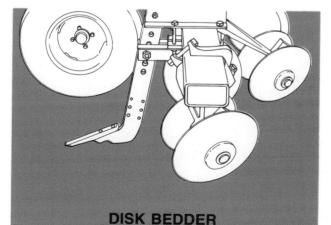

DISK BEDDER

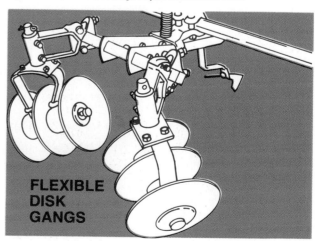

FLEXIBLE DISK GANGS

SOIL-ENGAGING TOOLS

Toolbars may be equipped with soil-engaging tools to match almost any soil, crop, or tillage condition (Fig. 38). They can use the full line of chisel-plow sweeps, shovels, furrow openers, reversible spike and chisel points, and twisted shovels. Also available, with different standards, are various sweeps, points, shovels, disk hillers, furrow openers, weed knives, etc., used for row-crop cultivation (See Chapter 19). In addition, there are bedder bottoms to match different soil types, and disk bedders.

Hundreds of combinations are possible when sweeps, shovels, and disks are used together for special bedding or cultivating operations provided tractor power, lift capacity, and stability are not exceeded.

MARKERS

Markers are used to guide the operator where there is a possibility of overlapping or skipping. Marker length is usually adjustable for a range of row widths or equipment sizes.

Markers may be raised automatically, by chain and sheave attachment on the tractor drawbar, or hydraulically (Fig. 39). Selective lowering may be accomplished automatically as the toolbar is lowered or by hand-releasing the proper latch. An optional electric release for tractors with cabs is available for some markers.

TOOLBAR TRANSPORT

Most toolbars up to about 15 or 16 feet (4.5 or 5 m) are transported on the tractor 3-point hitch without change. If length is much greater, some means of reducing width is usually required to permit safe transport. An endways transport attachment (Fig. 40) reduces width and permits easy movement with maximum tractor stability. Toolbar gauge wheels may sometimes be removed and used as transport wheels. The towing hitch for endways transport is equipped with a cross shaft to fit the tractor lower links or quick coupler for easy hookup and raising for transport.

Fig. 38—Almost Any Type of Soil-Engaging Tool Can Be Used with Toolbars

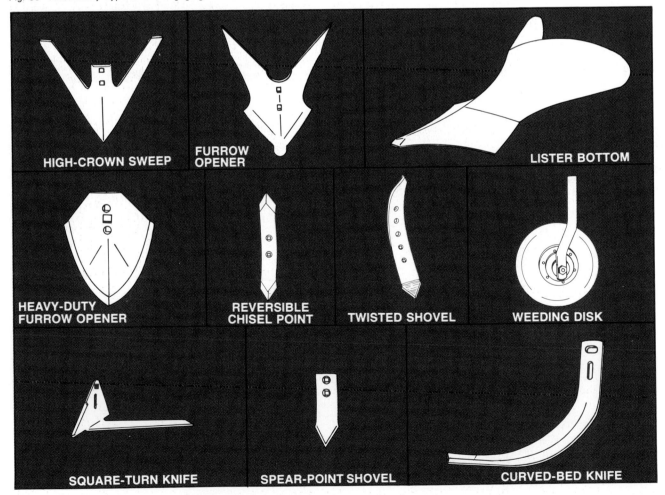

MARKERS MAY BE RAISED HYDRAULICALLY

Fig. 39—Markers May Be Raised Hydraulically or by Chain Attachment to Tractor Drawbar

Fig. 40—Endways Transport Attachment Reduces Width for Easy Movement

ENDWAYS TRANSPORT ATTACHMENT

Hydraulic cylinders mounted inside the toolbar control folding of some large toolbars (Fig. 41). For transport, a typical 32-foot (9.7 m) bar may be reduced to 21 feet (6.4 m), which still requires considerable care to avoid collisions or running tractor wheels too close to the road edge. Only a small area near each hinge restricts placement of attachments.

SUMMARY

To catalog all possible and practical combinations of soil-conditioning, cultivating, and planting combinations for hitch-carried toolbars would be a difficult task. In fact, just listing toolbar attachments from various manufacturers is almost impossible, for they include not only many types of soil-engaging tools but a wide variety of solid and box-beam bars, standards of various sizes and lengths, and such auxiliary equipment as various spacers, gauge wheels, lift-assist wheels, and markers.

Toolbars save money by providing a single basic frame for a wide choice of equipment. They are wellnigh an ideal basic unit for innovative farmers who want to try combinations not available in complete implements sold by dealers. Some farmers have two or more toolbars, permitting them to change operations without tearing down one hookup to use another.

Fig. 41—Hydraulically Folded Toolbar Saves Transport Time

But caution is necessary. Implement manufacturers carefully check strength, and keep size and weight within the power, hydraulic lift, and stability capacity of recommended tractors. They cannot do this for farm-assembled toolbar combinations. So it is up to users who experiment with toolbars to assemble combinations with sufficient overall strength, and to avoid overtaxing tractor power and stability and lift limits of 3-point hitches.

CHAPTER QUIZ

1. What is one of the major problems encountered in do-it-yourself implement design with toolbar components?

2. How can toolbar implement weight be reduced for some lighter small-bar applications?

3. Name two reasons for using multiple toolbars on one tool carrier.

4. List five general categories of commonly used toolbar attachments, one specific operation for each category and a crop on which the attachment could be used.

5
Moldboard Plows

Fig. 1—Plowing Buries Trash and Aerates the Soil

INTRODUCTION

For thousands of years, plows have been used as primary tillage tools to prepare seedbeds and rootbeds for crops. Modern moldboard plows, as well as the crude plows still used in some underdeveloped parts of the world, have evolved from forked sticks pulled by primitive farmers or their animals.

Such Colonial leaders as Thomas Jefferson and Daniel Webster helped adapt European plows to American conditions. Those early cast-iron plow bottoms worked well in sandy and gravelly soil, but they wouldn't scour properly, if at all, in sticky Midwest prairie soil.

That problem prompted development of steel bottoms, which became an important factor in the settlement and agricultural development of the Prairie states. Prior to that time, some people had predicted that the prairie soil would have to be abandoned for growing crops because of the difficulty in plowing.

Frontier blacksmiths and farmers continually modified plow design for localized soil and crop conditions. Modern farm equipment manufacturers are continuing this evolution as plowing speed increases and designers work to reduce draft and improve plow performance.

WHAT THE PLOW DOES

The moldboard plow cuts, lifts, and turns the furrow slice (Fig. 1), and in so doing it:

● *Buries some or all of trash and crop residue*

● *Aerates the soil*

● *Controls weeds, insects and crop diseases*

● *Incorporates fertilizer into the soil*

● *Provides good seedbeds for better germination*

Incorporation of plant residue (Fig. 2) and aeration of the soil by tillage also stimulates growth of microorganisms which decompose trash and other organic materials. Accelerating decomposition of organic matter increases the supply of available nitrogen, phos-

Fig. 2—Plows Incorporate Heavy Residue

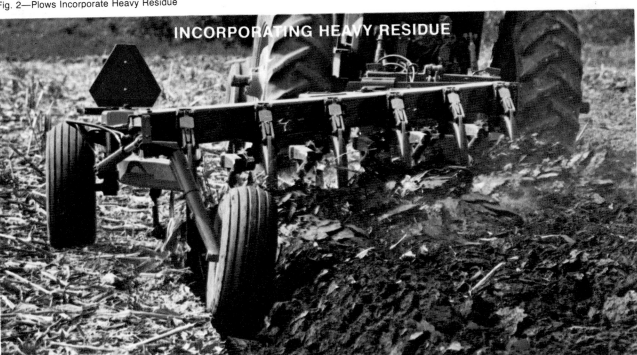

phorus, potassium, and other plant nutrients. Soil microorganisms require the same favorable conditions of warmth, moisture, and aeration as are needed for quick seed germination.

Good plowing sandwiches organic matter between furrow slices to form a "wick" for better water absorption and storage, and faster decomposition of residue. It also increases soil porosity and provides more air for faster, stronger root growth.

Good soil pulverization in plowing reduces the cost of later tillage, but unless planting will start immediately, each furrow should retain a distinct crown. This helps reduce runoff, reduces ponding and puddling in low spots, reduces wind erosion, increases moisture infiltration, and speeds surface drying.

Plowing to the same depth each year may produce a plow sole or hardpan, just below plowing depth, which can severely restrict root growth and water movement.Occasional chisel plowing, or plowing several inches deeper every few years, will help break the compacted layer.

Let's examine the moldboard plow in detail.

MOLDBOARD PLOW BOTTOMS

The "business end" of the moldboard plow is the bottom, normally from 12 to 20 inches (305-508 mm) wide. It is essentially a three-sided wedge, with a landside and share cutting edge as flat sides, and the moldboard as the curved side. Each bottom is attached to a standard, which in turn is fastened to the plow frame (Fig. 3).

The main plow-bottom parts are all attached to a common part called the frog (Figs. 4 and 5), and include:

- **Moldboard**
- **Share**
- **Shin**
- **Landside**

Fig. 3—Plow Bottom

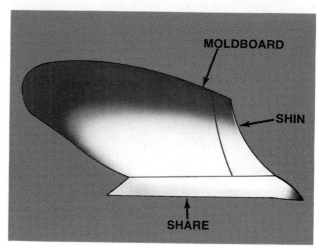

Fig. 4—Side View of Plow Bottom

SOIL ACTION ON THE PLOW BOTTOM

The wedging action of the plow bottom moving through the soil exerts pressure upward and toward the open furrow. This turning causes blocks of soil to be sheared at regular intervals (Fig. 6). This is the fracturing effect of a plow bottom under most soil conditions.

The blocks of soil rub and slip against each other as they move upward over the moldboard, causing granulation or crumbling of the furrow slice. As plowing speed increases, soil pulverization also increases, depending to a great extent on moldboard shape. Most of the granulation is done by the lower portion of the moldboard; the upper portion primarily turns the furrow slice.

LEVELNESS OF THE PLOW BOTTOM

Plow bottoms are designed to run level and exert uniform pressure on the furrow slice. If a plow bottom does

Fig. 5—Moldboard and Landside Attached to Frog

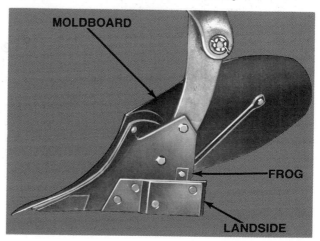

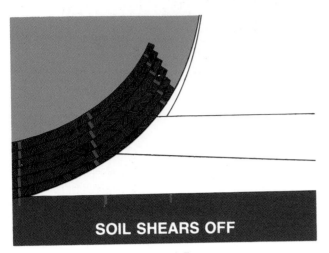

SOIL SHEARS OFF

Fig. 6—Soil Shears Off in Blocks as It Turns

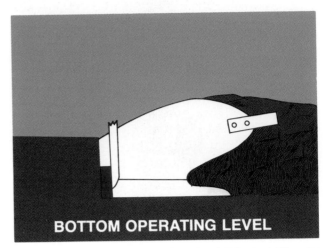

BOTTOM OPERATING LEVEL

Fig. 7—Furrow Slice Is Properly Turned and Pulverized When the Bottom Runs Level

Fig. 8—Pressure on Furrow Slice Is Released Too Quickly If Plow "Noses" Toward Unplowed Land

Fig. 9—Bottom Overcuts and Wastes Power If It "Wings" Toward Plowed Land

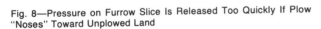

PLOW "NOSES" TOWARD UNPLOWED LAND

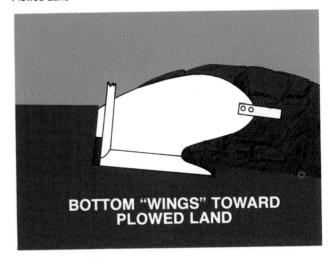

BOTTOM "WINGS" TOWARD PLOWED LAND

Fig. 10—Best Furrow Shape Is Formed at Correct Speed and Adjustment for the Bottoms Being Used

Fig. 11—Furrow Slope Is Too Steep From Plowing Too Slowly, Running the Plow on Its Nose, or Cutting Too Narrow

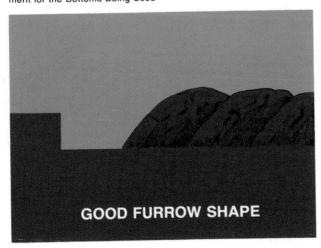

GOOD FURROW SHAPE

FURROW SLOPE TOO STEEP

54

FURROW SLOPE TOO FLAT

Fig. 12—Slope Is Too Flat from Excessive Speed, Winging Toward Plowed Land, or Cutting Too Wide

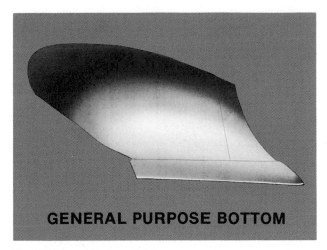

GENERAL PURPOSE BOTTOM

Fig. 13—General-Purpose Bottom

not run level, too much or too little pressure is exerted on the turning soil, resulting in poor plowing and poor granulation (Figs. 7, 8, and 9).

ANGLE OF THE FURROW SLICE

The angle or slope of the furrow slice is influenced by the speed of plowing, moldboard curvature, depth of plowing, and levelness of the plow. Figures 10, 11, and 12 show furrow slices in desired position, too steep, and too flat, respectively.

PLOW-BOTTOM DESIGN

Soil types and characteristics vary from light, abrasive sand to heavy, sticky clay. Depending on soil type, as well as climate and crops, soil moisture content when plowing may range from that of saturated rice paddies to extremely hard and dry soil, such as sun-baked sandy clay in irrigated desert areas. Too, soil physical condition may range from a loose, well-granulated structure to a hard, compact mass; this is affected by mineral, organic, and moisture content.

No single plow-bottom design can possibly do a satisfactory job in all soil conditions. For example, early plow bottoms were designed to be pulled slowly by animals, and didn't work well when pulled by tractors at higher speeds. Similarly, some current high-speed bottoms function poorly when operated at slower speeds. Draft varies widely between designs, and even between similar bottoms made by different manufacturers.

Hundreds of plow-bottom shapes have been made, each meant for a particular job. However, most present bottoms fall into six main types:

- **General purpose**
- **High speed**
- **Slatted**
- **Stubble**
- **Scotch**
- **Deep tillage**

Here's more about each type.

General-Purpose Bottoms (Fig. 13) have fairly slow-turning moldboards which work well in sod and for faster plowing of old land, stubble, ordinary trash, and stalk cover. These bottoms are usually designed for speeds of three to four miles an hour (4-6.5 Km/h).

High-Speed Bottoms (Fig. 14) have proved to be practical for plowing at four to seven miles an hour (6.5-11 Km/h). The moldboard has less curvature at the upper end than general-purpose bottoms. Reasonably priced throwaway shares can be discarded when worn out.

Slatted Moldboards (Fig. 15) scour better in difficult soil because about 50 percent of the moldboard has been removed. This concentrates the full pressure of the turning furrow slice on the remaining area, thus greatly improving scouring in waxy clay soils or loose, sticky soils.

Fig. 14—High-Speed Bottom

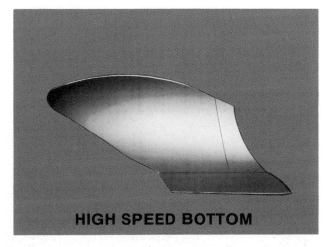

HIGH SPEED BOTTOM

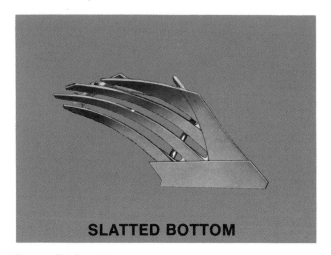

SLATTED BOTTOM

Fig. 15—Slat Bottom

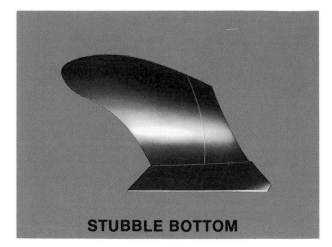

STUBBLE BOTTOM

Fig. 16—Stubble Bottom

Stubble Bottoms (Fig. 16) have an abruptly curved moldboard which turns the furrow slice quickly. These bottoms are used primarily for difficult scouring conditions such as stubble land. The sharply-turned moldboard thoroughly pulversizes soil, but doesn't lend itself to high-speed plowing. The limit is generally two and a half to three miles an hour (4-5 Km/h).

Scotch Bottoms (Fig. 17) are normally used in heavy, close-textured clay soils and tough sod. The long, curved moldboard sets each furrow slice on edge so it will catch rain and snow. The slice is not pulverized, but is exposed to air and weather. The narrow share on this bottom does not cut the full width of the furrow slice.

Deep-Tillage and **Semi-Deep-Tillage Bottoms** (Fig. 18) have high moldboards to permit plowing as deep as 16 inches (406 mm) in heavy soil. These bottoms are used most commonly in irrigated areas and in the south.

MOLDBOARDS

The curved moldboard and shin of the plow bottom receive soil from the share and lift, fracture, and turn the furrow slice. The sliding action of the soil generates a greal deal of heat and wear on the moldboard, and particularly on the shin, which therefore is replaceable on many bottoms.

Manufacturers try to use materials for moldboards best suited to the varying soils in which bottoms are intended to operate. Three basic moldboard materials are:

- **Off-center, soft-center steel**
- **Solid steel**
- **Chilled cast iron**

Let's look at these in detail.

Fig. 17—Scotch Bottom

Fig. 18—Semi-Deep-Tillage Bottom

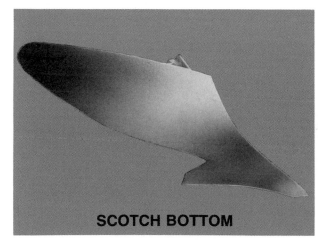

SCOTCH BOTTOM

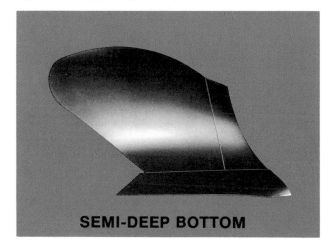

SEMI-DEEP BOTTOM

Off-Center, Soft-Center Steel

This is a highly polished, fine-textured steel for long wear and good scouring in sticky soil. The moldboard is made of three separate layers of steel which are hot-rolled together to form a single, tough sheet of metal (Fig. 19).

The outer layers are of different thicknesses, have a high carbon content, and are hardened by heat-treating. The thicker front layer provides good scouring and long wear. The inner layer, with low carbon content, provides a soft, ductile core to help absorb shocks.

Solid Steel

These moldboards are made of one sheet of steel specially treated to provide hard and soft layers similar to those of 3-ply steel. This steel has a slightly lower carbon content than the outer layers of soft-center moldboards, and is highly shock-resistant. However, these moldboards do not wear well in highly abrasive soils, and are used primarily where scouring is not a problem.

Chilled Cast Iron

Cast-iron bottoms were designed to withstand scratching and hard wear in sandy or gravelly soils. They have largely been replaced by more-modern types.

Teflon and other plastic coatings have been applied to moldboards to improve scouring in extremely difficult soils, particularly tropical and subtropical soils. Such coatings improve scouring and effectively reduce draft, but they are costly. They are not in widespread use, though considered worthwhile in certain problem areas.

Plow Shares

The share is the "business end" of the plow bottom. It gives suction and penetration and cuts the furrow slice loose. Some lifting and a slight turning action starts at the share, but little granulation takes place.

For many years plow bottoms had forged shares which could be resharpened when cutting edges and points wore dull. Now, however, many farmers find it economical to use throwaway or expendable shares which are simply discarded when worn. Several kinds of throwaway shares are manufactured to meet different plowing conditions (Fig. 20). Each type is usually available for several types of bottoms.

(Some plows are designed to permit tipping the points of the bottoms down to increase suction for better penetration in problem soils. However, this increases draft and is an expensive way of getting more life from worn shares, so it is recommended only where penetration is extremely difficult.)

Let's examine each type of share shown in Figure 20:

A. **Full-cut** shares extend the full width of the bottom and assure complete severing of heavy roots.

B. **Narrow-cut** shares penetrate better and pull easier than full-cut shares, and are best where roots

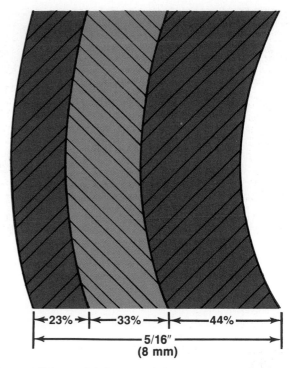

Fig. 19—Off-Center, Soft-Center 3-Ply Steel

are no problem. They also are recommended for rocky soil to reduce exposure to buried stones.

C. **Heavy-duty, deep-suck** shares are best in abrasive, rocky, or very hard soil. As with most throwaway shares, they are available in full- or narrow-cut.

D. **Hard-surfaced** shares are for extremely abrasive soils where regular shares wear out quickly.

E. **Extra-heavy-duty** shares are meant for extreme conditions.

F. **Gumbo** shares are for gumbo bottoms; the long point penetrates quickly and holds depth in sticky gumbo soils.

There also are various other shares, including chilled cast, for special, less-common conditions.

MOLDBOARD PLOW LANDSIDES

The landside is a flat metal piece bolted to the frog, and forms one side of the plow-bottom "wedge." It helps absorb side forces from the turning furrow slice, steadies the plow, and helps keep the plow straight behind the tractor. Varying plowing conditions and plow designs require different landsides (Figs. 21-24).

Steel landsides with reversible wear plates (Fig. 21) are used on high-speed bottoms in normal plowing conditions. The plate can be reversed for longer wear.

The long landside with adjustable heel is usually used on the rear bottom of integral moldboard plows (Fig.

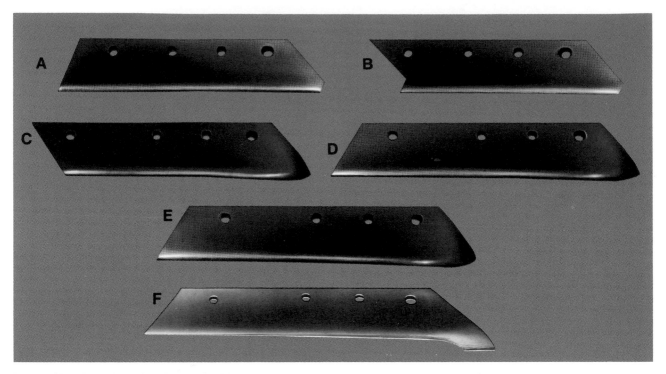

Fig. 20—Share Types for Various Plowing Conditions

22). The heel is adjusted vertically to mark the furrow bottom slightly and assist in controlling the rear of the plow.

An adjustable landside (Fig. 23) sometimes is used on the rear bottom of integral, semi-integral, drawn, and two-way plows. This landside reduces vertical movement of the plow, and helps maintain a smooth furrow bottom. The rear of the landside is adjustable vertically to contact the furrow bottom lightly to control plowing depth and steady the plow.

A rolling landside is used on the rear bottom of some integral plows (Fig. 24). Rear wheels of some drawn and semi-integral plows are used as rolling landsides. Rolling landsides on some plows may be adjusted to increase or decrease pressure on the furrow wall to help make the plow trail straight behind the tractor.

MOLDBOARD PLOW STANDARDS

Each bottom is bolted to a standard assembly, which in turn is attached to the plow frame. Some older plows

Fig. 21—Steel Landside with Reversible Wear Plate

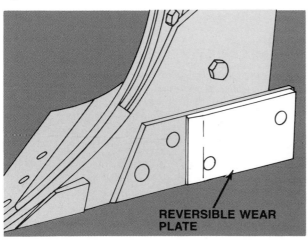

REVERSIBLE WEAR PLATE

Fig. 22—Long Landside with Adjustable Heel

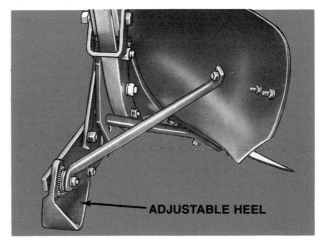

ADJUSTABLE HEEL

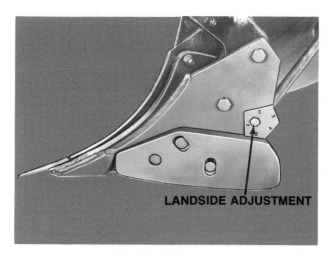

Fig. 23—Adjustable Landside

Fig. 24—Rolling Landside in Operation

(Fig. 25) had one-piece beams and standards, formed from a web or I-beam member, and bent and flattened to the desired shape. Others were rigidly assembled from rectangular bar stock and flat steel plates. No release mechanism was provided on these standards or beams to protect bottoms from obstructions in the soil, though a safety-release hitch was used on some drawn plows of this type to disconnect the plow from the tractor if a bottom hit an obstruction.

Release-type standards hold the bottoms in position on most current plows, and provide protection for bottoms and frame from damage by rocks and stumps in the soil. Design of standards varies greatly, but four basic types (Figs. 26-31) are commonly used.

- **Shear bolt**
- **Safety trip**
- **Hydraulic automatic reset**
- **Spring automatic reset**

Let's consider each type in detail:

Shear-bolt standards (Fig. 26) are economical protection where fields have very few buried obstructions. If a bottom strikes an obstacle the shear bolt fails and the bottom swings back. The operator must raise the plow, reset the standard by hand, and install a new shear bolt. (Note: use only shear bolts recommended by the plow manufacturer.) If bolts shear frequently for no apparent reason, check tightness of pivot bolt and bottom alignment. A loose pivot may cause excessive movement of the standard and premature shear-bolt failure.

When a bottom with a **safety-trip** standard (Figs. 27 & 31) strikes an obstruction, a release mechanism permits the bottom to trip back over the obstacle to prevent damage to the frame or the bottom. To reset the standard, raise the plow slightly, back the tractor until the standard locks in normal plowing position, then drive forward.

Fig. 25—One-Piece Beam and Standard Used on Old-Style Plows

Fig. 26—Shear-Bolt Standard

Fig. 27—Safety-Trip Standard

Most safety-trip standards may be adjusted for different tripping pressures (consult operator's manual), and provide simple protection for occasional obstructions.

Hydraulic, automatic-reset standards (Figs. 28-30) were designed for plowing in rocky fields. Bottoms quickly swing back and rise to clear obstructions, then automatically return to working position without interrupting forward travel.

Controlled hydraulic pressure holds the standard in working position, allows it to clear an obstruction, then returns it to working position. Hydraulic pressure may be provided either directly from the tractor remote-cylinder outlets or from a hydraulic accumulator system on the plow.

Tractors with sufficient remote-cylinder outlets and adequate pressure in a closed-center hydraulic system can operate automatic-reset bottoms directly. Manifold pressure is controlled by an adjustable valve and maintained by the tractor hydraulic system. When a bottom strikes an obstruction, pressure builds up in the manifold until it reaches a predetermined level and a relief valve opens. Oil is then returned directly to the

Fig. 28—Hydraulic-Reset Standard in Operation—Up, Over, and Back

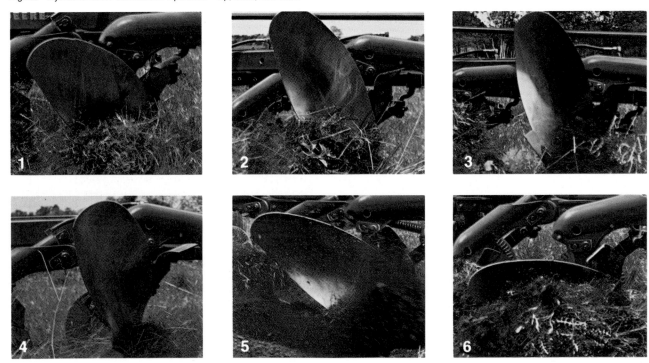

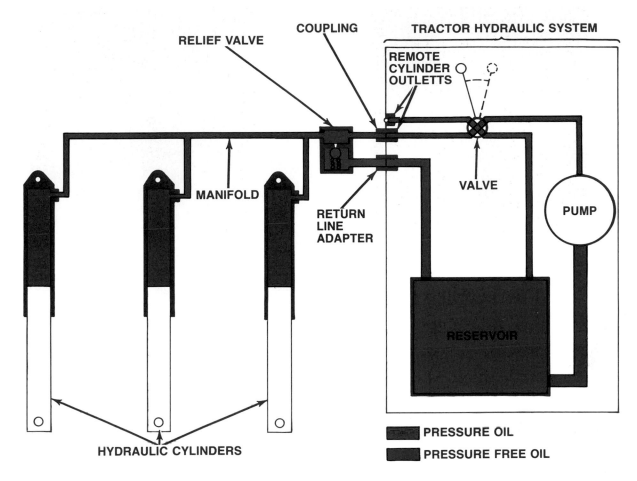

COUPLING

TRACTOR HYDRAULIC SYSTEM

RELIEF VALVE

REMOTE CYLINDER OUTLETTS

MANIFOLD

VALVE

PUMP

RETURN LINE ADAPTER

RESERVOIR

HYDRAULIC CYLINDERS

PRESSURE OIL

PRESSURE FREE OIL

Fig. 29—Schematic View of Tractor-Controlled Hydraulic Automatic-Reset Standard System

tractor hydraulic reservoir. This allows the cylinder to retract and the bottom to pass over the obstacle.

As soon as the obstruction is cleared, the relief valve closes and the tractor hydraulic system forces oil back into the manifold, the plow cylinder is extended, and the bottom returns to operating position (Fig. 29).

For tractors lacking sufficient remote outlets and a closed-center hydraulic system, or with insufficient hydraulic pressure, a hydraulic accumulator system controls the reset bottoms. A bladder-type accumulator is charged with an inert gas to maintain oil pressure in the plow manifold, instead of using direct pressure from the tractor system. After charging the accumulator with gas, the manifold is charged with hydraulic fluid to the proper pressure and disconnected from the tractor.

When a bottom strikes something solid, oil is forced into the accumulator, the gas is compressed, and the bottom rises over the obstacle. A one-way orifice allows free oil to flow into the accumulator, but restricts outward flow so the bottom returns to working position at a safe speed (Fig. 30). Specific instructions for charging the accumulator are given in the operator's manual.

Spring automatic-reset standards are replacing hydraulic-reset plows. The newer spring-reset plows also let you plow nonstop. Bottoms swing back over obstructions and then automatically return to working position (Figs. 32 and 33).

These standards are held in position by heavy-duty spring linkage which permits a bottom to pass over an obstacle. Parallel linkage also provides a vertical relief feature which allows the share to slide up and over sloping obstructions without swinging back.
Caution: automatic-reset bottoms operate fast and can injure anyone in their path of movement. Never allow anyone close to the bottoms or tripping mechanism while the plow is in operation.

MOLDBOARD PLOW ATTACHMENTS

Many attachments, each serving a specific purpose, are available to improve plow performance and the quality of plowing. Among them are:

- **Rolling coulters**
- **Disk coulters**
- **Moldboard extensions**
- **Root cutters**

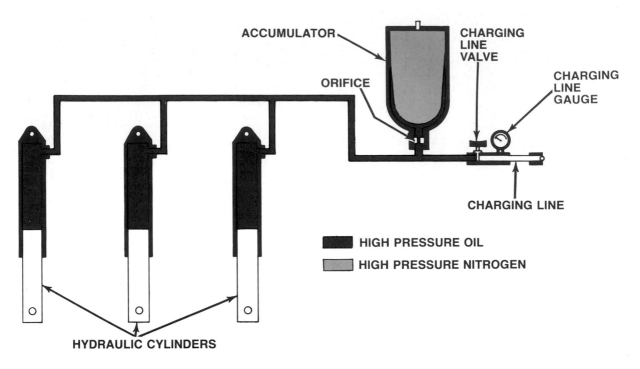

Fig. 30—Schematic View of Hydraulic Accumulator System

- **Trash boards**
- **Jointers**
- **Weed hooks**
- **Gauge wheels**

Let's discuss these attachments.

Rolling coulters improve plowing and help bury trash by:

1. Cutting trash into shorter lengths which are more-easily covered.

2. Cutting through trash which might otherwise drag on the shin or moldboard and plug the plow.

3. Cutting the furrow slice vertically to provide a clean, smooth furrow wall which reduces soil pressure and wear on the share and shin.

The three basic types of coulter blades (Fig. 34) are plain, notched or cutout, and rippled.

Plain coulters are recommended where ground penetration or scouring may cause problems. Blades are readily resharpened by grinding.

Notched or cutout coulters are used in heavy, loose trash where the notches will help pull the trash down for better cutting action. These blades are difficult to sharpen.

Fig. 31—Plow with Safety Trip Standards

SAFETY TRIP STANDARDS

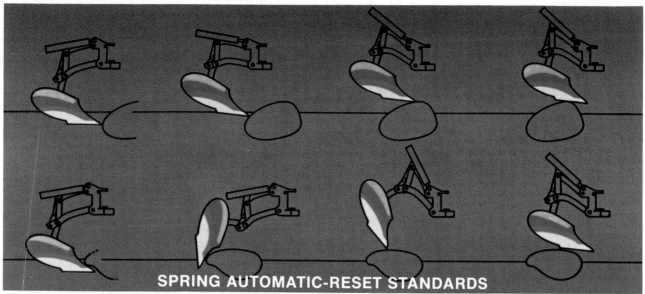

SPRING AUTOMATIC-RESET STANDARDS

Fig. 32—Spring Automatic-Reset Standard

Fig. 33—How Spring-Reset Standards Work

SPRING RESET STANDARDS

Fig. 34—A, Plain Coulter Blade; B, Notched Coulter Blade; C, Rippled Coulter Blade

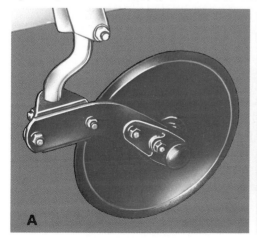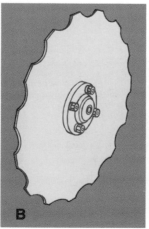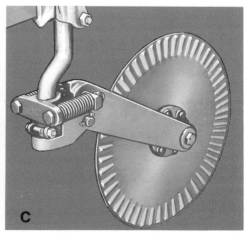

Fig. 35—Shear Bolt and Cushion Coulter Assembly

Ripple-edge coulters are very effective for cutting through trash because the ripple design "gears" the blade to the ground to keep it turning. Wear and cutting action tend to keep rippled coulters sharp. Penetration may be a problem in hard soil.

Coulter diameters range from 15 to 22 inches (381-559 mm), and must match the plow frame and operating conditions for optimum results. Small-diameter coulters penetrate better in hard soil, where large coulters tend to raise the plow out of the ground. On the other hand, large coulters are better for cutting heavy trash, and can be set deeper for cutting roots in heavy sod.

Coulters are mounted either on a single arm or on a yoke which supports both ends of the axle. They turn on cone-type bearings or anti-friction roller bearings. Single-arm mounting usually provides greater trash clearance around the coulter. Many coulter assemblies have shear-pin or spring-cushion (Fig. 35) blade protection from rocks or other obstructions.

Disk coulters have concave blades similar to a disk harrow (Fig. 35). These blades cut a ribbon of trash and soil from the edge of the furrow slice and turn it into the furrow bottom for better coverage. Disk-coulter assemblies may have either rigid or spring-cushion mountings, and are normally attached to the plow frame with the blade center about 4 inches (100 mm) ahead of the share point. Proper vertical positioning, blade angle, and pitch are very necessary for correct disk-coulter functioning.

Trash boards (Fig. 36) are mounted just above and at the leading edge of the plow moldboard, and deflect trash into the furrow bottom. When providing equal trash coverage to jointers, trash boards usually have less draft. Steel trash boards wear longer but may be harder to scour in sticky soils than plastic trash boards.

Removing trash boards permits some residue to be left on the surface to protect soil from wind and water erosion.

Jointers are shaped like miniature plow bottoms, and cut a small ribbon of soil just ahead of and above the share point. They are particularly effective in deflecting standing trash or heavy manure from the edge of the furrow slice into the furrow bottom for good coverage. However, plugging may be a problem in loose trash, such as cornstalks, where trash boards usually are more effective.

Standing jointers (Fig. 36) are mounted on the plow frame, but combination jointers are attached to and swing with coulters.

Moldboard extensions (Fig. 36), attached to the rear of the moldboard, provide increased turning action. They are especially helpful when plowing on hillsides

Fig. 36—Trashboard, Jointer, And Moldboard Extension

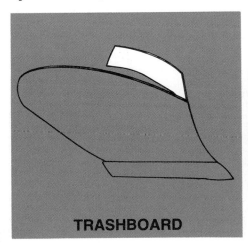

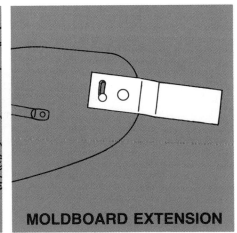

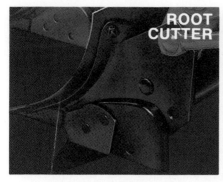

Fig. 37—Root Cutter, Weed Hook, and Gauge Wheel

or in heavy sod which turns over in ribbons. Adjustment is provided to permit applying extra pressure to the desired portion of the furrow slice for best turning.

Root cutters are small blades attached to the landside (Fig. 37). They provide complete cutting of the roots of alfalfa and similar plants which might slip around the end of the next plowshare. Suction in the blade aids plow penetration in gumbo-type soils.

Weed hooks (Fig. 37) bend weeds or standing trash down to improve coverage by the furrow slice.

Gauge wheels (Fig. 37) are not intended to carry the rear of the plow, but steady the plow and control plowing depth in varying soil conditions and on rolling land.

Gauge wheels are used most commonly on large integral and semi-integral plows, but may be used on smaller models in adverse conditions. They also may be desirable on drawn plows if there are extreme soil variations within the field.

MOLDBOARD PLOW FRAMES

The frame is the backbone of the moldboard plow. It holds the standards in position and they, in turn, support the bottoms.

As trash volume has increased with higher crop yields, so has the frame clearance of most plows. Modern backbone-style frame design provides large trash tunnels for free flow of crop residue to permit non-stop plowing. Wider bottoms, up to 20 inches (508 mm), and larger-diameter coulters also aid plowing in heavy trash.

Important frame dimensions are the vertical distance from share cutting edge to the lower side of the main backbone or frame members, and the fore-and-aft distance from share point to share point (Figs. 38 and 39). Another common measurement for trash clearance is the direct diagonal distance from share point to share point. However, on some plows, extremely wide standards may reduce diagonal clearance and restrict heavy trash flow.

Increasing fore-and-aft clearance requires greater plow length, but this does not increase plow draft. Longer hitches permit shorter turning of semi-integral and

drawn plows, and a flatter line of draft for more-uniform plowing depth—but, again, do not increase draft. However, any increase in integral-plow length requires more tractor front-end weight, because the center of plow weight and soil load are shifted farther to the rear during lift and transport.

Three basic plow-frame types are available:

- **Fixed frame**
- **Combination**
- **Adjustable**

Fixed-frame plows (Fig. 40) have simple construction and provide maximum strength at least cost. However, each size of plow and width of bottom requires a specific frame; for example, a 6-bottom 16-inch (406 mm) plow can only be used with six 16-inch (406 mm) bottoms.

Combination-frame plows permit conversion to different sizes. For instance, a basic 4-bottom plow may be converted to 5- or 6-bottom by adding frame extensions and bottoms. This permits matching of plow size to tractor power and traction and draft in differing soil types and conditions. Frame spacing on some models also is convertible, for instance from 14- or 16-inch to 18-inch (356 or 406 to 457 mm) bottoms.

Fig. 38—Vertical Plow Clearance

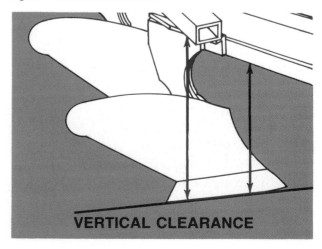

VERTICAL CLEARANCE

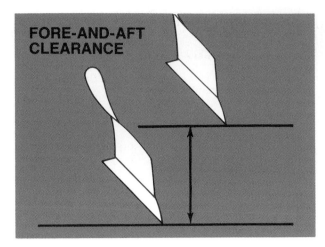

Fig. 39—Fore-and-aft Plow Clearance

Fig. 41—Adjustable-Frame Plows

Adjustable-frame plows (Fig. 41) permit varying the width of cut of each bottom to match tractor power and traction to field conditions. Some plows are adjusted manually by moving standards or shims and others are remotely controlled from the tractor seat by the operator actuating a remote hydraulic cylinder. Plowing productivity is increased by adjusting plowing width as desired on hillsides, contours, and in finishing lands and fields.

Reducing width of cut of adjustable-frame plows reduces soil inversion and leaves more residue exposed to help control erosion. Increasing width of cut leaves larger slabs and more open spaces to catch and hold moisture.

Fig. 40—Fixed-Frame Plow

CHOOSING A MOLDBOARD PLOW

Picking the right plow is more than selecting a basic frame or buying the largest plow a particular tractor can pull. It requires making a choice from integral, semi-integral, and drawn types, and in some areas considering the respective merits of one-way or two-way plowing. Plow types include:

- **Integral**
- **Semi-integral**
- **Drawn**
- **Reversible or two-way**

The following comparison of plow features, and later detailed discussion of the various types, will aid in making a suitable selection.

INTEGRAL PLOWS

Integral plows (Fig. 42) are attached to the tractor 3-point hitch or implement quick-coupler, and the entire plow is carried by the tractor during transport. The tractor lower draft links are attached to the plow-hitch crossbar and the upper link to the plow mast. Use of a quick-coupler permits hitching by backing to the plow, lifting, and then latching the coupler automatically or manually, depending on the design.

Integral plows are limited in size, due to tractor hydraulic-lift capacity and front-end stability, though models are available with up to six bottoms. The tractor-plow combination has excellent maneuverability for

COMPARISON OF PLOW TYPES							
Plow type	Maneuver- ability	Uneven terrain, contours	Uniform depth	Weight transfer	Tractor stability	Operator skill required	Cost per bottom
Integral	Best	Good	Least	Best	Least	Most	Least
Semi- integral	Good	Best	Good	Good	Good	Medium	Medium
Drawn	Least	Good	Best	Least	Best	Least	Most

short, quick turns at field ends, and easy backing into corners or plowing small, irregular fields. Integral plows require no transport wheels or axles, so they cost less than semi-integral or drawn models.

SEMI-INTEGRAL PLOWS

Semi-integral plows (Fig. 43) are attached to hitch links of tractors with lower-link draft sensing. On double-pivot hitch plows, the upper hitch link is used to stabilize the plow on turns. The upper hitch link is also required with an implement quick-coupler or if the tractor has top-link draft sensing. For top-link sensing, an A-frame (Fig. 44) is added which places the plow hitch pins below the tractor draft links. This provides leverage on the A-frame for proper activation of the tractor weight-transfer system.

Fig. 42—Integral Moldboard Plow

INTEGRAL MOLDBOARD PLOW

SEMI-INTEGRAL MOLDBOARD PLOW

Fig. 43—Semi-Integral Moldboard Plow

Fig. 44—A-Frame Coupler

A-FRAME COUPLER

The front end of a semi-integral plow is carried and controlled by the tractor linkage, and the rear of the plow rides on a furrow transport wheel. This rear wheel is steered by a rod from the front hitch point or through a closed circuit hydraulic system between the plow hitch crossbar and the tailwheel. When the tractor turns, oil is forced out of one end of a hydraulic cylinder mounted on the crossbar and extends the tailwheel cylinder and steers the plow. This lets the plow follow the tractor closely on turns.

A remote hydraulic cylinder raises and lowers the rear of semi-integral plows. This independent raising and lowering of front and rear permits more-uniform headlands, particularly with larger plows.

Fig. 45—Semi-Integral Plow with In-Furrow Hitch

IN-FURROW HITCH

Fig. 46—Semi-Integral Plow with On-Land Hitch

Semi-integral design permits longer plows, with more bottoms and better trash clearance between bottoms, than is possible with fully integral models. The tractor-plow combination is highly maneuverable when transporting, and also functions well on contours. Semi-integral plows may be operated with the right rear tractor wheel in the furrow (Fig. 45), or with an on-land hitch (Fig. 46) to permit use of dual rear tractor wheels and reduce compaction in the furrow.

Semi-integral plows with an on-land hitch have a front furrow wheel for transport and depth control of the front bottoms. Not much weight transfer is obtained with an on-land hitch, because the tractor carries only one front corner of the plow, which has little weight compared to tractor size, total plow weight, and draft.

With an on-land hitch, the edge of the tractor tire should be kept at least 4 inches (100 mm) from the furrow edge to prevent crumbling of the furrow wall. Clods and loose dirt knocked into the furrow may cause bouncing of the front furrow wheel and uneven plowing.

In-furrow hitching is normally recommended for 5-bottom and smaller semi-integral plows, due to difficulty in obtaining proper wheel tread and hitch settings for on-land operation.

DRAWN PLOWS

The drawn or pull-type plow (Fig. 47) is a complete unit in itself attached to the tractor drawbar. Most current models are raised for transport or lowered for plowing by remote hydraulic cylinders. Older models have a clutch-type lift on one wheel, activated by the tractor driver with a trip-rope.

Drawn plows have front and rear furrow wheels and one land wheel, which transport the plow and provide accurate control of plowing depth.

The rear wheel usually is held rigid during plowing, but is allowed to caster for easier turning when the plow is raised. However, many larger plows have steerable front and rear furrow wheels (similar to semi-integral plows with on-land hitch) for better maneuverability and sharper turning.

On most large drawn plows, the hitch may be adjusted or interchanged to permit on-land or in-furrow operation of the right rear tractor wheel.

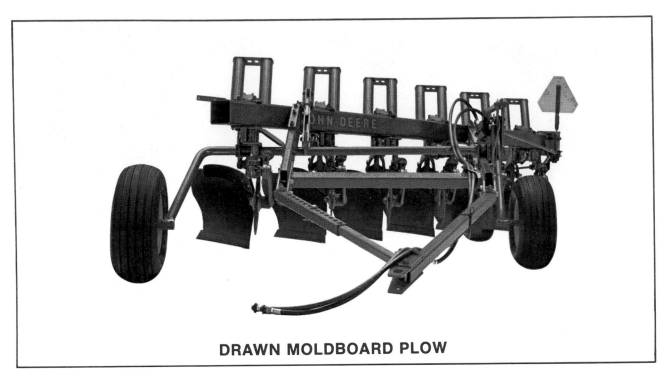

DRAWN MOLDBOARD PLOW

Fig. 47—Drawn Moldboard Plow

Fig. 48—Two Drawn Plows with Tandem Hitch

TANDEM HITCH

Pulling two plows with a tandem hitch helps utilize full power of large wheel or crawler tractors (Fig. 48). Two smaller plows also provide greater flexibility for plowing uneven ground, compared to one very large plow. When two plows are tandem-hitched, all tractor wheels operate on land.

Some newer large plows have a flexible frame instead of using tandem plows to permit uniform plowing over uneven surfaces.

REVERSIBLE OR TWO-WAY PLOWS

Two-way plows have both right- and left-hand bottoms, which are alternated at each end of the field (Fig. 49), so all furrows are turned in the same direction. They are used primarily in irrigated land where dead furrows and back furrows would impede water flow. They are also used to reduce travel time when plowing point rows in contoured fields, and to throw all furrows either uphill (preferred) or downhill, as desired.

Integral Two-Way Plow

Integral two-way plows are attached to the tractor 3-point hitch and are fully carried by the tractor in transport. They are generally limited in size, due to tractor hydraulic-lift limitations and front-end stability. Integral mounting provides excellent maneuverability for transport and short, quick turns at row ends.

Most integral two-way plows are indexed (reversed) at each end of the field (Fig. 50), but some are indexed automatically by mechanical linkage when the plow is raised to transport position (Fig. 51).

Fig. 49—Integral Two-Way Moldboard Plow

Fig. 50—Two-Way Plows Are Rotated at Each End of the Field

Fig. 51—Most Two-Way Plows Are Indexed Hydraulically

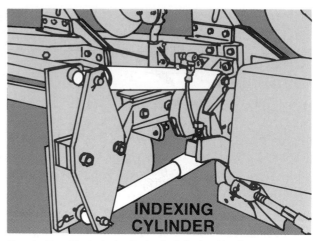

Fig. 53—Two Single-Acting Hydraulic Cylinders May Be Used to Index Semi-Integral Two-Ways

Integral two-way plows may be operated with one tractor wheel in the furrow or with both tractor wheels on land (Fig. 52).

Semi-Integral Two-Way Plow

Compared to integral two-way plows, semi-integral mounting permits use of longer plows with more bottoms and greater trash clearance.

Semi-integral two-ways are attached to lower hitch links of tractors with bottom-link draft sensing. The top link is not used. Tractors with top-link sensing require a special A-frame coupler on the hitch, which places

the plow-hitch crossbar several inches below the tractor lower links. This provides leverage for actuation of the top-link draft-control system.

The front of semi-integral two-ways is supported by the tractor hitch, and the rear is carried on a transport wheel controlled by a steering rod so the plow will follow the tractor closely on turns. This provides maneuverability and good trailing characteristics for plowing contours and for transport.

Two single-acting (Fig. 53) or one double-acting hydraulic cylinder (depending on design) are used to index semi-integral two-way gangs of bottoms. These

Fig. 52—On-Land Plowing with Integral Two-Way Plow

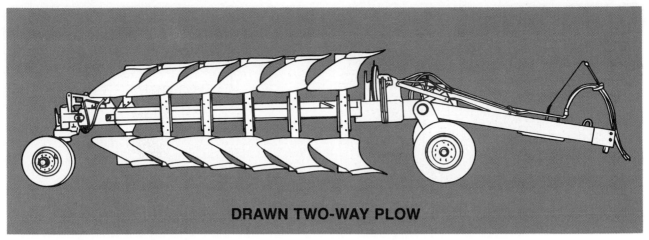

DRAWN TWO-WAY PLOW

Fig. 54—Heavy-Duty Drawn Two-Way Plow

plows may be operated with one tractor wheel in the furrow, or with both tractor wheels on land.

Drawn Two-Way Plow

Heavy-duty drawn two-way plows (Fig. 54) have size and strength to match the power of large 4-wheel-drive and crawler tractors, and in some cases have bottoms capable of plowing 20 to 24 inches (508-610 mm) deep.

HITCHING—IN FURROW OR ON LAND?

The operator must consider several factors before deciding whether to use in-furrow or on-land hitching. Here are the advantages of each method.

ADVANTAGES OF IN-FURROW HITCHING

1. Draft is reduced because soil pressure against the landside is reduced.

2. Driving is easier; the furrow wall serves as a guide, particularly for less-experienced operators.

3. It's easier to maintain uniform width of cut. Tractor wheel tread largely controls width of cut of the front bottom, so plowing is more uniform and level.

4. Weight transfer from integral and semi-integral plows is better. The front of the plow is controlled by the tractor, so a change in plow draft automatically results in weight transfer from the plow and tractor front wheels to the tractor drive wheels for more traction. Traction can be increased even more by maintaining proper vertical adjustment of the hitch, depending on tractor model and soil conditions. However, the on-land tractor wheel must be adequately weighted to prevent excessive slippage.

ADVANTAGES OF ON-LAND HITCHING

1. Reduces soil compaction. Taking the tractor wheel out of the furrow prevents compacting the furrow bottom and mashing the furrow wall and some plowed soil with oversized tires.

2. Improves tractor pull. Equalizing weight on tractor drive wheels through level operation provides equal traction for both wheels and reduces slippage.

3. Permits use of dual wheels. In some conditions duals provide better traction and flotation and less compaction than single wheels.

4. Improves operator comfort, for keeping all wheels on-land provides a more-level sitting position. This may be somewhat offset by the need for more-accurate driving to maintain proper width of cut of the front bottom. (A guide rod attached to the front of the tractor and extending to the furrow edge will help, especially for inexperienced operators.)

5. Increases tire life. With all tractor wheels on land, sidewall wear and scuffing is reduced. There's less slippage for tires that have equal footing and weight.

PRINCIPLES OF MOLDBOARD-PLOW OPERATION

Principles of plow hitching and operation are the same, regardless of plow type.

Understanding the "hows" and "whys" of correct plow operation makes it easier to get top plow performance, and to recognize and correct problems.

Principles to be discussed include:

- **Center of load**
- **Center of pull**
- **Line of draft**

Let's examine each of these.

CENTER OF LOAD

The theoretical center of load for a single plow bottom is assumed to be located one-quarter of the bottom width from the landside, and vertically at half the plowing depth. Thus the center of load for one 16-inch (41 cm) bottom, plowing 8 inches (20 cm) deep, is 4 inches (10 cm) from the landside and 4 inches (10 cm) above the cutting edge of the share.

Center of load for a multiple-bottom plow is found by measuring one-half the total width of cut plus one-fourth of the width of one bottom from the plowed-land furrow wall. Thus on a 5-bottom 16-inch (406 mm) plow the center of load is approximately 44 inches (1,118 mm) to the left of the furrow wall.

$$\frac{5 \times 16}{2} + \frac{16}{4} = 44 \text{ inches } (\frac{5 \times 406}{2} + \frac{406}{4} = 1,118 \text{ mm})$$

However, actual location of the center of load is affected by soil conditions, bottom type, and plowing speed. Plow and hitch adjustments have been designed to permit compensation for these variables.

CENTER OF PULL

Tractor horizontal center of pull is on the tractor center-line just ahead of the rear axle, and is vertically positioned according to tractor drawbar or the convergence of top and lower hitch links. With integral plows, the center of pull is at the convergence of the links (Fig. 55). Horizontal center of pull with semi-integral plows is where the two lower links converge and for drawn plows is at the front pivot of the tractor drawbar.

LINE OF DRAFT

For optimum tractor-plow performance, the center of load on the plow, the hitch point on the tractor, and the tractor's center of pull must all fall on a straight line of draft. This results in minimum side-draft on the tractor, and reduces wheel slippage, fuel consumption, and plow draft. It also reduces wear on plow-wheel bearings and landsides.

To perform properly, the plow must trail straight behind the tractor, and the plow frame must be parallel to the ground-line both fore-and-aft and laterally. Proper width of cut on the front bottom is determined by proper driving, tractor wheel tread, and hitch settings.

On small or very-large plows, it may be impossible or impractical to adjust tractor wheel-tread for the ideal line of draft parallel to the direction of travel. So a compromise is required and the hitch must be adjusted to match wheel settings to make the line of draft as nearly parallel as possible to the line of travel.

Line-of-draft setting for integral plows is established by plow designers. Tractor wheel-tread must be set according to the manufacturer's recommendations to achieve proper width of cut of the front bottom. In many cases designers have considered such factors as common spacings for row crops, and have designed plow hitches to reduce or eliminate need for changing tread setting between plowing and cultivation.

On semi-integral plows, proper lateral adjustment of hitch assembly and hitch crossbar places the hitch pivot (Figs. 56-57) on the line of draft between center of load and center of pull for in-furrow as well as on-land hitches.

Similarly, to reduce side-draft and improve performance with drawn plows, the hitch pin (Figs. 58-59) must be as nearly as possible on the line of draft, which passes through the center of pull and the center of load. Instructions for locating hitch points for various wheel-tread settings and either semi-integral or drawn plows are given in the plow manual.

Accuracy of these settings may be checked by stretching a string from the center of load to the center of pull, and verifying that the hitch point falls on this line of draft.

Normal soil forces tend to rotate a plow clockwise (looking down from above). Therefore, if the hitch cannot match the true line of draft, setting the tractor center of pull to the right (toward the furrow) of the plow center

Fig. 55—Center of Pull with Integral Plow

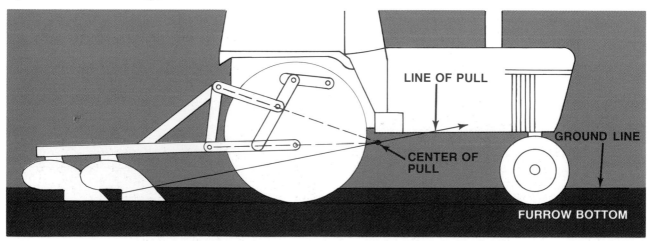

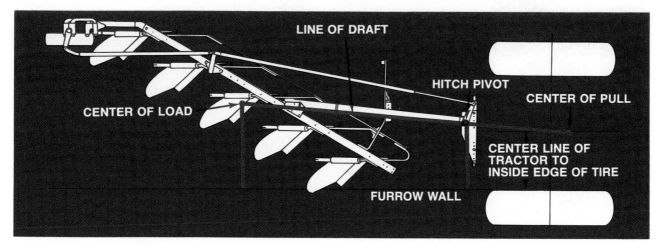

Fig. 56—Line-Of-Draft Hitching for In-Furrow Operation of Semi- Integral Plow

Fig. 57—Line-Of-Draft Hitching for On-Land Operation of Semi-Integral Plow

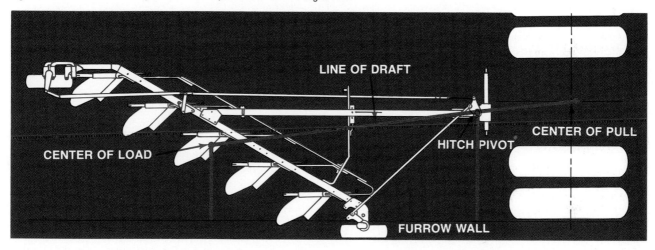

Fig. 58—Line-Of-Draft Hitching for In-Furrow Operation of Drawn Plow

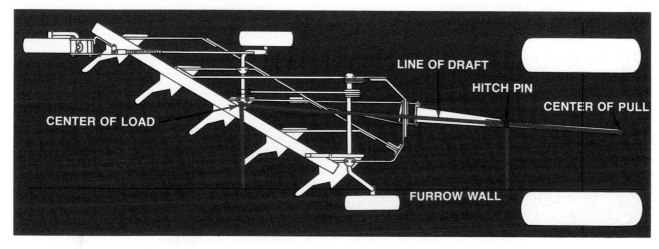

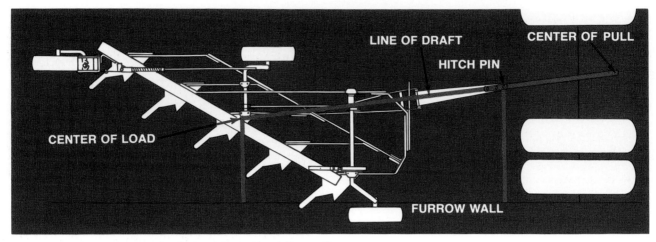

Fig. 59—Line-Of-Draft Hitching for On-Land Operation of Drawn Plow

of load will help offset the rotational tendency and reduce lateral pressure on tail wheel and landside.

BASIC MOLDBOARD-PLOW ADJUSTMENTS

Basic principles of moldboard-plow adjustments are similar regardless of plow make. Here, adjustments of one-way plows will be discussed; later, the differences of two-way plows will be described. Specific details and initial settings depend on design and are explained in the manufacturer's operator's manual for each plow.

Typical scope of adjustments are:

- **Vertical hitch**
- **Leveling**
- **Width of cut**

Fig. 60—Adjusting Length of Tractor Lift Links for Integral Plows

- **Rolling coulters**
- **Disk coulters**
- **Trash boards**
- **Jointers**
- **Gauge wheels**

Here's more about each of these.

VERTICAL HITCH ADJUSTMENT

Vertical hitch adjustment of integral plows is simpler than adjustment of semi-integral and drawn plows. Here are the differences.

Integral Plows

Vertical hitch alignment of integral plows is built into basic plow design. Therefore, the main factor to consider on most such plows is adjusting length of the tractor-hitch lift links (Fig. 60), according to tractor and plow operating instructions. Fore-and-aft leveling also is needed, and will be discussed later. A few integral plows are designed to permit mounting the hitch cross-shaft either above or below the main plow frame to compensate for variations in tractor hitch height, tire size, and plowing depth.

Semi-Integral and Drawn Plows

Holes in the hitch-plate assembly of most semi-integral and drawn plows permit vertical hitch adjustment to match soil conditions and tractor hitch dimensions. (Fig. 61). Proper vertical hitch adjustment results in more uniform plowing depth, smoother furrow bottoms, and improved scouring.

If the vertical hitch point is too high, the front of the plow is pulled down and the plow "runs on its nose." Shares have more tendency to dig in on striking an obstacle, and the plow may "bob" in operation. Running with the hitch too high also increases plow draft, and at the same time reduces weight transfer because the tractor keeps trying to pull the plow down, instead of up. This also reduces weight on the rear furrow wheel and throws extra weight on plow front wheels, causing needless plow wear and strain.

On the other hand, if the vertical hitch point is too low the front of the plow is pulled up, which may cause slow or difficult penetration in hard soils. Extra weight is shifted to the rear furrow wheel, making it difficult to level the plow in operation.

Variations from true line-of-draft hitching will be dictated by specific plowing conditions. Realizing the problems and consequences of improper hitching, and taking corrective action, will save time, fuel, and wear and tear on equipment, and result in a better plowing job.

LEVELING THE PLOW

Keeping a plow running level maintains even moldboard pressure on the furrow slice, uniform plowing depth, and furrow configuration for all bottoms. Width-of-cut of the front bottom is affected by running the front bottom too deep or too shallow. *Therefore, the plow must be leveled before width-of-cut adjustments are made.*

Measuring furrow depth immediately behind front and rear bottoms (Fig. 62) tells if both are cutting equal depth. Next, sighting across the plow frame from both side and rear (Fig. 63) tells if the plow is operating parallel to the ground line.

Leveling An Integral Plow

Integral plows are leveled fore-and-aft by adjusting length of the tractor's top hitch link (Fig. 64). Shortening the link raises the rear of the plow, while lengthening the link allows the rear of the plow to run deeper.

Tipping the plow forward can increase suction and improve penetration in hard soil, but is not a cure-all for worn shares. To avoid penetration problems never allow the rear of the plow to run lower than the front.

Fig. 61—Vertical Hitch Adjustment

Fig. 62—Measuring Furrow Depth

Fig. 63—Plow Level Fore-and-aft

Fig. 64—Adjust Top Link

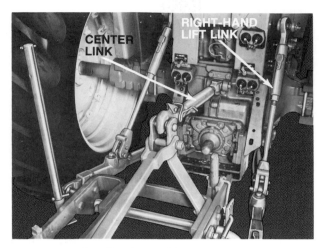

Fig. 65—Lateral Leveling of Integral Plow

Fig. 66—Fore-and-aft Leveling of Semi-Integral Plow

Fig. 67—Adjusting Rear Furrow-Wheel Stop

Lengthening or shortening the tractor right-hand lift link controls side-to-side (lateral) leveling of integral plows (Fig. 65). Improper leveling will cause the plow to "nose" on share points toward unplowed land, or "wing" toward plowed ground, causing uneven plowing.

Leveling A Semi-Integral Plow

Fore-and-aft leveling of semi-integral plows is controlled by adjusting the rear furrow-wheel stop (Figs. 66-67). Only enough weight should be carried on the rear landside to leave a clear mark in the furrow bottom. All remaining weight is carried by the rear wheel.

Plowing depth and lateral leveling of semi-integral plows with in-furrow hitch are controlled by the lift links of the tractor 3-point hitch. Shortening the right-hand link (Fig. 68) tips share points down; lengthening this link raises the points.

When the plow is leveled fore-and-aft by adjusting the rear-wheel stop, it will automatically seek a new level operating position following any change in rockshaft control setting.

For on-land plowing with semi-integral plows, adjust the tractor 3-point hitch for the desired plowing depth and level the front furrow wheel by adjusting the remote cylinder stop or (depending on plow design) the depth stop on the wheel. Rear of the plow is leveled in the same way as for in-furrow operation.

Leveling A Drawn Plow

Drawn plows are leveled laterally by adjusting the front furrow-wheel linkage with devices such as the "stop" screw (Fig. 69), and then adjusting the hydraulic remote-cylinder depth stop for desired plowing depth. The front furrow wheel and front bottom must run at the same depth.

For drawn plows with steerable front and rear furrow wheels, adjust the depth stop on the remote cylinder controlling each wheel.

After leveling the drawn plow laterally, check fore-and-aft level and adjust the stop for the rear furrow wheel as needed (Fig. 70). Allow enough weight on the landside to mark the furrow bottom, but leave most of the weight on the furrow wheel.

ADJUSTING WIDTH OF CUT

Each plow bottom must cut the same width and depth as the others to provide uniform plowing. To assure this, wheel tread, hitch adjustment, and coulter spacing must be correct.

Integral plows generally lack means for significant lateral hitch adjustment, so tractor wheel tread must be set within rather close limits. Plow manuals show proper tread setting, normally expressed in inches from center of tractor to inside of rear tires. This method avoids problems with variations in tire size and tractor models.

Tractor wheel tread is usually expressed in the same way for in-furrow operation of semi-integral and drawn plows. However, these plows normally have built-in

LATERAL LEVELING

Fig. 68—Lateral Leveling of Semi-Integral Plow (Tractor Shown With Implement Quick-Coupler)

FORE-AND-AFT LEVELING

ADJUSTABLE COLLARS

Fig. 70—Fore-and-aft Leveling of Drawn Plow

lateral hitch adjustments, which allow more leeway in tread setting compared to integral plows.

In some soil conditions, integral and semi-integral plows may trail improperly or not cut the correct width with the front bottom, even with plow and tractor correctly adjusted. This may be overcome by landing adjustment of the plow, as will be explained. However, the landing adjustment should not be used to compensate for improper wheel-tread or hitch adjustments. It should only be the last resort to obtain proper plow performance, and should always be returned to center posi-

tion before making initial plow adjustments when changing tractors or meeting changed soil conditions.

Integral Plows—Width of Cut

Width of cut of the front bottoms of integral plows is adjusted by changing the angle (landing) of the offset hitch crossbar (Fig. 71). If the front bottom overcuts with wheel tread properly adjusted, rotate the crossbar counterclockwise (viewed from the right side of the plow). This angles the whole plow toward the furrow and reduces cut of the front bottom.

Conversely, rotating the crossbar clockwise leads the plow toward the land and increases width of cut. An optional hydraulic landing attachment (Fig. 72) is available for many integral plows to control width of cut of the front bottom for hillside and contour work. This hydraulic landing can vary front-bottom cut up to 4 inches (100 mm) to either side of the true width of cut. Crank and lever-type landing attachments are available for some integral plows.

Fig. 69—Lateral Leveling of Drawn Plow

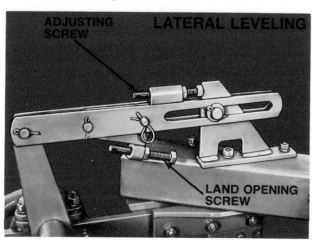

LATERAL LEVELING

ADJUSTING SCREW

LAND OPENING SCREW

Fig. 71—Landing Adjustment for Integral Plow

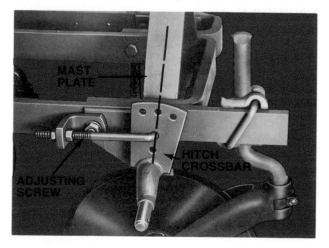

MAST PLATE

ADJUSTING SCREW

HITCH CROSSBAR

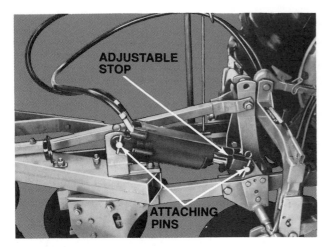

Fig. 72—Hydraulic Landing of Integral Plow

Fig. 73—Horizontal Adjustment for Semi-Integral Plow

HORIZONTAL ADJUSTMENT

Fig. 74—Hydraulic Landing Attachment for Semi-Integral Plow

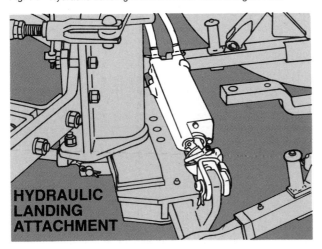

HYDRAULIC LANDING ATTACHMENT

Semi-Integral Plows—Width of Cut

Lateral movement of the hitch assembly on the front frame of semi-integral plows, plus lateral adjustment of the hitch crossbar, provide correct width of cut with various tractor-wheel settings (Fig. 73). Hydraulic or screw-type landing attachments are available for many semi-integral plows (Fig. 74).

Drawn Plows—Width of Cut

As to drawn plows, after setting the tractor wheel tread locate the proper line of draft between tractor center of pull and plow center of load. If the line passes through the tractor drawbar and plow hitch points, proper width of cut for the front bottom will be achieved. Minor changes in hitch-attaching points on the plow or tractor drawbar can be used to compensate for soil variations.

ADJUSTING ROLLING COULTERS

Rolling-coulter brackets permit vertical and lateral coulter adjustment. Set the rear coulter first, to obtain a well-defined furrow wall; then set others so each bottom is cutting the same width.

Set all coulters the same distance from shins of the moldboards to provide equal soil volume in each furrow slice. For instance, if the coulter on one bottom cuts a half-inch too wide, that bottom will turn more soil than the following bottom. Unevenly adjusted coulters cause:

• *Ridges and valleys in the plowed surface*
• *Rough riding in later tillage operations*
• *Need for more leveling*
• *An unsightly plowing job*

For average conditions set the center of the coulter blade approximately 2 to 4 inches (50 to 100 mm) ahead of the share point (adjust all coulters to same setting) and ½- to ⅝-inch (12 to 16 mm) to the left of the share point (Figs. 75-76).

Coulters must never undercut the bottom width. In hard ground, move coulters rearward to reduce their tendency to lift the plow. For trashy conditions, shift coulters forward to provide more clearance for cutting and clearing trash to reduce plugging.

Set coulters deep enough to cut trash, about 3 inches (75 mm) in ordinary conditions. If they are too deep, trash will be pushed ahead of the blades instead of being cut. For heavy sod, set coulters just deep enough to cut roots. If the plow tends to ride out of the ground in extremely hard soil, it may be necessary to raise the coulters.

ADJUSTING DISK COULTERS

Set disk-coulter blades just deep enough to cut through trash, about 2¼ inches (57 mm) in most conditions, and with proper angle to turn the desired amount of soil and trash into the furrow. Too-much angle in firm soil increases plow side-draft, but a sharper angle may be needed to keep blades turning in loose or sandy soil.

Fig. 77—Adjusting Gauge Wheel

Fig. 75—Fore-and-aft Coulter Position

ADJUSTING JOINTERS

Jointers should be set to run from 2 to 2½ inches (50 to 64 mm) deep, about ⅝-inch (16 mm) to the land from the furrow wall, and about ⅛-inch (3 mm) from the coulter blade. Setting jointers too deep prevents proper trash coverage and keeps the furrow slice from turning and laying properly. Normal jointer pitch is about 45 degrees, but may be increased to improve scouring if necessary.

ADJUSTING TRASH BOARDS

Positioning trash boards too low increases draft with little improvement in trash coverage. The front lower point of the trash board must always be in contact with the top edge of the shin or moldboard to prevent buildup of trash and dirt. Trash boards are normally used in conjunction with rolling coulters for optimum trash handling.

ADJUSTING GAUGE WHEELS

Gauge wheels (Fig. 77) are normally set to run on the land adjacent to the rear bottom, and are adjusted for average desired plowing depth. To maintain proper tractor load-and-depth system operation, gauge wheels should not carry appreciable weight most of the time.

Fig. 76—Recommended Horizontal Coulter Position

POWER REQUIRED FOR PLOWING

To realize the magnitude of work involved in moldboard plowing, and the necessity for proper plow adjustment and operation, consider the soil moved per acre (hectare). Assuming average soil density of 60 pounds per cubic foot (960 Kg/m³) and 6-inch (152 mm) plowing depth, 1,306,800 pounds of soil will be moved per acre (1,464,900 Kg/ha). That is 653.4 tons (1465 tonnes/ha).

Viewed another way, on the average a 14-inch (356 mm) bottom travels approximately 7 miles to plow one acre (27.8 Km/ha), or a 7-bottom 14-inch (356 mm) plow covers one acre for each mile traveled. (A 5-bottom 410 mm plow covers one hectare for each 5 kilometers traveled.)

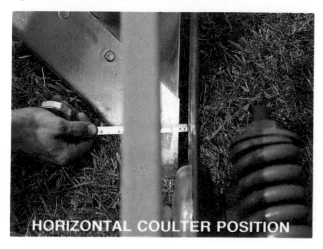

(To determine the miles traveled per acre in plowing, or with any other implement, simply divide 99 by the operating width in inches. To determine kilometers traveled per hectare, divide 10 by the operating width in meters.)

Moldboard-plow draft ranges from about 3 to 20 pounds per square inch (2 to 14 N/cm^2) of furrow cross-section. Thus a 16-inch (40 cm) bottom plowing 8 inches (20 cm) deep has (8 × 16) a 128-square-inch (800 cm^2) furrow section. In a moist clay loam with 7 pounds resistance per square inch (5 N/cm^2), draft of such a bottom would be (7 × 128) 896 pounds (4000 N). Eight bottoms would have a total draft of (8 × 896) 7,168 pounds (32,000 N).

As plowing speed increases, draft also goes up, but usually at a higher rate of speed. This poses a problem with a given tractor in choosing an economical balance between using a small plow at high speed or pulling a larger plow more slowly. However, most current tractors are built with less weight per horsepower for higher-speed operation.

It is normally best to plow as shallow as possible and still meet desired objectives of covering trash, aerating the soil, and providing suitable seedbed and rootbed (Fig. 78). In most areas only minor yield differences have been found to favor deep plowing. This is particularly true where soil freezes below plowing depth during the winter. Shallow plowing also saves time and fuel. Horsepower-hours per acre needed for plowing may range from 6.6 to 24.2 (12.2 to 44.6 KW-h/ha), according to a North Dakota study. Soils ranged from silt to compacted clay loam which had been irrigated.

High horsepower required for moldboard plowing has recently caused some farmers to seek other tillage tools for seedbed preparation. However, use of other implements may require more chemicals for control of weeds, insects, and diseases. Generally speaking, for most soil and crop conditions, the moldboard plow remains the most consistently reliable tillage tool for row-crop seedbed preparation.

TRACTOR LOAD-AND-DEPTH AND DRAFT-CONTROL SYSTEMS

Tractor draft control maintains constant draft load on the tractor. If draft increases, draft sensing in either the

Fig. 78—Do Not Plow Deeper Than Necessary To Do a Good Job

PLOW ONLY DEEP ENOUGH TO DO A GOOD JOB

tractor lower links or the top link activates the lift-control valve. Then the hitch automatically rises just enough to reduce the draft load to the preset level. As the plow or other implement is lifted, weight is transferred to tractor rear wheels for added traction to pull the increased draft load.

The lifting action on the implement also tends to lift the tractor front end, and additional weight is shifted to tractor drive wheels, providing still more traction. As draft decreases the hitch automatically lowers until the chosen draft load is again obtained.

Tractor load-and-depth control combines the benefits of draft control, as described above, with depth or position control which holds the hitch at a fixed height. Thus by changing the control setting the desired system response can be obtained, from entirely draft-responsive to controlled hitch height. This reduces variations in working depth caused by soil changes.

Most tractors with top-link draft sensing respond to changes in tension and compression of the top link as implement load varies. Thus increased suction on the plow activates the control system just as an increase in soil resistance tends to rotate a smaller plow about the crossbar and push on the top link. (See John Deere's FMO Manual, TRACTORS, for a more-detailed explanation of hitch-control system functioning.)

Some sacrifice must be made in uniformity of plowing depth with draft or load-and-depth control, but extreme changes are unlikely except in cases of major soil change — clay to sand, for instance. For such cases a gauge wheel on the plow will help maintain reasonably uniform plowing depth. Occasionally, if soil conditions are extremely variable, changes in load-and-depth setting may be required to hold the desired plowing depth.

Properly adjusted plows may still tend to "bob" in some soil conditions due to frequent draft changes. This usually can be overcome by slowing the response of the draft or load-and-depth control system. This tends to average system response to draft changes, and the tractor ignores extreme draft changes of short duration.

Fig. 79—Load-and-depth Control with Integral Plow

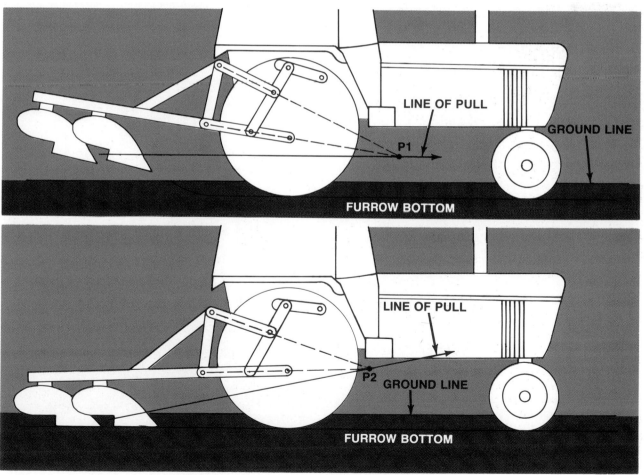

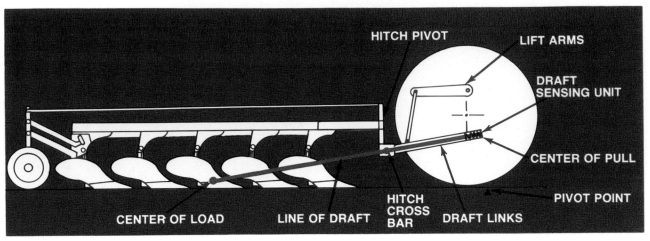

Fig. 80—Load-and-depth Control with Semi-Integral Plow

LOAD-AND-DEPTH CONTROL WITH INTEGRAL PLOWS

When an integral plow enters the ground, the lift links are dropping freely and the line of pull passes through point P_1 (convergence of tractor upper and lower hitch links) as shown in Fig. 79, top. This provides a very low line of pull, which aids quick plow penetration.

When the integral plow reaches its working depth, as determined by the load-and-depth control setting, it is held by the tractor hitch and hydraulic system. This results in a higher line of pull through P_2 (Fig. 79, bottom). Plow draft, weight and furrow slices, plus suction of the bottoms all tend to make the line of pull steeper. This transfers weight from the plow and tractor front wheels to tractor rear wheels for better traction.

The steeper the line of pull at the center of the rear wheel, the greater is the tendency for tractor front wheels to rise, and the greater is the amount of weight transferred to drive wheels. Ballast normally must be added to the tractor front end to maintain stability with integral plows, and to rear wheels to provide adequate traction.

LOAD-AND-DEPTH CONTROL WITH SEMI-INTEGRAL PLOWS

The line of pull (draft) of semi-integral plows also passes through the line of the tractor lower links. Thus, as the plow is lowered it is pulled into the soil until it reaches the depth and load level established by the tractor load-and-depth setting (Fig. 80).

Draft changes are sensed by the tractor, and the hitch is raised or lowered accordingly to maintain uniform draft. When draft increases, the front of the plow is raised and weight is transferred from the plow and tractor front wheels to the tractor drive wheels. Most draft variations are of short duration, with the exception of those caused by changes in soil type, and normally cause only minor changes in plowing depth.

ADJUSTMENTS FOR TWO-WAY PLOWS

Adjustments for two-way plows are basically the same as for one-way moldboard plows. However, some procedures are peculiar to two-ways.

For instance, both right- and left-hand hitch lift links must be set to the same length (Fig. 81), and for most two-ways should be short enough to obtain maximum ground clearance for rotation of the gangs of bottoms.

Tractor rear wheels must be equidistant from the tractor centerline (Fig. 82), with the exact setting to be found in the plow manual. Set front-wheel tread the same, center to center, as rear wheels; or, for in-furrow operation, at least 2 inches (50 mm) wider than rear wheels from center of tractor to inside of tire wall.

Consult the tractor operator's manual to be certain the tractor has sufficient lift capacity and stability for operation of the particular plow model considered. Provide

Fig. 81—Set Tractor Lift Links to Equal Length

additional front and rear weight as recommended to provide tractor stability in operation and transport and eliminate excessive wheel slippage. Do not exceed tire or tractor manufacturers' recommended maximum weight for any wheel.

Integral two-way plows are heavier than one-way plows of the same capacity, and are closer-coupled to the tractor. This may require removal of the tractor drawbar and PTO master shield (but keep PTO guard in place) to provide adequate clearance. Check plow manual and be extremely careful when first lifting and rotating the plow.

Hitching and fore-and-aft leveling of integral two-way plows is the same as for one-way plows, as previously explained. However, adjusting screws are provided on the hitch frame (Fig. 83) to level the plow laterally. Turning the right-hand screw levels the plow for right-hand plowing; turning the left-hand screw levels the plow for working the other way.

Similar adjusting screws are used to level semi-integral two-way plows.

Proper width of cut on the front bottom of semi-integral two-ways is obtained by adjusting the tractor rear-wheel tread according to the plow manual. Some plows have provision for hydraulically sliding the hitch crossbar in the hitch to secure the correct cutting width on the front bottom (Fig. 84). Shifting the plow on the crossbar toward plowed land helps reduce tractor sidedraft.

TRACTOR PREPARATION FOR PLOWING

The tractor must be matched to the plow from the standpoints of drawbar pull, hydraulic lift capacity, and tractor stability. Some preparations depend on tractor and plow design, and details are explained in each operator's manual. However, many preparations apply to all tractors and plowing situations, and should be covered before going into the field.

Because plowing is normally the greatest power-consuming operation on most farms, it pays to have the tractor properly tuned and serviced. A good tune-up can save fuel, increase power, and reduce plowing time. University studies have shown that poorly-tuned tractors may be wasting 10 to 15 percent of the fuel consumed, and losing 10 percent or more of their possible power output.

Wheel slippage of 25 percent, which is not too uncommon but 10 percent more than the recommended maximum, wastes more fuel and time — 10 gallons (liters) of fuel for 9 gallons' (liters') worth of work, or 10 hours of plowing that should have been done in 9 hours. Plowing also makes heavy demands on tractor hydraulics, so these must also be in top condition. Tires, electrical system, engine cooling, and lubrication are other major areas requiring careful preparation.

Important tractor-preparation steps include:

Fig. 82—Tractor Rear Wheels Must Be Equidistant from the Centerline

Fig. 83—Lateral Leveling of Integral Two-Way

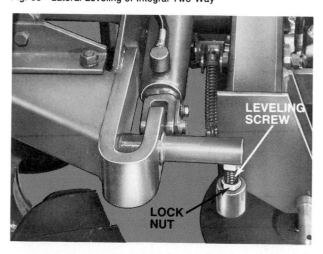

Fig. 84—Hydraulic Landing for Semi-Integral Two-Way

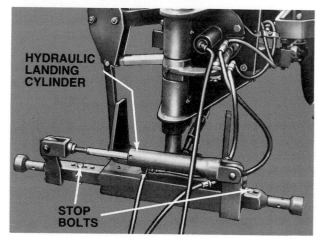

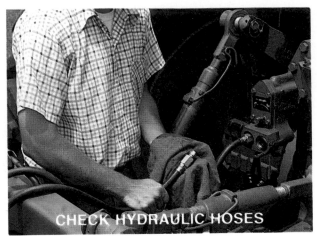

Fig. 85—Check Hydraulic Hoses, Couplings, and Cylinders

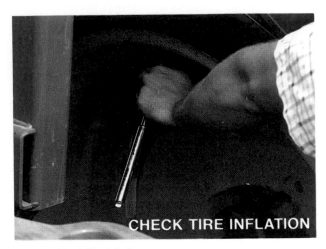

Fig. 86—Check Tire Inflation

1. Engine tuneup, including timing, ignition system (on gasoline engines), air filter, cooling system (flush and refill if necessary), and engine oil and filter.

2. Complete hydraulic check. Clean or replace filter, maintain proper oil level, check compatability of pump capacity and pressure (see operator's manual), and examine all hoses, couplings, and cylinders (Fig. 85) for leaks, repairing or replacing as necessary.

3. Determine if tractor hydraulic-system pressure and capacity are adequate for automatic-reset plows. Be sure hoses are properly connected.

4. Set front and rear tread as specified in manual, as nearly as possible on true line of draft. For in-furrow plowing set all wheels equidistant center-to-center, or set front wheels 2 inches (50 mm) wider than rear from center of tractor to inside of tires. For on-land operation, allow at least 4 inches (100 mm) between edge of tire and furrow wall. Shift duals to narrowest possible setting.

5. Inflate tires to pressure recommended in operator's manual (Fig. 86). Hard pulling and weight transfer can seriously damage underinflated tires.

6. Add tractor front-end ballast as required for stability in transport and operation. For instance, with integral plows some manufacturers recommend that the tractor front end be weighted so that at least 25 percent of the static front-wheel weight is retained on front wheels when the plow is raised. Added weight may be required for operation (Fig. 87), especially with semi-integral plows using the lower-range gears, but weight should never exceed tractor and tire manufacturers' recommendations. *Front wheels must remain on the ground at all times to maintain safe steering control.*

As a starting point for drawn plows, the recommended minimum front-wheel weight is approximately one-fourth of the total tractor shipping weight without ballast. More weight can be added for stability.

7. Add sufficient rear-wheel weights or tire fluid to eliminate excessive slippage. Do not exceed tractor or tire makers' maximum weight recommendations, or add enough weight to eliminate all slippage. Such weighting could cause severe damage to the tractor drive train.

Normally, slippage should not exceed 15 percent, at which point soil between lug marks will be slightly broken and shift (Fig. 88) when the tractor is pulling its rated load. If too much weight is added, tracks will be clear and distinct (Fig. 89). With too little weight, tread marks are completely wiped out (Fig. 90).

Fig. 87—Proper Front-End Weighting Is Important

Fig. 88—Tread Marks Show Proper Slippage

Fig. 89—Too Much Weight Has Been Added to Rear

8. Adjust tractor lift links and top link to suggested starting lengths for integral and semi-integral plows (Fig. 91). Raise 3-point hitch as high as possible to clear drawbar when using drawn plows.

9. Adjust tractor drawbar to recommended height for drawn plows, and so it will clear integral and semi-integral plows in operation.

10. Select load-and-depth or draft-control setting on tractor hydraulic system (except for drawn plows).

11. For integral plows, set sway blocks or stabilizers to permit draft-link sway in operation and sway lockout in transport (Fig. 92). For semi-integral plows, eliminate draft-link sway in both transport and plowing (Fig. 93). Lower links must be held laterally rigid for quick steering response and good semi-integral plow maneuverability, except for articulated (center-steered) tractors, which must have some sway in the links to prevent plow damage.

MOLDBOARD PLOW PREPARATION AND MAINTENANCE

A little extra care in preparing a plow for operation can be time well spent. Clean bottoms and a properly adjusted plow can save time and fuel in the field, reduce downtime for repairs and field adjustments, and do a better plowing job. Here are some general suggestions; see the plow operator's manual for specific details.

Fig. 91—Adjust Tractor Lift Links

Fig. 90—Not Enough Weight Has Been Added to Rear

Fig. 92—Allow Links To Sway for Operation of Integral Plows

Fig. 93—Eliminate Link Sway for Semi-Integral Plows

Before Each Season:

1. Lubricate according to instructions (Fig. 94). Clean grease fittings to avoid forcing dirt into bearings. NOTE: Some safety-trip standards should not be lubricated.

2. Inflate tires to recommended pressure. Underinflation damages tires and throws off adjustments.

3. Clean, inspect, lubricate, and tighten wheel and coulter bearings. Replace as needed.

4. Remove any rust, paint, or other protective coatings from moldboards, shares, and landsides to assist scouring.

5. Replace dull or broken shares and sharpen coulters if necessary. After replacing shares, always re-tighten bolts again after one or two rounds.

6. Check all bolts for tightness (Fig. 95), especially on bottoms and standards. Replace badly worn or broken parts.

7. Check bottom alignment—shares pointing up or down, to left or right. Measure vertical distance from share point to frame on all bottoms; distance between top corners of moldboards and between moldboard tips (Fig. 96); also the distance from share point to share point. Each of the measurements should be essentially equal for all bottoms. If any measurement is off more than ½ inch (13 mm), consult plow manual for proper corrective action. If any bolts have loosened, remove dirt from between frog and moldboard, share, landside or standard before retightening.

8. Replace bent or cracked frogs, standards or beams.

9. Hand-release each safety-trip standard (Fig. 97) to be sure it functions freely. Refer to the operator's manual for proper method of checking spring-trip or hydraulic-reset standards.

Fig. 94—Lubricate According to Manual

Fig. 95—Check All Bolts for Tightness

Fig. 96—Check Moldboard Alignment

10. Check nitrogen supply in hydraulic accumulator on all new automatic-reset plows, and at least once prior to each plowing season thereafter.

Daily Before Operation:

1. Check for loose, broken, or missing bolts, nuts and parts. Replace as needed.

2. Lubricate as recommended.

3. Check for badly worn or broken shares, especially if plowing in abrasive or rocky soil.

4. When the plow is not in use—even overnight—protect bottoms and coulters with oil to prevent rust. For long periods use soft, black paint.

Before Storage At End Of Season:

1. Clean soil, trash, and accumulation of grease from the plow and repaint spots which are scratched or worn.

2. Protect bottoms and coulters from rust with heavy grease or plow-bottom rust-preventive paint.

3. List any parts or repairs needed. Buy and install them prior to the next plowing season.

4. Relieve pressure in hydraulic reset system—see operator's manual.

5. Store plow inside and remove weight from tires.

6. Place boards under bottoms to prevent ground contact and rusting.

7. Remove hydraulic cylinders or fully retract cylinders to protect cylinder rods from rust.

FIELD OPERATION
FOR ONE-WAY PLOWS

Careful planning of field layouts and just a bit of extra time spent adjusting the plow can save hours of plowing time, save fuel, and provide a better plowing job.

Fig. 97—Do Trips Function Freely?

PLANNING FIELD LAYOUT

The best plowing pattern for any particular field must be consistent with good soil-conservation goals and avoid excessive travel on the ends. The pattern should also avoid extra turning and unnecessary point rows and dead furrows. The optimum number of lands, back furrows, and dead furrows will depend upon field width, travel speed on the ends, and the relative amount of time spent on the ends compared to plowing time. If soil compaction from repeated travel on the ends is a problem, reduce land width.

A typical plowing pattern used for a rectangular field (Fig. 98) is: Mark scratch furrows (A) across each headland parallel to field boundaries and far enough into the field for easy turning. To save time in finishing headlands, make each headland a multiple of the plow width (two, three or four times the plow width, etc.). For convenient turning the headland should be at least as wide as one-half the total length of tractor and plow.

To mark headlands, use rear bottom only and cut just deep enough for the furrow to be readily seen. When plowing, lower and raise the plow at these marks on each end.

Next, lay out the first back furrow (B) and continue to plow on each side of it (C) until the land is plowed to the field boundary. Then start a new back furrow (D) and plow a new land (E) equal in width (L) to the first land. When the unplowed strip between lands equals the width of the plowed lands, plow out that strip (F) leaving

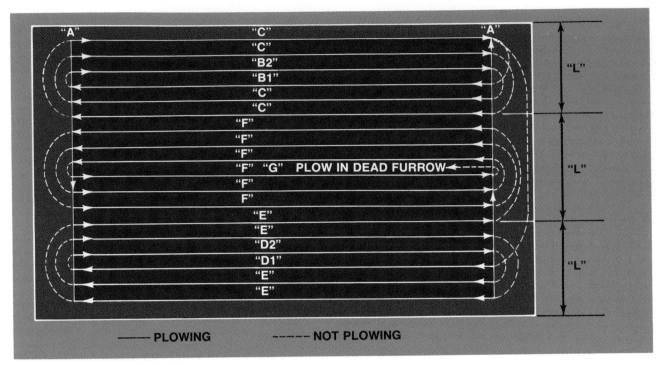

The diagram labels include: "A", "C", "C", "B2", "B1", "C", "C", "F", "F", "F", "F" "G" PLOW IN DEAD FURROW, "F", "F", "E", "E", "D2", "D1", "E", "E", "A", "L", "L", "L"

———— PLOWING   ----- NOT PLOWING

Fig. 98—Typical Plowing Pattern for Rectangular Field

a dead furrow (G) in the center of the field. After this land is finished, set the plow shallow and turn some soil back into the dead furrow to help level the field.

Finish the field by plowing headlands. This may be done by turning all furrows toward the plowed field and leaving a dead furrow next to the border, or by plowing furrows out toward the border and leaving a furrow in the field which is relatively easy to level with later tillage. Many farmers alternate plowing headlands toward and away from field borders each year to help keep headlands level.

In a variation of this pattern, some farmers leave strips down each side of the field the same width as headlands and finish the field by plowing all the way around the field instead of backtracking on each headland.

Irregularly shaped fields (Fig. 99) may be easier to plow by working around the field, throwing soil to the outside. A combination of rounded corners and turn strips are made to permit easier plowing. Divide the angles of sharp corners and make scratch furrows for lowering and raising the plow. If angles change rapidly, plow only short scratch furrows each time. After the field is finished the turn strips are plowed out leaving a dead furrow in each strip. If erosion is likely to start in the dead furrows it may be best to simply leave the turn strips in grass and follow the same pattern for planting.

CONTOUR PLOWING

Contour plowing requires more turning and driving than straight fields, but helps reduce erosion on sloping land. Always seek professional help from the USDA Soil Conservation Service or private engineers in laying out contours, terraces, or strip-cropping patterns to help reduce unnecessary turning and point rows.

The contour pattern used will depend upon slope and shape of the field. Less energy is required to turn furrows downhill, and this method is sometimes recommended. However, this will cause a gradual downhill shift of topsoil and eventually expose considerable subsoil at the upper end of the slope. Therefore, it is generally best to turn furrows uphill whenever possible.

To start plowing a contoured field, begin about midslope (Fig. 100) and lay out a back furrow. Plow the first land around the back furrow and then plow the second land, leaving a turn strip of uniform width in the center to avoid turning on plowed ground. When the rest of the land is finished, plow out the turn strip, leaving a dead furrow in the center. Complete third and fourth lands in the same manner. Do not plow the headlands or leave any open furrows up and down slopes in which erosion could start.

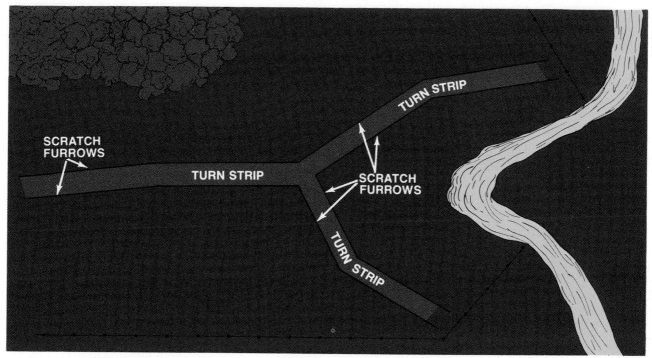

Fig. 99—Plowing Pattern for Irregularly Shaped Field

PLOWING IN HEAVY TRASH

A carefully chosen plow, properly equipped and adjusted, can operate in heavy trash with relatively little difficulty (Fig. 101). Select the widest bottoms possible and maximum vertical and fore-and-aft frame clearance for extremely trashy conditions. Add large rolling coulters and keep them sharp. Use trash boards to help turn trash from the edge of the furrow slice into the furrow for better coverage. Keeping shares sharp improves penetration and helps avoid dragging trash when the plow is lowered.

Set coulters farther forward than normal and just deep enough to cut the trash and to keep turning. This provides the best "scissor action" between blades and soil. Coulters set too deep may push trash ahead of the blade or drag it on the coulter hub. Properly adjusted coulters are very important in reducing plugging.

To improve trash coverage, raise rear of trash board and, if possible, tilt blade forward at the top. But keep the trash board point flush with and tight against the shin or upper edge of the moldboard to prevent catching trash. Tipping trash boards too far forward increases draft and improves trash coverage very little.

Shredding or disking trash before plowing frequently can save more than enough plowing time (and permit better coverage) to justify the extra operation. Both

disking and shredding of trash seldom are justified economically (do one but not both). Avoid plowing when trash is extremely wet, and do not plow so deep that trash clearance is seriously reduced. Reducing speed can also improve trash flow in most instances, but working too slow can cause plugging because soil is not moving through the plow fast enough to keep trash flowing.

OPERATING SEQUENCE

The following procedures are general in nature — consult the sections on Principles Of Plow Operation, Basic Moldboard-Plow Adjustments, and the plow operator's manual for details.

1. Drop the front of semi-integral plows first (Fig. 102), then the rear, to make straight headlands.

2. Start back furrow with front bottoms set slightly shallower than the rear for the first round. Set rear bottom to desired plowing depth.

3. Clean bottoms frequently, most adjustments should be made after bottoms are scouring well.

4. When starting second round, readjust front and rear bottoms to desired depth. Set rear coulter to leave a clean furrow wall (Fig. 103).

Fig. 100—Typical Plowing Pattern for Contoured Field

WORKING IN HEAVY TRASH

Fig. 101—Good Work in Heavy Trash

Fig. 102—Drop the Front of Semi-Integral Plows First, Then the Rear

LOWERING SEMI-INTEGRAL PLOWS

CLEAN FURROW WALL

Fig. 103 — On Second Round, Set Rear Coulter to Leave Clean Furrow Wall

ALL BOTTOMS CUT EQUAL DEPTH

Fig. 104—Level Plow Fore-and-aft for Uniform Plowing Depth

5. Level plow fore-and-aft so that all bottoms are cutting equal depth (Fig. 104). Measure furrow depth immediately behind front and rear bottoms and sight across plow frame.

6. Sight across plow frame from the rear and adjust tractor linkage or plow furrow wheel as necessary to level the plow laterally (Fig. 105).

Fig. 105—Level Laterally for Proper Furrow Shape

LEVEL PLOW LATERALLY

Fig. 106—Check Width of Cut of Front Bottom

Fig. 107—Adjust Rear Furrow Wheel

7. Check width of cut of front bottom (Fig. 106) and adjust if needed. Landing adjustment can change width of cut and help set the plow straight behind the tractor. The top hitch-link must be parallel to the direction of travel.

8. Check all coulters for proper width of cut. Start with rear coulters and set others to match.

9. Adjust rear furrow-wheel of semi-integral and drawn plows to lead away from furrow wall to relieve some load on landsides (Fig. 107). Never lead furrow wheel toward furrow wall or it may tend to climb and lift the plow from the ground.

10. Quickly recheck all adjustments in sequence—plowing depth, fore-and-aft leveling, lateral leveling, and width of cut.

11. When properly adjusted and operated at the correct speed, all furrows will be uniformly shaped and the surface even (Fig. 108). It should be impossible to detect each round of the plow by noting a repetitive pattern in furrow shapes.

12. To scour new or rusted bottoms, remove paint or as much rust as possible. Set coulters wide, or take them off to provide maximum pressure on moldboards. Plow shallow and fast. Frequently stop and clean sticking soil from moldboards to keep soil moving.

13. After bottoms are scouring well, return coulters to proper position and return to normal plowing depth. Recheck for level operation, width of cut, etc.

FIELD OPERATION FOR TWO-WAY PLOWS

Mark headlands for two-way operation the same as for one-way plows. Then start plowing at one side of the field and reverse the plow at each end to permit plowing adjacent passes until finished. If the field has been leveled for drainage or irrigation, headlands may be left unplowed or plowed toward the plowed field as desired. This will leave a single furrow next to field borders on each end and along one side for minimum interference with water movement.

When starting to plow contoured fields first mark off a shallow scratch furrow on previously determined contour lines (1, 2, 3, 4, 5, Fig. 109). Start plowing at the top of the slope (A, Fig. 109) and plow approximately on the contour until reaching Contour Line (1). Now plow parallel to Line (1) until reaching Line (2). After plowing all land above Line (2) repeat the process until reaching Line (3), etc.

When all land on one side of the waterway is plowed go to point (B) at the top of the opposite slope and follow the same procedure in plowing that land. To minimize soil erosion, do not plow headlands on contoured fields. They should be left in grass as turn strips for tillage, planting, etc. Also, leave furrow ends slightly staggered by lifting the plow at different times on adjacent passes to reduce water flow downslope at the edge of the plowed land.

TROUBLESHOOTING

Improper plow adjustment causes most plowing problems. If difficulties persist after following the assembly, operating, and adjusting instructions in the plow manual, use this guide to help locate and solve the problems.

UNIFORM FURROWS

Fig. 108—When Plow Is Properly Adjusted, All Furrows Are Uniform

Fig. 109—Typical Plowing Pattern for Contoured Field

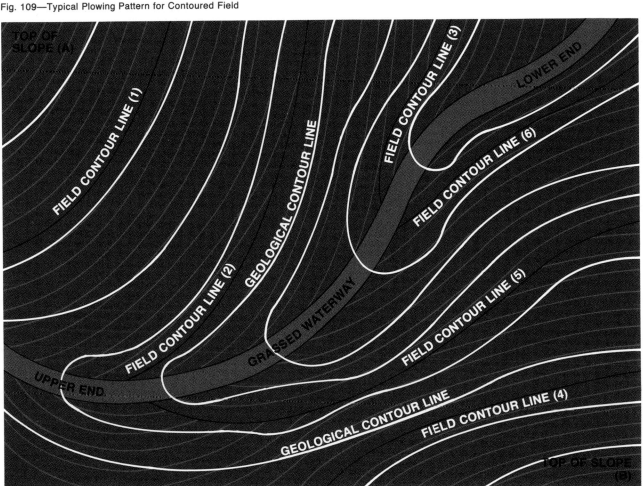

TROUBLESHOOTING CHART

PROBLEM	POSSIBLE CAUSE	POSSIBLE REMEDY
SLOW GROUND ENTRY	Bottoms not scouring.	Clean bottoms frequently until land polish is obtained.
	Worn shares.	
	Improper coulter settings.	
	Dull coulters.	
	Coulters too deep.	
	Integral	
	Insufficient plow bottom suck.	Shorten center link slightly.
	Too much load or draft response.	Reduce sensitivity.
	Rear landside heel too low.	
	Gauge wheel set too low.	
	Semi-Integral	
	Improper rear-wheel setting.	Set rear wheel so rear landside contacts bottom of furrow and so plow is level fore-and-aft.
	Hitch crossbar too low.	Raise hitch plate on front of plow.
	Improper tractor rockshaft control-level setting.	Adjust stop so front bottom cuts desired depth—see tractor manual.
	Rear landside too low.	
	Drawn	
	Improper rear-wheel setting.	Set rear wheel so rear landside contacts bottom of furrow and so plow is level fore-and-aft.
	Improper hitch adjustment.	
	Insufficient plow bottom suck.	Loosen bottom attaching bolts and force share points down. Re-tighten bolts securely.
	Worn shares.	
POOR DEPTH CONTROL	**Integral And Semi-Integral**	
	Too much load response.	Reduce sensitivity.
	Varying soil conditions.	Set gauge wheel to carry part of weight.
	Rockshaft control lever in wrong position.	Refer to tractor manual for proper operation.
	Semi-Integral And Drawn	
	Rear of plow won't stay down.	Lower hitch at hitch plate assembly on front of plow.

PROBLEM	POSSIBLE CAUSE	POSSIBLE REMEDY
	Rear of plow goes too deep.	Lower rear wheel or lower rear landside, or both.
		Raise hitch on front of plow.
		Level plow laterally.
	Rear wheel not set correctly.	Set wheel so rear landside contacts furrow bottom. Wheel must run on furrow bottom with slight lead away from furrow wall.
	Improper remote cylinder operation.	Reset control lever—see tractor manual.
PLOW NOT RUNNING STRAIGHT	Bottoms not scouring.	Clean bottoms frequently until land polish is obtained.
	Improper leveling.	Set front bottom to cut proper depth. Make sure plow is level laterally.
	Improper tractor wheel tread.	
	Hitch not set properly.	
	Improper coulter adjustment.	
	Integral	
	Improper lateral adjustment.	If plow overcuts, rotate crossbar counter-clockwise. If plow undercuts, turn crossbar clockwise (Fig. 65).
	Plow running on nose.	Lengthen center link and/or shorten right-hand lift link.
	Insufficient rear-landside pressure.	
	Semi-Integral	
	Sway blocks set wrong.	
	Semi-Integral And Drawn	
	Improper rear-furrow wheel adjustment.	Give rear wheel slight lead toward plowed ground and set rear landside to run on furrow bottom and against furrow wall.
RIDGING	Front bottom cutting too wide or too narrow.	Check tractor wheel-tread settings.
	Front bottom cutting too deep.	Raise front bottom and relevel plow.
	Front bottom cutting too shallow.	Lower front of plow and relevel.
	Plow not running straight.	See preceding section.
	Improper coulter setting.	Set coulters to same spacing as frame size. Check distance with plow in soil.

PROBLEM	POSSIBLE CAUSE	POSSIBLE REMEDY
HIGH DRAFT	Bottoms not scouring.	Clean bottoms frequently until land polish is obtained.
	Worn shares.	
	Insufficient traction.	Use differential lock. Increase tractor ballast—front and rear.
	Integral	
	Too little load response.	Adjust control system—see tractor manual.
	Plow running on nose.	Lengthen center link (or leveling screw on two-way plows).
	Semi-Integral And Drawn	
	Plow running on nose.	Check plow-hitch setting. Level plow fore-and-aft. May be necessary to put tractor drawbar in high position (drawn).
	Improper rear-wheel adjustment.	Lower rear wheel so landside contacts furrow bottom and wheel carries part of plow weight.
POOR TRASH COVERAGE	Bottoms not scouring.	Clean bottoms frequently until land polish is obtained.
	Improper coulter adjustment.	Set coulters just deep enough to cut trash and keep turning. Keep coulters sharp.
	Plow running crooked.	See "Plow Not Running Straight".
	Trash boards not used.	Add trash boards.
	Trash boards improperly set.	When plowing more than 7 inches (18 cm) deep, raise rear of trash board slightly. Keep point of trash board tight against upper edge of the moldboard.
RAGGED FURROW WALL	Improper coulter adjustment.	Set rear coulter 3 to 3½ inches (7.5 to 9 cm) deep and far enough toward unplowed land to leave a clean furrow wall.
	Wrong rear-coulter angle.	Be sure coulter collar doesn't keep coulter from running straight.
	Dull coulters.	
	Semi-Integral And Drawn	
	Rear wheel improperly set.	Give rear wheel slight lead toward plowed ground—allow rear landside to contact furrow wall and bottom.

PROBLEM	POSSIBLE CAUSE	POSSIBLE REMEDY
BOTTOMS DON'T SCOUR	New bottoms.	Remove paint; clean bottoms frequently until land polish is obtained.
	Plow running on nose.	Level plow fore-and-aft.
	Coulters not set properly.	Set coulters shallower and farther back.
SAFETY-TRIP STANDARDS TRIP TOO EASILY	Tripping resistance not great enough.	Adjust safety trip—see plow manual.
	End of standard worn or chipped.	
	Throwout roller damaged.	
	Pivot stud loose.	
	Excessive pivot-stud wear.	
	Safety-trip spring broken.	
	Dirt packed under throwout roller.	
	Excessive lubrication.	
SHEAR-BOLT STANDARDS TRIP TOO EASILY	Using wrong shear bolts.	Replace with proper bolt—see manual.
	Pivot stud too loose.	
	Excessive pivot-stud wear.	
SHEAR-BOLT STANDARDS TRIP TOO HARD	Pivot stud too tight.	
	Excessive pivot-stud wear.	
	Using wrong shear bolt.	
	Pivot stud loose.	
	Rust between standard and side plates.	
SAFETY-TRIP STANDARDS TRIP TOO HARD	Standard seldom tripped.	Trip standard periodically by hand.
	Tripping resistance too great.	
	Trip not lubricated.	
	Standard binding at pivot stud.	
	Parts of trip mechanism broken or lost.	

PROBLEM	POSSIBLE CAUSE	POSSIBLE REMEDY
HYDRAULIC RESET STANDARDS "FLOAT"	Insufficient hydraulic pressure at manifold.	Check pressure valve setting. Check for proper hose hook-up. Check for air in hydraulic system. Check tractor hydraulic-pump pressure. Check nitrogen pressure in accumulator.
HYDRAULIC RESET STANDARDS FAIL TO TRIP	Insufficient lubrication. Faulty valve.	
SPRING-RESET STANDARDS TRIP TOO EASILY	Too much suck in bottoms. Pivot stud loose or worn.	Check bottom and plow settings. Tighten or replace stud. Be sure standard can be rotated by hand when spring pressure is released.
	Inadequate spring tension. Excessive lubrication.	
SPRING-RESET STANDARDS TRIP TOO HARD	Pivot stud too tight. Accumulation of paint, dirt or rust on standard.	
EXCESSIVE LATERAL LOOSE-NESS OF STANDARD	Standard pivot stud loose.	Tighten nut until standard will just move without binding.
	Pivot stud worn or broken.	Replace stud and lubricate pivot.
EXCESSIVE VERTICAL PLAY AT SHARE POINT (SAFETY TRIP PLOWS)	Safety trip improperly adjusted. Pivot stud worn or broken. Throwout roller damaged.	
LOSS OF NITROGEN FROM RESET ACCUMULATOR	Loose core valve.	Have dealer tighten valve and recharge accumulator.
	Faulty core valve.	Have dealer replace valve core and recharge accumulator.
PLOW RAISES TOO SLOWLY	Pump delivering low pressure.	Check pump pressure, repair as needed.
	Using small hoses and cylinders with low pressure system.	Match hoses and cylinders to tractor and load.
	Tractor low on hydraulic fluid.	

TRANSPORT AND SAFETY

Many accident studies show that hurrying and human error are responsible for or involved in a vast majority of equipment accidents. An operator with a complete understanding of the function, operation, and limitations of the equipment reduces the chances of increasing the "statistics" of farm accidents. There must also be a determination not to be hurried into an accident.

Strict observance of the following precautions and the safety tips in plow and tractor manuals will help make plowing safer and prolong equipment life as well.

1.   Provide adequate front-end weight for tractor stability in transport and operation, particularly with integral and semi-integral plows. Never pull from any point higher on the tractor than the recommended hitch point.

2.   Use extreme caution and reduce speed when transporting the plow and the tractor over rough ground.

3.   Avoid sharp turns at high speeds, especially on slopes.

4.   On tight turns, avoid swinging rear of plow into fences or other obstacles.

5.   Turning stops on semi-integral plows limit turning radius. Shorter turns may severely damage plow frame and tractor hitch.

6.   Never carry passengers on the tractor or permit others to ride on the plow—particularly plows with automatic reset.

7.   Always lower the plow when not in use or left unattended.

8.   Lower the plow and securely pin the parking stand before detaching the plow from the tractor.

9.   Always use proper lighting, reflectors, SMV emblem, and other safety devices for road travel as required by state and local laws (Fig. 110).

10. When hitching drawn plows, always use a hitch pin with adequate strength for the tractor-plow combination.

SUMMARY

Dating from sharpened sticks drawn by men or animals back in the dawn of agriculture, plows were—and are today—among primary tillage tools for making seedbeds and rootbeds for domesticated crops.

Those forked sticks have evolved into several types of tractor-drawn plows, each with a special purpose . . . carrying various types of bottoms to meet local soil and crop conditions . . . with numerous precise adjustments to assure best work with lowest draft . . . and with improvements not dreamed of only a few decades ago.

Moldboard plows remain, for most soil and row-crop conditions, the most consistently reliable tool for seedbed preparation. Despite comparatively new tillage im-

Fig. 110—Use Proper Lighting and SMV Emblem for Road Transport

plements, such as disk tillers, the moldboard plow is still the best for burying surface trash and crop residues. It also mixes fertilizer throughout the plowed depth, increases soil porosity, aerates the soil, promotes growth of valuable microorganisms, helps control weeds, insects and crop diseases, and improves seed-oil contact for better germination.

But moldboard plowing is normally the most power-consuming operation on farms. Therefore it is important to understand every facet of plow choice, adjustment, operation, and special equipment for special conditions.

CHAPTER QUIZ

1.   Name three purposes of moldboard plowing.

2.   Name four key parts to a plow bottom.

3.   What are four of the most common bottom types now available?

4.   (Fill in blanks) Three basic moldboard materials are _____, _____ and _____.

5.   (Choose one) The greatest advantage of off-center, soft-center steel is (long wear and easy scouring) (abrasion resistance) (low cost).

6.   How do coulters improve plow performance?

7.   What usually limits the maximum size of integral plows?

8.   Name two advantages of semi-integral plows.

9.   What are two advantages and two disadvantages of plowing with tractor wheels in the furrow?

10. List two advantages and two disadvantages of plowing with on-land hitch.

11. Through what points must the line of draft pass for efficient operation of plows?

12. What happens if the hitch on semi-integral or drawn plows is set too high? Too low?

13. At what percent is wheel slippage considered excessive, and what is appearance of tracks at that point when pulling rated load?

14. What are two main advantages of two-way plows?

15. Where are two-way plows most commonly used?

6
Disk Plows

Fig. 1—Integral Four-Bottom Disk Plow

INTRODUCTION

Disk plows are used for primary tillage and the work performed is similar to that of the moldboard plow. They consist of a series of individually mounted, frame-supported, concave, rotating disks, with working depth of the disks controlled by one or more wheels, or the tractor hydraulic systems (Fig. 1).

Disk plows are best suited for such conditions as:

● *Hard, dry soils where a moldboard has difficulty penetrating.*

● *Sticky soils (waxy muck and gumbo) where a moldboard will not scour.*

● *Hardpan and highly abrasive soils where the cost of moldboard plow-bottom wear would be prohibitive.*

● *Soils containing heavy roots.*

● *Loose, "push-type" soils, such as peat land.*

● *Soils where deep plowing, to 12 to 16 inches (305 to 406 mm) is desired.*

This better performance in difficult conditions costs an approximate 10 percent increase in draft per square inch (soil cross-section turned) compared to moldboard plows. Because of this high draft, time spent assuring top tractor performance can reduce fuel cost, save time, and result in better all-around operation.

TYPES AND SIZES

Disk plows generally have from one to seven concave disks which cut 7 to 12 inches (178 to 305 mm) of width per disk. Disk blades are available in 24- to 38-inch (610 to 965 mm) diameters.

The disks are tilted backwards from the vertical at an angle of 15 to 25 degrees (tilt angle), and operated with the plane of the disk face at a horizontal angle of 42 to 45 degrees from the direction of travel.

Four basic disk plows are in current use:

● **Integral One-Way**

● **Integral Reversible**

● **Semi-Integral Reversible**

● **Drawn Reversible**

Each type will be covered later in more detail.

PRIMARY COMPONENTS

Primary components of a disk plow are (Fig. 2):

Fig. 2—Primary Components of a Disk Plow

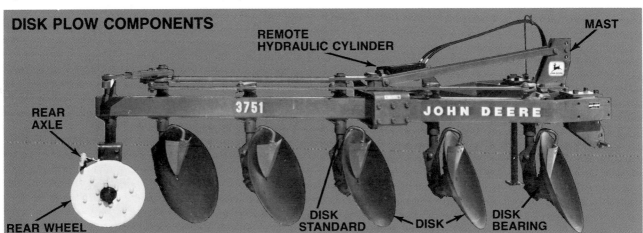

Fig. 3—Disk Plows Turn and Mix Soil

- **Disk**
- **Disk Bearing**
- **Rear Wheel**
- **Rear Axle**
- **Disk Standard**
- **Mast**
- **Support Stand**
- **Remote Hydraulic Cylinder**

PRINCIPLES OF OPERATION

When plowing with a disk plow, soil and trash are cut and moved with a rolling action. Used without scrapers, the disks produce a mixing action of the soil rather than inversion. If scrapers are used, the soil is turned in a similar way to the action of a moldboard plow, but usually not as well (Fig. 3).

A disk blade creates no suction; therefore, obtaining the desired plowing depth requires proper disk-angle setting and ample plow-frame weight (usually 400 to 1200 pounds (180 to 545 Kg) per blade). Even then, additional weight sometimes must be added.

Disk plows must be operated at a fairly slow, uniform speed for the best cutting action and width-of-cut control. Because of the tendency to pitch soil unevenly, they generally do not perform as well as high speeds. Higher speeds also tend to reduce depth.

Disk penetration and speed of rotation relative to ground speed are governed largely by disk position on the plow frame. Faster rotation during plowing improves trash flow through the plow.

The key to good plow trailing is proper adjustment of the rear furrow wheel (Fig. 4). Because there are no landsides on a disk plow, this wheel must absorb the

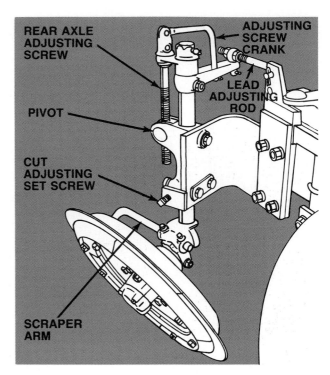

Fig. 4—Adjustable Rear Wheel

side thrust of the soil on the disks. Proper setting assures a steady plow, maintains an even cut on each blade, and keeps the plow operating straight.

Let's look at other factors that affect disk-plow performance:

- **Disk shape and design**
- **Cutting width**
- **Disk angle**

DISK SHAPE AND DESIGN

The heart of the disk plow is the concave disk. It cuts, lifts, and rolls the furrow slice.

Larger-diameter disks can take a wider cut, permit deeper plowing, and cut trash better. However, smaller disks will penetrate better in hard soils.

Disks are made from high-carbon, heat-treated steel to obtain the best wear and strength characteristics. Common blade thicknesses are 3/16 and ¼ inch (5 and 6.4 mm)

Disk-plow blades are bolted to a hub which is mounted on sealed, tapered bearings (Fig. 5). These bearings permit disks to rotate freely by action of the soil forces. The bearing housing is attached to the standard, and vertical disk angle can be adjusted to match soil conditions (Fig. 6).

Disk edges are available for various soil conditions. Three basic types of cutting edges are (Fig. 7):

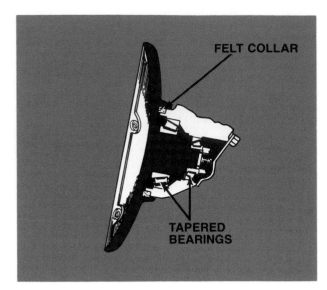

Fig. 5—Sealed Disk Bearing

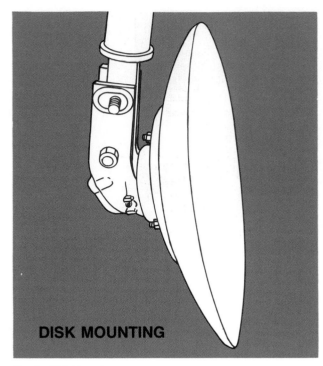

DISK MOUNTING

Fig. 6—Disk Mounting

- **Outside-beveled edge**
- **Inside-beveled edge**
- **Notched edge**

Disks with *OUTSIDE-BEVELED EDGE* adapt well to a wide variety of soil conditions. The edge can be re-sharpened by grinding on the convex or back side.

The *INSIDE-BEVELED EDGE* is more aggressive and penetrates better than the outside-beveled disk in very hard and dry soil. The edge can be resharpened by rolling or grinding the concave or front side.

NOTCHED-EDGE disks perform well in most heavy trash conditions, where the notches aid in rotation and pull trash down for better cutting.

CUTTING WIDTH

Some disk plows are built to permit reducing the cutting width per disk or reducing the number of disks to obtain optimum tractor-plow performance in various soils. The width-of-cut per disk can be reduced by moving disk standards forward on the main frame, or by altering the frame angle in relation to the direction of travel, depending on plow design. In very hard soils, it may be desirable to reduce the width of cut or number of disks (Fig. 8).

Fig. 7—Disks With Outside-Beveled Edge, Inside-Beveled Edge, and Notched Edge

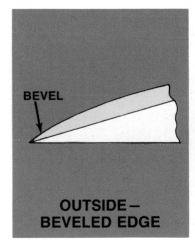

OUTSIDE— BEVELED EDGE

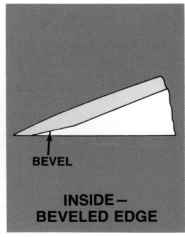

INSIDE— BEVELED EDGE

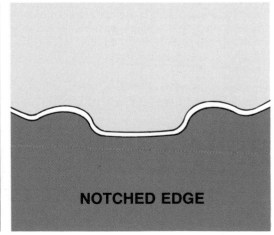

NOTCHED EDGE

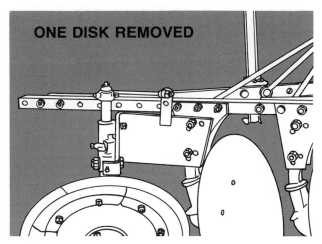

Fig. 8—Combination Frame Allows Removal of One Disk

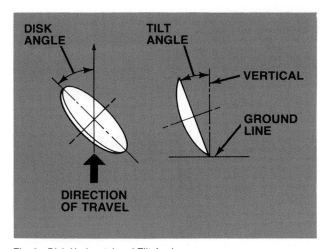

Fig. 9—Disk Horizontal and Tilt Angles

In tough, hard soils, a narrow width-of-cut per disk reduces draft and clod size and helps stabilize the plow. A wider cut per disk increases plowing capacity in light, sandy soils.

DISK ANGLE

Provision is made in the plow standard for adjustment of the **horizontal** disk angle and **vertical** tilt angle to obtain optimum disk operation in different soil conditions (Fig. 9).

HORIZONTAL DISK ANGLE is normally 42 to 47 degrees from the direction of travel. Reducing the angle (less abrupt setting) increases disk rotation with respect to ground speed, and reduces the tendency of the plow to overcut due to furrow-wall pressure on the heel or back side of the disk. Increasing the disk angle (more abrupt setting) improves disk penetration.

The *vertical angle* normally ranges from 15 to 25 degrees. Increasing the tilt angle (farther from vertical) improves disk penetration in heavy, sticky soils that have a tendency to roll under the cutting edge in the bottom of the furrow. Decreasing the tilt angle (closer to vertical) improves disk performance in loose or brittle soils. Setting disks in the steeper position puts greater soil pressure on the disk, resulting in faster disk rotation, greater soil pulverization, and better cutting and coverage of trash.

Now let's look at the different types of disk plows in detail.

INTEGRAL ONE-WAY DISK PLOWS

Integral one-way disk plows are attached to the tractor 3-point hitch and are fully carried by the tractor when transporting. The tractor lower draft links are attached to the plow hitch pins and the tractor upper link is attached to the plow mast.

The integral plow is generally limited in size to two to five disks due to tractor front-end stability and hydraulic lift capacity. The tractor-plow combination has excellent maneuverability in transport, for plowing irregular fields, and for quick turns at field ends.

TRACTOR PREPARATION

The tractor used with an integral disk plow must be matched to the plow from the standpoint of hydraulic lift capacity, tractor front-end stability, and drawbar pull. The following typical instructions for preparing a tractor for a given plow are shown in the plow operator's manual:

1. Adjust front and rear wheels to specified tread setting (Fig. 10).

2. Provide adequate front-end ballast for tractor stability in transport and operation.

Fig. 10—Adjust Front and Rear Wheels

3. Provide proper rear-wheel weighting (a disk plow does not have suction to transfer weight from the soil, plow, and tractor front end to the rear wheels for traction). The ideal amount of weight can be determined by observing tracks of rear wheels. When a tractor is pulling its rated load, soil between lug marks should be broken or shifted. If too much weight is used, tread marks will be clear and distinct. If too little weight is used, tread marks are entirely obliterated.

4. Adjust lift links and center link to suggested starting lengths (see operator's manual).

5. Select load-and-depth control setting according to tractor operator's manual.

6. Set sway blocks or sway chains to provide link sway during plowing and sway lockout in transport (Fig. 11).

OPERATION AND ADJUSTMENT

The integral disk plow is attached to the tractor 3-point hitch in the normal manner. The tractor load-and-depth control system regulates working depth of the plow.

For optimum performance the plow must trail straight behind the tractor (center link must extend straight rearward), all disks must operate at the same working depth, and all furrow slices must be the same width. Width of cut of the front disk is determined by the tractor wheel-tread setting.

Most integral disk plows are designed for the frame to run parallel to the ground, both fore-and-aft and laterally. The fore-and-aft level of the plow is controlled by the tractor center link. Lateral leveling is controlled by length of the tractor lift links (Fig. 12). These adjustments must be made with the plow in operating position.

The rear furrow wheel can be adjusted vertically and horizontally, and turned to lead the plow toward or away from the plowed land (Fig. 4). In normal soil conditions,

operate the wheel in the bottom of the furrow against the furrow wall with the wheel leading slightly away from the wall. The bottom of the wheels should be 12-inch (13 mm) or more below the bottom of the disks. If the furrow wall crumbles, or soil is soft and loose, lower the furrow wheel enough for it to "bite in" and stabilize plow operation. Keep furrow wheel clean of soil build-up for good depth control and proper width of cut on front blade.

In extremely hard soil, if the furrow wheel cannot grip the furrow bottom, adjust it to run higher against the furrow wall.

If the plow has a tendency to overcut (plow swinging toward unplowed land) reduce width of cut by increasing the lead angle in the wheel, lowering the wheel, or moving the wheel toward the furrow wall. To increase width of cut (plow not trailing straight) reverse this procedure.

If the plow does not respond to hitch or rear-wheel adjustments, adjust the disk angle or tilt angle as described earlier.

Note: Reducing horizontal disk angle increases speed of disk rotation and reduces tendency to overcut. Increasing disk angle improves penetration.

In heavy soils it may be necessary to increase tilt angle for improved penetration and greater width of cut.

Decreasing tilt speeds disk rotation, causes better pulverization and cutting and coverage of trash, and can improve performance in loose soils.

Fig. 12—Leveling Adjustment

Fig. 11—Sway Block Position for Plowing

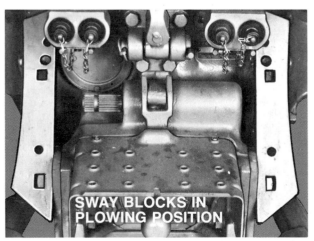

HORIZONTAL ANGLE ADJUSTMENT

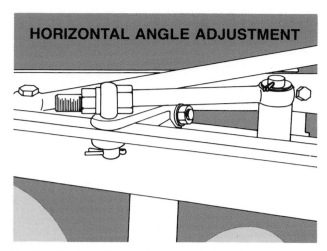

Fig. 13—One Screw Changes Horizontal Angle of All Disks

Provision is made in the frame for changing disk angle by turning wedges or adjusting screws, depending on the plow make and model (Fig. 13). Disk-tilt angle can be changed by reversing adjusting blocks (Fig. 14) or by using a series of adjusting holes, again depending on plow design.

Set disk scrapers so they barely touch near the center of the disk and just clear the edge of the disk. If set too close, they will cause wear and friction and slow disk rotation; if set too far, dirt will build up on the blade.

Fig. 14—Tilt-Angle Change By Adjusting Blocks

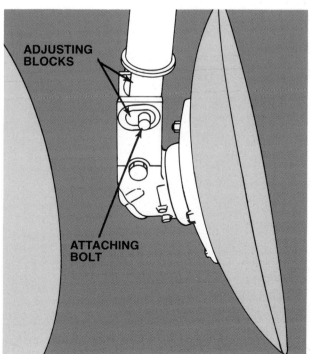

ADJUSTING
BLOCKS

ATTACHING
BOLT

MANUALLY INDEXED PLOW

Fig. 15—Manual Plow Indexing

Lubricate the plow at the intervals recommended in the operator's manual. When the plow is not in use protect blades from rust with a layer of grease or special plow-bottom paint.

REVERSIBLE DISK PLOWS

Reversible disk plows may be integral, semi-integral, or drawn. They generally are used in irrigated areas to keep land level for good water control, and in soils where moldboard plows do not perform well. All furrows are turned the same way, so there are no ridges, dead furrows, or back furrows.

INDEXING

Some small, integral, reversible disk plows can be indexed manually from right- to left-hand plowing and back by a hand lever reached from the tractor seat (Fig. 15). The subframe is held in position by soil pressure on the blades. Other plows index automatically when the plow is raised to full transport position.

Larger integral, reversible disk plows are indexed hydraulically (Fig. 16). The disk subframe is held in plowing position by the remote cylinder or an automatic locking device. Width of cut per disk on some plows can be reduced by limiting subframe travel. This automatically increases horizontal disk angle from direction of travel and improves penetration.

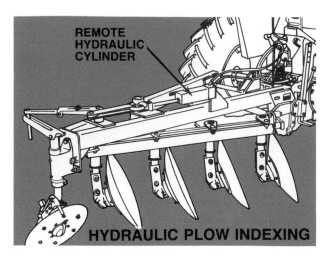

Fig. 16—Hydraulic Plow Indexing

INTEGRAL, REVERSIBLE DISK PLOWS

These plows are attached to the tractor 3-point hitch and are fully carried by the tractor in transport, providing maneuverability (Fig. 17). Integral, reversible plows usually are limited in size to two to four disks due to tractor front-end stability and hydraulic-lift capacity.

Tractor Preparation

Tractor preparations are similar with reversible plows and integral disk plows, except that hitch lift links on reversible plows must be set at the same length because the plow turns right- and left-hand furrows on alternate passes through the field. (See Tractor Preparation for integral disk plow.)

Hitching and Adjusting

The integral, reversible disk plow is attached to the tractor 3-point hitch in the normal manner, and the tractor load-and-depth control system regulates plowing depth.

Fig. 17—Integral Reversible 4-Disk Plow

For optimum tractor-plow performance the plow must trail straight behind the tractor.

The tractor center link and rear furrow wheel must be adjusted to level the plow fore-and-aft and permit all disks to work the same depth. These adjustments must be made with the tractor wheel operating in the furrow after the first pass across the field has been completed.

Adjustment of the rear furrow wheel is the key to proper trailing of the plow. Correct adjustment assures uniform operation for both right- and left-hand plowing (Fig. 18).

The rear wheel illustrated in Fig. 18 reverses automatically (approximately 200 degrees rotation) when the disk subframe is indexed from right- to left-hand plowing or vice versa. The wheel is fully adjustable for height, lead, and lateral setting. In normal soil conditions, operate the wheel in the bottom of the furrow against the furrow wall with the wheel leading slightly away from the wall. The bottom of the wheel should be approximately ½-inch (13 mm) below the bottom of the disks.

If the plow does not respond to hitch or rear-wheel adjustments, change the disk angle or tilt angle as described under Disk Angle (Fig. 9) and Operation and Adjustment. These adjustments are covered in each plow operator's manual and must be made correctly so that blade angle is the same when furrows are turned either right or left.

Disk tilt-angle change is essentially the same as for integral disk plows shown in Fig. 13.

Fig. 18—Rear-Wheel Adjustments

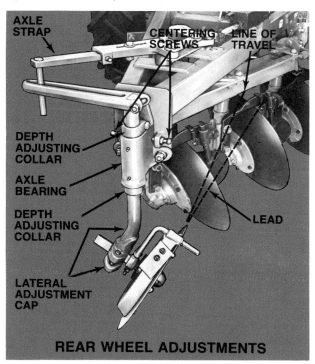

SEMI-INTEGRAL, REVERSIBLE DISK PLOWS

Semi-integral, reversible disk plows are attached to the tractor lower draft links and are raised and lowered by the tractor hitch and a remote hydraulic cylinder on the rear transport wheel (Fig. 19). The rear wheel on some of these plows operates on the land and serves as a gauge wheel when plowing and as a transport wheel (free to caster) when the plow is raised. Actually, the rear wheel is an attachment for converting a fully integral, reversible disk plow to semi-integral operation.

The semi-integral concept permits use of larger, longer plows with more space between blades, while still retaining tractor front-end stability.

This provides more trash clearance and greater plowing capacity, compared to integral plows. Common plow sizes are four to six disks.

Tractor Preparation

Tractor preparations are similar for using semi-integral reversible plows and fully integral models. Hitch lift links must be set the same length for uniform plowing on alternate passes through the field. (See *Tractor Preparation* for integral disk plow.)

Hitching And Adjusting

Hitching and adjusting the semi-integral, reversible disk plow is essentially the same as for integral reversible models. Care must be taken that adjustments made when plowing in one direction do not adversely affect operation in the opposite direction. (See *Hitching and Adjusting* for integral, reversible disk plow.)

The rear transport wheel of the semi-integral plow also serves as a gauge wheel for plowing. Vertical adjustment of the gauge wheel helps level the plow and controls working depth of the rear disk.

The subframe is indexed from right- to left-hand plowing and back by a hydraulic cylinder and is held in plowing position by this same cylinder.

The plow in Fig. 19 can be set for 9- or 11-inch (23 to 28 cm) cut per disk by adjusting travel of the subframe. The plow width may be altered by adding or removing one or two diskframe extensions. This permits easy matching of plow size to tractor power and soil conditions.

DRAWN REVERSIBLE DISK PLOW

Many drawn, reversible disk plows are built to match the performance of regular farm tractors, while others are built for the power and capacity of large crawler or 4-wheel-drive tractors. These plows range from three to six disks, and some models use up to 38-inch (97 cm) diameter blades for plowing 18 to 20 inches (457 to 508 mm) deep.

Drawn reversible disk plows are attached to the tractor drawbar and trail behind the tractor. To provide uniform plowing depth, the three transport wheels also serve as gauge wheels during operation. The plow is raised to transport or lowered to plowing position by a hydraulic cylinder. A rear furrow wheel stabilizes side thrust on the plow.

Note: Never make sharp turns toward the plowed land when plowing, because this forces the rear furrow wheel against the furrow wall and could seriously damage the plow. Raise the plow before making turns.

All furrow slices are turned in the same direction, thus eliminating ridges and dead furrows and leaving the field level for good water control.

The center of load of a disk plow is generally considered to be located at the center of the total width of cut. For a 6-disk plow cutting 12 inches (305 mm) per disk, the center of load would be 36 inches (914 mm) from the furrow wall.

The attaching point of the plow tongue to the plow must be lower than the tractor hitch point. If the hitch is too high, the tractor will pull down on the front of the plow, resulting in loss of penetration, excessive width of cut, undue wear on wheel bearings, and poor plow control. If

Fig. 19—Semi-Integral Reversible Disk Plow

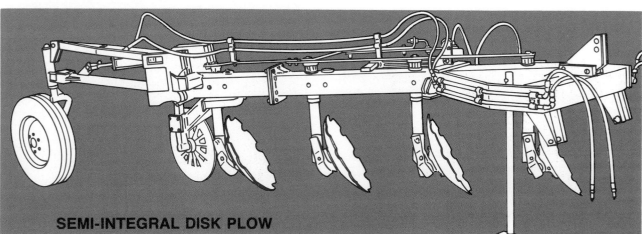

SEMI-INTEGRAL DISK PLOW

the hitch is too low, the front of the plow is raised by the tractor, causing uneven penetration.

These plows are easy to adjust because they are heavy and relatively stable in most soil conditions. Stops on the wheel cylinders control working depth; the hitch can easily be adjusted for height and proper line of draft, and the disk-angle and tilt-angle adjustments are similar to that previously discussed for other disk plows.

TRANSPORT AND SAFETY

Integral plows are transported completely raised, and all weight is carried by the tractor 3-point hitch. Adequate tractor front-end weights are required to offset the plow weight.

When transporting on a road or highway, always display SMV emblem and use lights and reflectors as required by state and local regulations.

Semi-integral plows are quite long and caution must be used when turning to prevent swinging the plow into fences or irrigation ditches.

Reduce speed when transporting over rough ground, and avoid quick, sharp turns at high speeds.

When transporting semi-integral or **drawn plows,** always install cylinder locks to prevent accidental lowering of the plow. Relieve the load on hydraulic cylinders before starting to transport.

Lower the plow to the ground or install hydraulic cylinder locks when the plow is not in use.

Watch for other people when raising, lowering, or indexing the plow.

Never permit anyone to ride on the plow, and allow only the driver on the tractor. Do not permit children to play on or near the plow either when parked or in operation.

Lower the parking stand and securely pin it in place before detaching integral or semi-integral plows from the tractor.

FIELD OPERATION

Operating disk plows in the field is essentially the same as for moldboard plow operation. Plowing patterns are the same as moldboard plowing patterns, depending on whether the plow is one-way or reversible.

DISK PLOW ATTACHMENTS

The three attachments discussed here are:

- **Disk scrapers**
- **Rear-wheel or frame weight**
- **Rear-wheel flange and scraper**

DISK SCRAPERS

Three types of disk scrapers are available to clean disk blades and aid in turning and burying trash:

- **Hoe**
- **Moldboard**
- **Reversible**

Let's look at why each is used.

Hoe Scraper

The hoe scraper (Fig. 20) is a flat blade set well to the outer edge of the disk blade and low enough to catch the furrow slice before it falls away from the disk surface. Two hoe scrapers are required for each disk on reversible disk plows. Spring tension holds the scraper against the blade for good cleaning and permits the scraper to clear itself of trash and return to work.

Moldboard Scraper

The moldboard scraper (Fig. 21) gives better coverage and furrow-slice control than the hoe scraper. However, in sticky, gummy soils the hoe-type keeps disks cleaner. Two scrapers are required per disk for reversible plows.

The point of the spring-loaded scraper is positioned against the disk and the outer edge just off the blade to permit trash to work out. Scrapers are set low enough to catch and turn the furrow slice before it falls away from the disk.

Fig. 20—Hoe Scraper

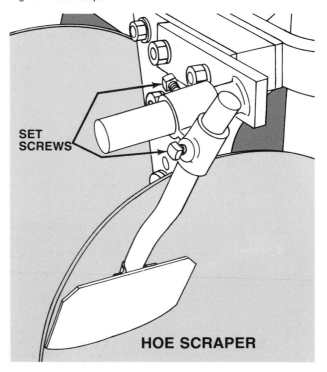

SET SCREWS

HOE SCRAPER

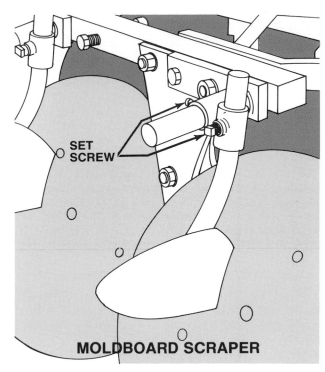

MOLDBOARD SCRAPER

SET SCREW

Fig. 21—Moldboard Scraper

Reversible Scraper

Reversible scrapers are formed like a shield and positioned on the centerline of the disk (Fig. 22). They are used only to keep the disk clean. Soil forces hold the scraper against the disk for scraping action. With this design, only one scraper is required per disk for reversible disk plows.

Fig. 22—Reversible Scraper

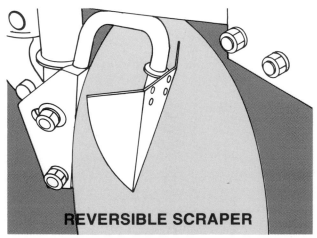

REVERSIBLE SCRAPER

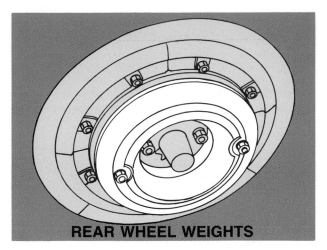

REAR WHEEL WEIGHTS

Fig. 23—Rear-Wheel Weight for Better Penetration and Plow Control

REAR-WHEEL OR FRAME WEIGHTS

In tough, hard soil conditions where penetration is difficult, rear-wheel weights or frame weights are frequently used. Wheel weights are bolted to the outer surface of the rear wheel (Fig. 23). Positioning of frame weights will depend on plow design and construction. Extra weight not only helps force disks into the ground, but enables the rear furrow wheel to keep the plow operating in a stable condition.

Fig. 24—Rear-Wheel Flange and Scraper for Soft Sticky Soil

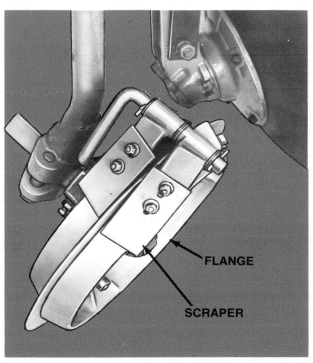

FLANGE

SCRAPER

REAR-WHEEL FLANGE AND SCRAPER

In loose ground or sandy conditions, an auxiliary flange and scraper are used to prevent the rear wheel from penetrating too deeply into the furrow bottom (Fig. 24). The scraper cleans soil from the wheel, and should be adjusted barely to clear the wheel surface.

TROUBLESHOOTING

Improper adjustments account for most plowing difficulties. If problems persist after you have followed the operating and adjusting instructions in the plow operator's manual, and checked for proper plow assembly, use the following chart to locate possible additional solutions. If you are still unable to remedy the trouble, ask your dealer or company representative for assistance. Where possible remedies to problems are obvious, based on the possible cause, a blank space is left in the possible remedy column.

TROUBLESHOOTING CHART		
PROBLEM	**POSSIBLE CAUSE**	**POSSIBLE REMEDY**
SLOW GROUND ENTRY	Improper center-link adjustment (integral plows).	Shorten center link until front disk cuts about the same depth as rear disk.
	Improper rear-wheel adjustment.	Bottom of rear wheel should run about ½-inch (13 mm) below bottom edge of disk. Rear wheel should have slight lead toward plowed land.
	Improper disk-tilt angle.	Tilt blade back (increase angle) for wet, sticky soil. Medium angle for average to hard ground. Steepest setting in loose soils.
	Improper horizontal disk angle.	Increase blade angle (from direction of travel) for hard soil. Reduce blade angle in loose, sandy soil for faster rotation.
	Dull disks.	
PLOW CROWDING	Crowding toward unplowed land.	Lengthen center link with integral plow. Recheck length of hitch lift links for integral and semi-integral plows. Plow rear wheel should be deep enough to hold plow straight. Wheel should have slight lead toward plowed land and be properly adjusted laterally.
	Incorrect tractor rear-wheel setting.	Set wheel tread as recommended in plow operator's manual.
	Crowding toward plowed land.	Shorten center link with integral plows and readjust right-hand lift link with integral or semi-integral one-way models. Adjust front-wheel cylinder stops on drawn plows. Rear wheel may be too deep and have too much lead toward plowed land, or may have soil build-up.
	Improper rear-wheel adjustment.	Make lateral wheel adjustment to correct width of cut on front disk.

PROBLEM	POSSIBLE CAUSE	POSSIBLE REMEDY
UNEVEN PLOWING DEPTH	Load-and-depth control too active (integral or semi-integral plows).	Change center link to less sensitive setting and check position of tractor selector lever (see tractor operator's manual).
	Plow won't stay at desired depth.	Shorten center link with integral plows. Change disks to hard-ground lateral setting. Change disks to steeper vertical setting. Raise rear wheel slightly. Check blade sharpness and rear-wheel scraper.
	Plow goes too deep.	Lengthen center link on integral plows. Adjust wheel-cylinder stops on drawn plows. Lower rear wheel.
	Plow runs too shallow	Shorten center link on integral plows. Adjust wheel-cylinder stops on drawn plows. Raise rear wheel. Rear wheel should lead slightly toward plowed ground. Check rear-wheel scraper.
	Improper disk-angle settings.	See Possible Remedy under "Slow Ground Entry."
	Dull disks.	
RIDGING	Front disk too deep.	Lengthen center link for integral plows. Readjust right-hand lift link on integral or semi-integral one-way plows. Adjust front-wheel cylinder stops on drawn plows.
	Front disk too shallow.	Shorten center link on integral plows. Readjust right-hand lift link with integral or semi-integral one-way plows. Adjust wheel-cylinder stops on drawn plows.
	Improper rear-wheel adjustment.	Adjust wheel to have slight lead toward plowed ground.
	Front disk cutting wrong width.	Check tractor rear-wheel setting; see plow operator's manual. Make sure plow is level.
	Improper scraper setting.	Set disk scrapers alike for uniform furrows. See "Disk Scrapers" under DISK PLOW ATTACHMENTS.
	Disks not scouring.	Clean disks frequently until land polish is obtained. Make sure scrapers are properly set.
POOR TRASH COVERAGE	Plow not level.	Level plow both fore-and-aft and laterally (see operator's manual).
	Improper scraper setting.	Set scrapers to catch and turn furrow slice. (moldboard scrapers, Fig. 22, provide best trash coverage.)

PROBLEM	POSSIBLE CAUSE	POSSIBLE REMEDY
	Disks not scouring.	Clean disks frequently until land polish is obtained.
	Disks not turning fast enough.	Swing disks laterally to loose ground setting. In loose or sandy soils, set disks as steep as possible.
DISKS DON'T SCOUR	New disks.	Clean surface with gasoline, kerosene, diesel fuel, or paint remover.
	Old disks.	Remove paint or heavy grease used to protect blades, or remove rust and smooth blade surface by sanding or polishing.
	Improper scraper setting.	Set moldboard-type scrapers close to disks, with slight opening at outer edge.
	Wrong type of scrapers.	Moldboard-type scrapers control furrow slice better than hoe-type. However, in very sticky soil hoe-type scrapers may be required to keep disks clean.
PLOW PULLS HEAVY	Front disk too deep.	Level plow fore-and-aft and laterally.
	Front disk cutting too wide.	Check tractor-wheel setting—see operator's manual. Level plow fore-and-aft and laterally. Check rear plow wheel setting. Wheel should lead slightly toward plowed ground and have correct vertical setting.
	Excessive tractor wheel slippage.	Be sure tractor load-and-depth setting is correct for integral and semi-integral plows and that the system is functioning properly. Add rear-wheel weights or ballast in tires if necessary.

SUMMARY

Disk plows are used for primary tillage. They are good tillage implements in conditions such as hard, dry soil, waxy muck and gumbo soil, hardpan and highly abrasive soil, and soils containing heavy roots.

Unfortunately, this good working potential in such conditions has its price. Disk-plow draft is about 10 percent more than that of moldboard plows per square inch (cm²) of soil cross-section turned.

Moldboard-plow bottoms invert the furrow slice and surface trash. Disk blades have more of a mixing action as they cut, lift, and roll the slice.

Disk plows perform best at fairly slow, uniform speed. They tend to lose depth and pitch soil unevenly at higher speeds.

Moldboard-plow bottoms have "suction," which aids penetration. Disk-plow blades have no suction, so proper disk-angle setting and ample added frame weight are necessary for good penetration.

Disk plows in common use include integral one-way, integral and semi-integral reversible (2-way), and drawn reversible (2-way). They range in size from one to seven blades, usually 24- to 38-inch (609 to 965 mm), cutting 7 to 12 inches (178 to 305 mm) per blade.

CHAPTER QUIZ

1. List three conditions where disk plows usually perform better than moldboard plows.

2. (True or false) Draft of disk plows per square inch of furrow cross-section is less than for moldboard plows.

3. Increasing or decreasing (choose one) disk-blade angle in relation to direction of travel improves disk-penetration.

4. Increasing or decreasing (choose one) vertical disk angle improves disk-plow performance in loose or brittle soil.

5. Name three types of blades used on disk plows.

6. Where is the center of load of a disk plow?

7
Chisel Plows

Fig. 1—Wide Chisel Plows Can Work Hundreds of Acres a Day

INTRODUCTION

The basic function of a chisel plow has changed little from that of the forked stick pulled through the soil by primitive man thousands of years ago. Alloy steels have replaced the wood and tractors have replaced animal and human muscle power, but the purpose still is to stir and aerate the soil with little inversion.

The primitive farmer, with his limited ability to scratch up the soil, would be amazed at the number of acres a man can cover today. Now one man can work 250

acres (100 ha) in a day with a chisel plow up to 45 feet (14 m) wide (Fig. 1) and at depths and speeds which early man couldn't dream of accomplishing with his crude tools.

CHISEL PLOWS VERSUS FIELD CULTIVATORS

Although chisel plows are referred to in some areas as field cultivators, we are classifying them here as distinct machines. Chisel plows have heavier construction and are basically used for primary tillage. Field cultivators are used principally for secondary tillage,

Fig. 2—Shanks are Arranged in Staggered Rows

SHANKS ARRANGED IN STAGGERED ROWS

NARROW CHISEL POINTS

Fig. 3—Narrow Points and Twisted Shovels Shatter and Break Soil

weed control, and seedbed preparation. Field cultivators are much lighter in construction than chisel plows, and are designed for shallower operation.

Modern chisel plows normally have two or three rows of curved spring-steel shanks attached to a rugged box-steel frame. The shanks are arranged in staggered rows (Fig. 2) to permit better trash flow and laterally balance the draft load of the machine.

CHISEL PLOWS VERSUS MOLDBOARD PLOWS

Draft of a chisel plow is perhaps half that of a moldboard plow per foot of width — both working the same depth. Therefore, chiseling is faster and more economical than moldboard plowing where complete trash coverage is not required. However, if soil is chisel plowed twice to prepare a seedbed, more fuel may be used than would be needed for moldboard plowing. Chisel plows are also frequently used to break up the hardpan or plow sole formed from years of plowing at the same depth with a moldboard plow.

Fig. 4—Sweeps Kill Weeds and Pulverize Surface Soil

SWEEPS KILL WEEDS

Because chisel plows break and shatter the soil, they perform best when soil is dry and firm. When too wet, soil is merely split open by the shank with no shattering or pulverizing. In fact, if soil is chiseled when it is too wet, large clods may be formed that are difficult or nearly impossible to break up with subsequent tillage to form a suitable seedbed.

THE VERSATILE CHISEL PLOW

Chisel plows may be operated to just scratch the soil surface, or worked down to 15 inches (380 mm) or more, depending on machine design, trash conditions, and desired results. They may be equipped with narrow chisel points which dig, stir, and break the soil (Fig. 3), or with a variety of other points—from shovels for erosion control to wide sweeps for seedbed preparation and weed control (Fig. 4). Some points are hard-faced to resist wear in highly abrasive soils.

OPERATING SPEED

Operating speed depends on chisel-plow size, power available, soil conditions, and the results desired. Faster speed causes more breaking and pulverizing of the soil and more ridging, which may be desired for holding water and reducing wind erosion. However, for seedbed preparation, a somewhat slower speed will leave the soil smoother and require less additional work before planting (though work of very flat sweeps is little affected by speed).

CHISELING METHODS

Soil is normally left loose and rough after chiseling with some trash mixed under, but most crop residue remains exposed on the surface.

Some university studies indicate that approximately 25 percent of the crop residue is covered each time the soil is chiseled (Fig. 5). This, of course, depends on the amount and kind of residue and the depth of chiseling.

If the soil is to be chiseled twice before planting, it is best to work the second time diagonally to the first to break any ridges left between chisels and to prevent chisels from following the same slots in the soil.

STUBBLE-MULCH TILLAGE

The chisel plow is an ideal tool for stubble-mulch or mulch-tillage farming; it helps prevent wind erosion and water runoff, and promotes water infiltration. If soil slopes 2 percent or more, it is advisable to chisel plow on the contour to reduce water runoff and erosion.

Chisel plows used for stubble-mulching in a summer-fallow operation are usually equipped with wide sweeps (12 to 30 inches; 305 to 760 mm) and operated just deep enough to cut off weeds with a minimum of surface disturbance.

Fig. 5—Crop Residue Remains on Surface after Chisel Plowing

The primary goals of a stubble-mulch system are:

• *Retaining maximum surface residue to control erosion*

• *Encouraging infiltration and storage of the maximum amount of moisture*

• *Limiting surface evaporation*

• *Killing weeds which deplete stored water and plant nutrients*

If initial residues are quite heavy, narrow chisel points may be used to work the soil from 5 to 8 inches (127 to 200 mm) deep and leave the surface rough and open.

FALL CHISELING

Research and on-farm experience have shown that fall disking or moldboard plowing leave soil quite subject to wind and water erosion. However, fall chiseling leaves the surface rough and tends to anchor the crop residue and bring larger, more stable clods to the surface to help resist erosion. If fertilizer is broadcast before chiseling, a light disking in the spring is all that's needed to prepare the land for planting corn.

To save time, reduce operating costs, and control winter erosion, some Corn Belt farmers chisel-plow cornstalks in the fall after broadcasting fertilizer (Fig. 6). To reduce trash problems, stalks are usually chopped or disked before chiseling. In the spring a heavy tandem disk is used once or perhaps twice prior to planting corn.

Due to the large amount of surface residue it is often impractical or virtually impossible to incorporate pre-emergence herbicides on chisel-plowed corn stubble. This forces dependence on post-emergence herbicide applications.

EFFECTS OF SURFACE RESIDUE

In the main Corn Belt, surface residues left on chiseled soil insulate the surface and slow soil warmup in the spring. This may delay seed germination and early plant growth. The greatest disadvantage lies in possibly delaying pollination into hot, dry weather, which could severely reduce yields. The insulating, chilling effect which surface residues have on the soil is particularly severe in wet, cool springs, but may have little effect in warmer, drier seasons.

There is no soil inversion with a chisel plow (Fig. 7), so broadcast fertilizers are simply stirred into the upper few inches by chiseling. Some scientists insist that phosphorous and potassium must be mixed deeper into the soil with a moldboard plow. Other scientists and some farmers have obtained adequate yields without plowing down these elements. If a chisel plow is used regularly in a row-crop program, usually the best routine is to incorporate a heavy broadcast fertilizer application with a moldboard plow every few years, with lesser amounts chiseled in each year.

Like crop residue and fertilizer, weed seeds also tend to remain in the upper layer of chisel-plowed soil, where they pose a greater problem than if the land is

CHISEL-PLOWING CORNSTALKS

Fig. 6—Fall-Chiseled Cornstalks Help Control Erosion

Fig. 7—Chisel Plows Do Not Invert Soil

turned regularly with a moldboard plow. Therefore, more herbicides are generally required when a chisel plow is used for primary tillage in row-crop operations.

The large amount of surface residue left by chisel plowing may absorb some herbicide, and keep part of the chemical from reaching the soil, requiring still higher applications. Over a period of time, there may also be a change in the dominant weed species present. This may necessitate further changes in weed-control practices.

VIBRATING CHISELS

Much research has been conducted since the early 1950's on vibrating or oscillating tillage tools. Primary goals have been reduction of draft, improved use of energy, and better control of the degree of soil pulverization. Results have varied widely and frequently have been contradictory; but, in general, total energy input has not been significantly reduced.

CHISEL-PLOW TYPES AND SIZES

Most chisel plows are composed of a basic center frame, usually with one shank per foot (305 mm) of width. Rigid-frame extensions of up to 6 feet (1.8 m) may be added to each end to match tractor power available and capacity needed. Folding outriggers or wings provide even greater width to match the power of big 4-wheel-drive and crawler tractors (Fig. 8).

CHISEL PLOW WIDTH

Chisel plows are available in sizes and styles to match almost any tractor size and field condition. Integral models are available from 5 to 20 feet (1.5 to 6 m) wide, but are limited in size by tractor power, lift capacity, and front-end stability. Drawn models start at about 10 feet (3 m) in width, with larger models introduced to match increasing tractor power.

Large plows can chisel up to 270 acres (112 ha) a day at 6 miles an hour (10 Km/ha). Approximate area cov-

Fig. 8—Winged Chisel Plows Match Big-Tractor Power

WINGED CHISEL PLOW

OUTRIGGERS ON CHISEL PLOW

Fig. 9—Outriggers Flex Over Uneven Ground

THREE-BAR FRAME

Fig. 11—Three-Bar Frame Provides Maximum Trash Clearance

ered in 10 hours equals operating width in feet times speed in miles per hour (45 × 6 = 270); or, width in meters times speed in kilometers per hour, times 0.80 (80 percent field efficiency). A 14 m chisel plow traveling 10 Km/h would cover approximately 112 ha per day (14 × 10 × 0.80 = 112 hectares).

Most chisel plows less than 20 feet (6 m) wide have rigid frames. Those more than 20 feet (6 m) wide are usually equipped with flexible outrigger sections. The outriggers follow ground contours (Fig. 9) and can be folded for transport.

A tongue jack permits easier hitching of drawn chisel plows, and may be available as standard or optional equipment, depending on machine design and price (Fig. 10).

CHISEL-PLOW CLEARANCE

Many early chisel plows offered only 18 to 20 inches (457 to 508 mm) of vertical clearance from frame mem-

Fig. 10—Hitching and Unhitching are Easier with a Tongue Jack

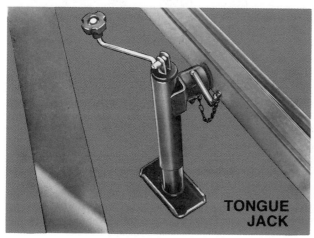

TONGUE JACK

bers to chisel points. As crop yields and residue have increased, so has trash clearance. Many plows now have 28 to 32 inches (711 to 813 mm) of vertical clearance and as much as 3 feet (1 m) center-to-center between rows of shanks. A few frame members are 44-inches (1,118 mm) apart for even greater trash clearance.

To reduce overhanging weight on the rear of the tractor, some integral chisel plows have only two shank bars, with shanks spaced 24 inches (610 mm) apart on the bar. Other integral models and most drawn chisel plows have three ranks of staggered chisels which provide 3-foot (1 m) lateral intervals between shanks for better trash flow (Fig. 11).

Shanks may be arranged in different patterns for maximum trash flow or to work extremely hard ground. However, wheel location on some models limits choice of shank placement, particularly when using 12- or 14-inch (305 or 356 mm) sweeps.

SHANK TYPES

Semi-rigid shanks are standard equipment on many current chisel plows. These shanks (Fig. 12) are clamped directly to the frame bar. They are recommended for economy, but only in soils which are free of such obstructions as rocks or stumps.

Various types of spring-cushion, spring-reset, and spring-trip shank mountings are available (Fig. 13). All are designed to protect the shank and frame when the point or sweep strikes an obstruction.

127

Fig. 12—Semi-rigid Shanks for Obstruction-Free Soil

Shank clamps allow the shank point to lift anywhere from 6 to 14 inches (152 to 356 mm), plus any deflection of the shank, as it passes over stones or stumps. Some also allow rearward movement of the shank point before rising, which permits more uniform depth operation in extremely hard soil.

The spring-cushioning effect of this type of mounting can produce a vibrating action in firm, dry soil, which helps break and shatter the crust.

Most current shanks are made of 1 × 2-inch (25 × 50 mm) heat-treated alloy steel for strength and durability. These shanks are naturally spring-acting because of the material and shank shape. Shanks on some heavy-duty plows are 1¼ × 2 inches (32 × 50 mm) or even 1½ × 2 inches (38 × 50 mm).

SOIL-ENGAGING TOOLS

The wide variety of available sweeps, spikes, chisels, and shovels makes it easy to match equipment to soil conditions and desired tillage results. A representative sample of these tools is shown in Fig. 14, and may be described as follows:

A. The **regular wheatland sweep** takes land polish quickly and the bottom is beveled to maintain sharpness. It kills weeds while keeping ridging to a minimum. It is available in 8- to 20-inch (203 to 508 mm) sizes.

B. **High-crown sweeps** lift and stir soil more than regular sweeps. They are usually available in 12- to 20-inch (305 to 508 mm) sizes from ¼ to ⁶⁄₁₆ (6 to 16 mm) thick.

C. **Heavy-duty, low-crown sweeps** kill weeds with less soil stirring. They are available in 12- to 18-inch (305 to 457 mm) sizes.

D. **Chisel sweeps** break up the soil for better water absorption, and anchor stubble to retard wind erosion. Ribbing beneath the points provide added strength on some brands. The sweep shown is 6 inches (152 mm) wide.

NOTE: Tools shown in A through D wrap around the shanks for greater strength and reduced twisting of the tool.

Fig. 13—Typical Spring-Cushion Chisel-Plow Shanks

E. **Heavy-duty, regular sweeps** are designed for maximum strength and durability. Sizes range from 6 to 14 inches (152 to 356 mm).

F. **Furrow openers** in 6-, 8-, 10-, 12-, 15- and 18-inch (152, 508, 635, 762, 965 and 1168 mm) widths feature heavy points for good penetration. The soil is grooved to catch and hold water and help control wind erosion.

G. **"Beavertail" shovels** function similarly to furrow openers and are also recommended for dry conditions. The beveled bottoms retain a sharp edge. They are available in 4-, 5- or 6-inch (100, 130, 150 mm) widths.

H/I. **Reversible chisel points** kill weeds and open packed or hard soil for better water penetration. The beveled edges improve cutting. They are available in sizes from 1½ × 11 inches to 2 × 16 inches (38 × 279 mm to 50 × 406 mm). To double the wear, they can be reversed.

J. **Double-point chisels** feature forged indentations to help keep the cutting edge sharp in tough, abrasive soil. They are especially good in dry conditions. Sizes range from 2 × 16 inches (50 × 406 mm) to 2 × 18 inches (50 × 457 mm).

K. **Spikes** kill weeds and rip up hardpan or plow-sole for better water infiltration. They are available in 2 × 12-inch to 2 × 16-inch (50 × 305 mm to 50 × 406 mm) sizes.

L. **Reversible, double-point shovels** kill weeds and roughen and groove the soil. This tool is an extra-wide, 4 × 14-inch (100 × 356 mm) shovel, reversible for double wear.

128

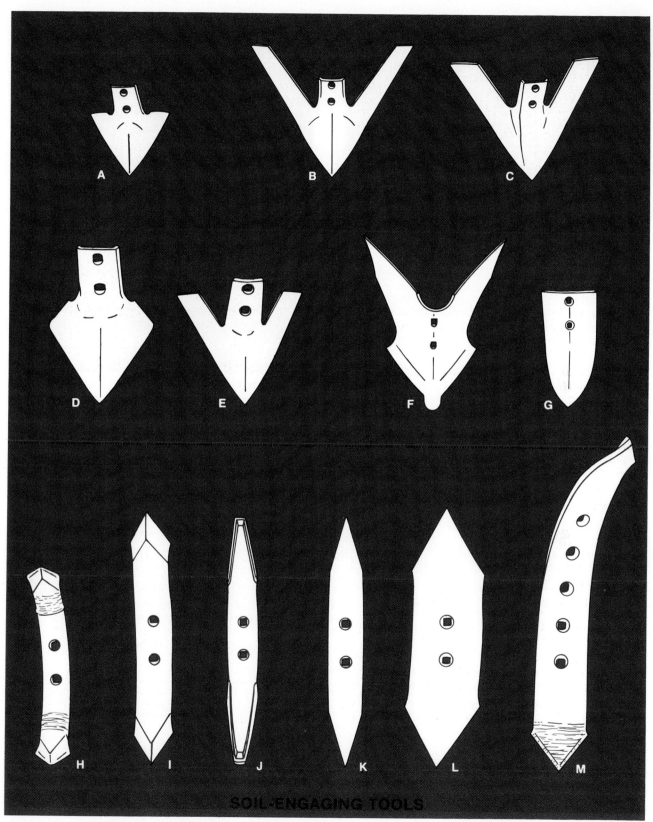

SOIL-ENGAGING TOOLS

Fig. 14—Various Chisel-Plow Soil-Engaging Tools

M. **Twisted shovels** feature side-throw action, some-what like a moldboard plow, and work deep, even in heavy cover. They are good for persistent weeds in summer fallow. Also, they are reversible for longer life and available with right- and left-hand turn to help balance machine draft. Common sizes include 3 × 22 inches (75 × 560 mm) and 4 × 22 inches (100 × 560 mm).

Many of the most-commonly used tool sizes and types are available with hard-facing for longer wear in abrasive soils.

Some farmers use chisel points or spikes for deep fall tillage. Then they switch to sweeps for fast, shallow seedbed preparation in the spring to kill more weeds and work all of the soil.

LIFTING AND FOLDING

Integral chisel plows are lifted and carried by the tractor 3-point hitch, and these plows usually do not have any outer folding "wing" or outrigger sections. Wider integral plows are too wide to transport, so they are usually equipped with special transport wheels and hitch to be pulled endwise.

Modern drawn chisel plows use remote hydraulic cylinders to raise and lower the plow and control operating depth. Older models have ratchet levers or hand-screw lift jacks for use with tractors without hydraulics.

On some folding-section models, a single remote hydraulic cylinder on the center section controls the entire

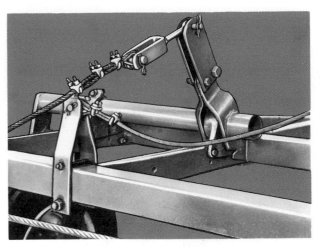

Fig. 15—A Single Remote Hydraulic Cylinder and Cables Control Raising, Lowering, and Operating Depth

machine, either through a rockshaft and mechanical linkage, or a cable arrangement (Fig. 15). Others use individual remote cylinders on each section. The cylinders are connected in series for synchronized raising and lowering (Fig. 16). Some models may use one or two remote cylinders nearly 3 feet (1 m) long to fold outriggers (Fig. 17).

TRANSPORT AND GAUGE WHEELS

As an aid in depth control, integral chisel plows may

Fig. 16—Three Remote Cylinders in Series Provide Full Machine Control

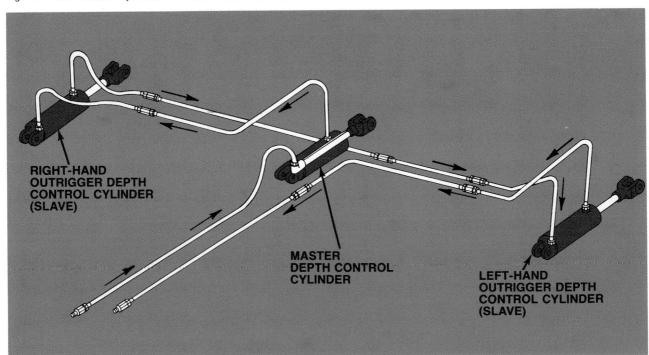

RIGHT-HAND OUTRIGGER DEPTH CONTROL CYLINDER (SLAVE)

MASTER DEPTH CONTROL CYLINDER

LEFT-HAND OUTRIGGER DEPTH CONTROL CYLINDER (SLAVE)

REMOTE CYLINDER FOLDS OUTRIGGERS

REMOTE CYLINDER

Fig. 17—Tractor Hydraulic Power Folds Outriggers for Easy Transport

GAUGE WHEEL

Fig. 18—Gauge Wheels for Integral Chisel Plows Help Control Depth

be equipped with gauge wheels, usually mounted ahead of the front frame bar to avoid interference with the normal shank pattern (Fig. 18). Some gauge wheels are relocated on the frame to transport the plow endwise on roads, through gates, etc.

Most rigid-frame drawn chisel plows have two transport wheels, which also gauge depth when the machine is in operation. On most plows, single wheels are also placed on each outrigger if it is more than 5 or 6 feet (1.5 to 1.8 m) long.

Some chisel plows have main frames equipped with

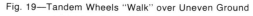

Fig. 19—Tandem Wheels "Walk" over Uneven Ground

TANDEM WHEELS

tandem wheels on opposite sides of a "walking" beam (Fig. 19) to improve flotation and support and maintain more uniform working depth. When one wheel rolls over a hump, or drops into a hole, its mate and the other wheels remain relatively stable. Tandem wheels stabilize the frame, distribute the weight for better shock absorption, and make transport safer. Single or tandem wheels are available for outriggers on these implements.

PRINCIPLES OF CHISEL-PLOW OPERATION

Chisel plows are designed to penetrate hard soil, shatter compacted layers, and break up large clods. The surface is left broken and open to catch and hold rainfall and resist wind erosion. Most crop residue is left on the surface, where it helps reduce evaporation and erosion.

Tests have shown that minimum draft on such tools as chisel plows and subsoilers occurs when the lift angle is 20 degrees between the face of the tool and horizontal. Shattering occurs with the least effort when the tool is applying a lifting force, rather than cutting horizontally or pushing vertically against the soil. Thus the common curved shank (Fig. 20) is ideally suited to provide optimum soil fracturing with reduced draft.

Operating a chisel plow so deeply that the upper curved portion of the shank is pressing down on the soil can increase draft unnecessarily. The alternatives are to get a machine with more clearance or work the land twice—once at a shallower depth, and then, at

right angles to the first pass, chiseling at the maximum desired depth.

Most tests indicate a moderate increase in specific draft (draft per square inch (cm²) of tilled cross-section) as depth increases. It is difficult to predict the effect of increasing depth on total chisel-plow draft because of the following reasons:

- *Variations in soil type.*
- *Moisture conditions.*
- *Shape of the sweeps and points used.*
- *The angle at which the points penetrate the soil.*

Draft also increases as speed increases, but again tests show varying results depending on depth, surface area of the sweeps or points, lift angle, and soil conditions. As speed increases, so does soil shattering (Fig. 21). Therefore, though more power is expended for fast chiseling, it may be offset later by reduced additional work required for seedbed preparation.

HYDRAULIC CYLINDERS

To provide uniform control when changing operating

Fig. 21—Soil Shattering Increases as Speed Increases

Fig. 20—Curved Shank Does Good Work with Least Draft

depth of wide chisel plows, take care to assure equal setting of hydraulic-cylinder stops on the center section and outriggers. On some large machines, the cylinders are connected in series. In one such system, the center cylinder is the master cylinder (Fig. 16) and outrigger cylinders are slave cylinders. As oil is pumped into the base of the master cylinder, oil is forced out of the ram end and into the base end of the first slave cylinder. In turn, this forces oil out of the first slave cylinder and into the base of the second slave cylinder at the other end of the chisel plow. Oil forced from the second slave cylinder returns to the tractor hydraulic reservoir.

To compensate for the smaller volume of oil which passes from the ram end of one cylinder to the base of another, the cylinders must be progressively smaller in diameter. For instance, in the system shown in Fig. 16, the master cylinder is 3½ inches (90 mm) in diameter; the first slave cylinder is 3¼ inches (82.6 mm), and the second slave cylinder, 3 inches (76 mm). All have an 8-inch (645 mm) stroke.

As the chisel plow is raised completely out of the ground, these specially designed cylinders are synchronized for uniform lifting by holding the tractor remote hydraulic control lever in the raised position for a few seconds. This allows a small volume of oil to bypass from the base of the master cylinder to the base of the first slave cylinder, to the base of the second slave cylinder, and back to the tractor. Thus all cylinders are fully extended simultaneously.

When working with more than one cylinder, arrange the cylinders in the proper sequence, attach all hoses in the correct order, and bleed all air from cylinders and hoses. Follow the synchronization procedure above for 10 to 15 seconds.

Check depth-control occasionally for internal leakage during field operation. Excessive leakage is occurring if cylinders leak down more than ¼-inch (6 mm) per hour, measured at the ram. Viewed another way, the loss of ½-cubic-inch (8.2 cm³) of oil from internal or external leakage can result in 1-inch (25 mm) difference in plowing depth between center and outrigger sections.

If single-acting cylinders are used instead of double-acting cylinders, they must be equipped with breathers and hoses according to the manufacturer's recommendations. Unless tractor valves are designed for either type of operation, the tractor must be equipped with one or more single-acting valves (depending on the number and arrangement of cylinders used), or dual-acting valves must be adjustable for one-way action.

On chisel plows using a single hydraulic cylinder and cables, inspect the cable for proper threading through guides and over pulleys, and check regularly for frayed cables or loose clamps.

HITCHING THE CHISEL PLOW

Integral chisel plows are attached to the tractor 3-point hitch in the usual manner (Fig. 22), with or without an implement quick-coupler. They are leveled laterally by adjusting the length of the lift links. To provide uniform penetration of all shanks, the machine

Fig. 22—Integral Chisel Plows are Carried on 3-point Hitch

3-POINT HITCH ATTACHMENT

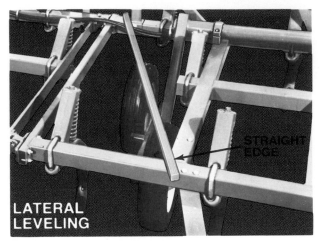

Fig. 23—Lateral Leveling of Drawn Chisel Plow by Changing Shims

sired depth in heavy soil when the load is highest. Gauge wheels carry more weight as draft is reduced in lighter soils. The tractor control system's usual reaction would call for increased operating depth to maintain uniform draft.

Drawn chisel plows are usually equipped with a rigid hitch which is attached to the tractor drawbar. Vertical hitch adjustment is provided for leveling the machine fore-and-aft to compensate for changes in operating depth, variations in drawbar height, tire size, and soil conditions.

Some drawn plows are leveled laterally by using a straight edge and shims (Fig. 23). The right-hand wheel is raised hydraulically until it touches the bottom of the straight edge. If the left-hand wheel also touches the bottom of the straight edge, the center frame is level. If not, use shims between the outer bearings and frame to level. Air pressure in all tires must be equal, and tires should be of equal size for level operation. Operating depth is controlled by one or more remote hydraulic cylinders, depending on machine size and design.

Determining the center of load, laterally, of a chisel plow simply requires location of the midpoint of the overall operating width. This point must be located on the centerline of the plow, directly behind the tractor hitchpoint, to keep the plow from veering to one side or the other in hard soil. Thus, when assembling a chisel plow, use care to provide symmetrical shank arrangement on each side of the machine (Fig. 24).

must also be leveled fore-and-aft by adjusting the tractor center link. Follow procedures and recommendations in operator's manual.

Tractor load-and-depth control is often used with integral chisel plows to regulate operating depth and provide weight transfer for better traction. However, in varying soil conditions it may be desirable to provide gauge wheels to limit operating depth and reduce the need of manually changing load-and-depth control frequently. When properly adjusted, gauge wheels should carry very little weight when plowing at the de-

Fig. 24—Symmetrical Shank Arrangement Provides Uniform Draft

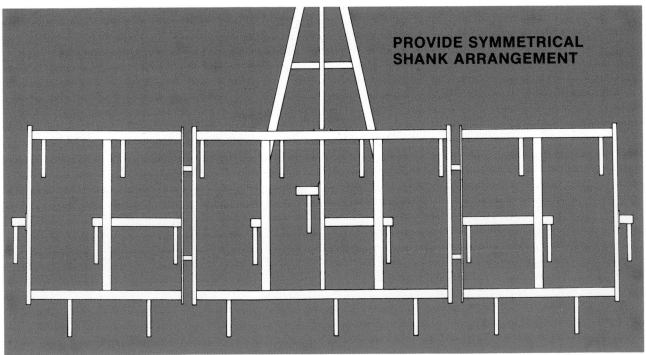

PROVIDE SYMMETRICAL SHANK ARRANGEMENT

If it becomes necessary to remove shanks to reduce draft in extremely hard soils, always remove the same number of shanks from each side of the plow to maintain a balanced load.

TRACTOR PREPARATION

Chisel plowing, like most other primary-tillage operations, requires relatively high power input. Therefore, the tractor must be in good mechanical condition and tuned for maximum power output and optimum fuel consumption. Tires must be inflated to recommended pressures and ballast added to prevent excessive wheel slippage, as outlined in the tractor operator's manual.

Change engine oil, oil filter, and hydraulic filter, clean or replace air filter, and lubricate according to instructions (Fig. 25). Make certain that the hydraulic system is functioning properly and has an adequate supply of oil, and that hydraulic cylinders and hoses are free of leaks.

If using new cylinders and hoses for the first time, extend and retract cylinders several times to fill them with oil and expel air from the system. Then recheck hydraulic-oil level in the tractor, particularly if the plow has several large cylinders.

Some operators prefer to leave the drawbar free to swing for easier turning with wide chisel plows, but pinning the drawbar in the center of the tractor provides stable operation, particularly in very hard soil.

For small integral chisel plows, adjust tractor wheel-tread so overall wheel width is less than operating width to keep wheels off of worked soil. When chiseling cornstalks or other row crops it may be desirable to adjust wheel-tread to keep tires off of rows for easier steering and control.

Fig. 25—Lubricate Tractor According to Instructions

MACHINE PREPARATION AND MAINTENANCE

Well-cared-for equipment reflects pride of ownership and concern for getting optimum performance. Properly serviced and maintained equipment is not only more reliable in the field, but is worth more when sold or traded.

Careful chisel-plow preparation for operation or storage should include the following items:

Before Each Season:

1. Tighten loose nuts and bolts to specified torque and replace broken, worn, or missing parts.

2. Check soil-engaging tools for excessive wear or broken points.

3. Lubricate entire machine as recommended in operator's manual.

4. Check tires for proper inflation and for damage that could cause failure in the field. Be sure all tires are of equal size.

5. Check lift linkage for proper travel and level operation side-to-side.

6. Check cables (if used for folding wings or lifting) for fraying or loose ends.

Daily Before Operation:

1. Lubricate as recommended in operator's manual.

2. Visually check for loose bolts, nuts, cotter pins, worn or broken parts, and under-inflated tires.

Before Storage At End Of Season:

1. Remove all trash, accumulated soil, and dirty grease from the machine.

2. Repaint spots where paint has been scratched or worn off.

3. Coat soil-engaging parts with plow-bottom paint or heavy grease to prevent rust.

4. Lubricate entire machine as recommended in operator's manual.

5. Store the plow inside, if possible, to prevent weathering.

6. Raise plow and block it up to remove weight from tires. Protect tires from sun if stored outdoors.

7. Install safety lock and relieve pressure in hydraulic cylinders or lower machine to the ground. Fully retract cylinders to protect cylinder rods from rust.

8. Place boards under the points if they are lowered to the ground for storage.

ONE CHISEL PLOW FIELD PATTERN

Fig. 26—One of Many Chisel Plow Field Patterns

9. For safe storage, lower folding sections to the ground.

FIELD OPERATION

Chisel plows may be operated satisfactorily in many field patterns (Fig. 26). Work may be started at one side of the field and adjacent passes made until the field is finished. Headlands, usually about twice the width of the machine, are worked last.

Chisel plows may also be worked in lands with headlands plowed last, or operated around the field until reaching the center.

Regardless of the plowing pattern, the plow must be raised from the ground when making sharp turns. This makes steering much easier and protects shanks and the frame from heavy, twisting side forces.

PLOWING ANGLE

To provide better leveling and maximum loosening of row-crop stubble, such as cornstalks, it is best to cross rows at an angle of 20 to 30 degrees. This assures that all roots are cut, and spreads the ridges that were formed during cultivation. If the crop was planted on ridges or furrowed for irrigation, it is usually necessary to follow the rows for the first pass. Subsequent operations are then made diagonally or at right angles for better leveling and more thorough working of the soil.

PLOWING DEPTH

When the soil is extremely hard, it may be necessary

first to loosen a shallow layer. Then make a second pass at the desired depth while working at an angle to the first pass.

If a relatively thin plow sole or hardpan has developed just below normal moldboard-plowing depth, chisel plows may be used to break it. However, the soil must be relatively dry to assure adequate shattering of the hard layer, or there will be little benefit from working to the greater depth.

Adding a rod-weeder attachment to a chisel plow combines many advantages of both machines—deep tillage, weeding, packing, and mulching (Fig. 27).

TRANSPORT AND SAFETY

Simplicity and safety are prime considerations in the design and manufacture of current farm equipment, but there is no safety device which can replace a careful operator. Follow the rules suggested here and the specific steps listed in the operator's manual to improve the safety and welfare of the operator and bystanders.

1. Reduce speed when transporting chisel plows over rough or uneven terrain.

2. Use lock-up straps or transport locks when transporting a chisel plow.

3. Be sure wings are locked in the folded position before traveling.

4. Use proper lights, reflectors, and a clean SMV emblem when transporting equipment on road or highway.

5. The transport width of most folding-section plows exceeds maximum width of normal vehicles. Therefore, use extreme caution when meeting other traffic, to avoid collisions and the possibility of transport wheels dropping into holes, drains, or ditches along the road edge.

6. Allow only the operator to ride on the tractor.

7. Never allow passengers to ride on the chisel plow.

8. Never allow anyone to stand or work near the chisel plow when it is in operation, particularly when raising or lowering outriggers.

9. Do not permit children to play on or near the chisel plow during operation or storage.

10. Provide adequate tractor front-end ballast for stability in transport and operation, especially with integral models.

11. Be particularly careful of escaping hydraulic fluid which can penetrate the skin and cause serious infection or reaction if not given immediate medical treatment.

12. Do not unhitch from the tractor or store a chisel plow when outriggers are in the raised position.

13. Make sure raised outriggers will safely pass under power and telephone lines.

ROD WEEDER

ROD WEEDER ATTACHMENT

Fig. 27—Chisel Plow with Rod-Weeder Attachment

TROUBLESHOOTING

Most chisel-plow performance problems stem from improper adjustment or servicing of the equipment. When difficulties arise, use the following checklist and make corrections as called for here and explained in the operator's manual.

Where possible remedies to problems are obvious, based on the possible cause, a blank space is left in the Possible Remedy column.

TROUBLESHOOTING CHART		
PROBLEM	**POSSIBLE CAUSE**	**POSSIBLE REMEDY**
UNEVEN PENETRATION FORE-AND-AFT	Machine not properly leveled.	With integral plow—extend tractor top link to increase penetration of rear shanks, shorten to increase depth of front row. Drawn chisel plow—raise tongue in relation to drawbar to increase depth of rear shanks, lower tongue for more penetration in front.
UNEVEN PENETRATION SIDE-TO-SIDE	Machine not properly leveled.	Level integral models laterally by adjusting 3-point-hitch lift link. Drawn plows—be sure all wheels have equal-sized tires equally inflated. Be certain all remote cylinders are adjusted for same

PROBLEM	POSSIBLE CAUSE	POSSIBLE REMEDY
		operating depth and working against stops. Adjust shims on rockshaft or check cable length and linkage adjustments.
TOO MUCH POWER REQUIRED	Machine set too deep.	Reduce operating depth. Make second pass at greater depth and different angle for best results.
	Machine too big for available power.	Remove some shanks from frame to reduce working width (remove equal number from each side of frame). Use larger tractor.
	Using improper tools for depth being worked.	Use spikes or points for deep penetration, sweeps or shovels for shallower work.
	Soil-engaging tools dull.	
PLOW YAWS SIDE-TO-SIDE OR SWINGS TO ONE SIDE IN HARD SOIL	Machine not leveled laterally.	Relevel machine, check while in operation.
	Uneven penetration.	Install gauge wheels toward the outer ends of plow frame.
	Operating too deep.	Reduce working depth on first pass, make second pass at desired depth at 30 degree angle to first pass.
	Soil too hard for sweeps.	Install chisels or spikes on front row to break hard soil, use sweeps on other shanks.
	Front Shanks breaking most soil.	Set chisel plow slightly lower at the rear than in front.
	Shanks not equally spaced on each side of frame.	
	Unequal number of shanks on each side.	
EXCESSIVE TRASH PLUGGING	Poor shank arrangement.	Place shanks with maximum lateral space between shanks and staggered so that trash flows through freely.
	Twisted shovels improperly installed.	Half of twisted shovels must be right hand—half left hand. All twisted shovels on front row turn soil same direction; entire middle row turns soil opposite way. Split rear row and turn soil toward the center of the plow.
	Too much long, heavy trash.	Disk or shred before chiseling.

SUMMARY

Chisel plows basically are used for primary tillage. Their main function— as opposed to disk or moldboard plows—is to break, stir, and aerate the soil with little inversion or coverage of trash. Thus, strangely, they are "sophisticated descendants" of the forked sticks the world's first farmers used!

Versatile modern chisel plows have two or three rows of staggered shanks (usually one shank per foot of width), attached to box-beam frame bars, which may be used for a wide variety of shovels, chisel points, spikes, and sweeps ranging up to 30 inches (76 cm) wide (for killing weeds or summer fallowing).

Chisel-plow draft is perhaps half that of a moldboard plow working to the same width and depth, which means that any given tractor can cover up to twice as many acres in a day when chisel-plowing.

Chisel plows leave a rough, trash-covered surface which is ideal for moisture retention and reducing wind erosion. They are excellent tools for stubble-mulch practices. They accomplish little in wet soil, but are very effective when soil is dry and firm, and are at their soil-shattering best at fairly fast speeds in such conditions.

They are available for almost any size tractor. Widths range from 5- to about 20-foot (1.5 to 6 m) integral models to drawn models with folding outriggers which can work 350 or more acres (140 ha) a day at 6 miles (10 km/h) an hour (Fig. 28).

CHAPTER QUIZ

1. Chisel-plow draft is about _____ that of a moldboard plow working at the same width and depth.

2. (True or false) Chisel plows perform best when soil is soft and moist.

3. (Choose one.) When used for summer fallowing, a chisel plow should be equipped with:

    A. Chisel points and worked as deeply as possible.

    B. Wide sweeps set just deep enough to kill weeds with minimum disturbance of surface residue.

4. Why is it particularly important to have chisel-plow shanks equally spaced on each side of the plow centerline?

5. What is the best way of getting deep tillage with a chisel plow if soil is extremely hard or power is limited?

6. List six safety requirements for transporting and using chisel plows.

Fig. 28—Wide Chisel Plows Increase Daily Capacity

WIDER PLOWS INCREASE DAILY CAPACITY

8
Stubble-Mulch Plows

WIDE-SWEEP PLOW

Fig. 1—Wide-Sweep Plow for Stubble-Mulch Tillage

INTRODUCTION

Wide-sweep or stubble-mulch plows cut off weeds at the roots and leave residue anchored to the surface with minimum soil disturbance, a real advantage where limited quantities of residue are present and erosion is a serious problem. Reduction of wind and water erosion and storage in the soil of all available moisture are primary objectives.

Wide-sweep plows are used for both primary and secondary tillage, either immediately after harvesting or just prior to planting. Sweep plows also are used to control weeds in summer fallow.

Wide-sweep plows have long V-shaped sweeps or straight blades which operate nearly at right angles to the direction of travel. Because of the wide shank spacing and relatively flat blade angle, these plows normally cover no more than 10 to 15 percent of the original residue with each pass (Fig. 1).

If the mulch is extremely heavy it may be desirable to use a wide-sweep plow for early tillage and then mix some residue into the soil with a disk tiller or disk harrow before planting.

For maximum weed kill, plowing should be done on a hot day when soil is dry enough to crumble well. Under most conditions, speeds of 4 to 6 miles an hour (6.5 to 10 Km/h) provide the best results and do the best job of loosening soil from weed roots.

In stubble, sweeps are normally operated just deeply enough to pass under the crown of the plant — about 3 to 4 inches (75 to 100 mm). If heavy accumulations of residue have been mixed previously with the soil, it may be necessary to operate deeply enough for blades to pass under that trash. Repeated plowings will work the trash back toward the surface. The shattering, crumbling action caused by the blade leaves a loose, porous surface for better water infiltration. It also helps level rough fields.

Improved weed control may be obtained by adding rod-weeder or stubble-treader attachments to wide sweep plows (Fig. 2), or using them as separate implements. Additional weeds are killed and brought to the surface by these units, and residue is more firmly anchored to reduce blowing and water runoff.

Additional machine versatility is provided by installing anhydrous-ammonia applicators behind the sweeps (Fig. 3). The ammonia tank can be mounted directly on the plow frame (Fig. 4), or on a trailer behind the plow.

An optional disk marker mounted on each end of the plow frame makes steering easier and reduces overlapping and unworked spots.

TYPES AND SIZES

Wide-sweep-plow sizes range from one to nine sweeps, with each sweep ranging from 4 to 8 feet (1.2 to 2.8 m) wide. Different combinations of blade size and number are available to match tractor power and field size. Fewer and wider sweeps may be used on some machines. The sweeps usually are staggered on the plow frame to permit smooth trash flow, and overlapped to assure complete

STUBBLE-TREADER ATTACHMENT

Fig. 2—Stubble-Treader Attachment for Wide-Sweep Plow

cutting of weed roots and to keep large weeds from slipping around the ends. Most common sweep sizes are in the 5-to 6-foot (1.5 to1.8 m) range.

Wide-sweep-plow frames are much like chisel-plow frames, but somewhat stronger, with extra reinforcement at stress points and for shank attachment.

Small wide-sweep plows may have one or two sweeps, and transport width is essentially the same as cutting width. Larger units have "wings" with one or two sweeps each. These wings are folded hydraulically for transport, and are attached to a center section with one or two sweeps. Some 6-sweep models have two wings on each side, with the outer sections folded flat across the top for transport (Fig. 5). This reduces transport height as well as width.

One or more remote hydraulic cylinders control working depth, raise and lower the machine, and fold wings for transport. Optional flotation tires on some machines provide increased stability in soft or loose soils.

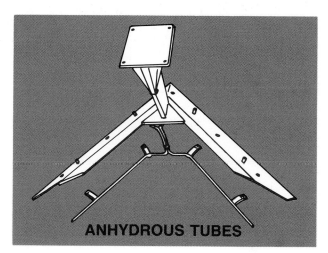

ANHYDROUS TUBES

Fig. 3—Anhydrous-Ammonia Attachment for Wide-Sweep Plow

Fig. 4—Ammonia Tank on Plow Frame

AMMONIA TANK

Fig. 5—Six-Sweep Plow Folded for Transport

PLOW FOLDED FOR TRANSPORT

Rolling coulters, usually equipped with rippled blades for more positive cutting and turning, slice through trash directly in front of each shank to prevent trash buildup and plugging. This means that more field time is spent plowing—less time spent unplugging trash.

PRINCIPLES OF WIDE-SWEEP PLOW OPERATION

Wide sweeps, operating from 3 to 6 inches (76 to 152 mm) beneath the soil surface, slice off weed roots, lift and shatter the soil (if it is dry enough), and leave residue on top (Fig. 6) The resulting mulch of stubble and dead weeds, plus the rough, shattered clods on the surface, reduce wind erosion and help catch and hold rainfall.

Basic performance and operation of wide-sweep plows is very similar to that of chisel plows, except that the wider sweeps leave more crop residue on the surface.

TRACTOR PREPARATION

Tractor preparations are similar for wide-sweep plows and chisel plows:

• *Provision of adequate drawbar horsepower for machine size*

• *Ample hydraulic capacity for depth control and wing folding*

• *Sufficient weight to eliminate excessive wheel slippage*

• *Good mechanical condition for optimum performance and minimum fuel use*

See Tractor Preparation in Chapter 7 on CHISEL PLOWS.

Fig. 6—Wide Sweeps Leave Residue Exposed on Top

WIDE SWEEP EXCESS RESIDUE

MACHINE PREPARATION AND MAINTENANCE

Extra time spent caring for equipment and properly preparing it for operation will usually be recovered several times over in better field performance, more economical operation, less down time, and fewer repairs.

BEFORE EACH SEASON;

1. Clean, repack and tighten wheel bearings. Replace worn bearings as needed.

2. Examine hydraulic hoses, couplings, and cylinders for wear, damage, or leaks. Repair or replace as needed.

3. Check for loose or missing bolts and nuts. Replace worn or broken parts.

4. Sharpen or replace dull sweeps for better cutting and reduced draft.

5. Make certain that all tires are the same size and inflated to the recommended pressure to provide level machine operation.

DAILY BEFORE OPERATION:

1. Lubricate the machine as recommended in the operator's manual.

2. Visually check for loose bolts, worn parts, or underinflated tires.

BEFORE STORAGE AT END OF SEASON:

1. Clean all trash, accumulated soil, and dirty grease from the machine.

2. Repaint spots where the paint has been scratched or worn off.

3. Coat sweeps with plow-bottom paint or heavy grease to prevent rust.

4. Lubricate the machine as recommended in the manual.

5. Store the plow inside if possible to prevent weathering. Block up the plow to remove weight from the tires.

6. Install the safety lock and relieve the pressure in the hydraulic cylinders, or lower the machine to the ground. Fully retract cylinders to protect the cylinder rods from rust.

7. Place boards under sweeps if sweeps are lowered to the ground for storage.

8. For safety, do not store the plow with wings in the folded position.

FIELD OPERATION

Operating characteristics of wide-sweep plows and chisel plows are similar, and both machines are handled in much the same manner. Sweep-plow performance is best and draft lowest when all blades are running level from side to side and at the same depth (Fig. 7). Sweep points should either be set level with the blade ends, or as much as a half-inch (13 mm) lower to aid penetration in hard soils.

Shims or other means of adjustment between the main frame and shanks permit leveling sweeps with the frame. Hitch adjustments are usually provided for fore-and-aft leveling. The hitch setting must be readjusted when changing operating depth or switching to a tractor with different height drawbar. Adjustable links between the center and wing sections permit lateral leveling.

Adjust the rolling coulters to work directly ahead of the sweep shanks and just deep enough to cut through trash and keep turning.

Sweep blades must be sharp to be able to penetrate hard soil. If penetration remains a problem after sweeps have been sharpened and leveled, it may be necessary to add additional weight to the frame.

Important: Always fasten weights securely to the frame to prevent shifting, bouncing, and possible equipment damage. NEVER raise wings unless weight is removed or firmly attached to the plow frame.

Work is normally started at one edge of the field, a 180-degree turn made at the end of the field, and the return pass made adjacent to the previously worked land. When the field is finished, headlands are plowed to complete the work.

If erosion is a problem it may be desirable to work at right angles to prevailing winds, or on the contour to reduce runoff from heavy rainfall.

Always lift the machine from the soil when making sharp turns to avoid twisting of shanks and sweeps.

TRANSPORT AND SAFETY

Most accidents are caused by the failure of some individual to follow simple and fundamental safety rules. Anticipating dangers and taking preventive action can prevent many accidents.

1. Use the SMV emblem, lights, and reflectors as required by law for transporting equipment on roads or highways.

2. Install the transport safety lock and relieve hydraulic pressure in the cylinders when transporting.

3. Be certain wings are securely locked in the folded position before transporting.

4. Limit transport speed as recommended—15 miles an hour (24 Km/h) for some machines, even less on rough or uneven terrain.

5. Never ride or allow others to ride on the machine during operation or transport. Allow only the driver on the tractor.

6. Do not unhitch from the tractor or store the machine with wings in the folded transport position.

7. Sweeps are sharp—watch out for them when wings are folded for transport or when servicing, adjusting, or repairing the plow. Always position wrenches to pull away from sharp edges or corners.

8. Never stand with feet under blades while making adjustments or during maintenance. Be extremely careful while working within the implement frame.

9. Do not stand or walk on the plow frame, or under wings when they are folded for transport.

10. Stand with both feet on the same side of the tongue when hitching or unhitching.

11. Do not allow children to play on or near the plow.

Fig. 7—Blades Run Level at Equal Depth

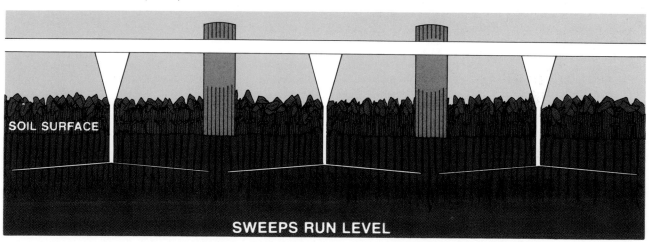

SOIL SURFACE

SWEEPS RUN LEVEL

TROUBLESHOOTING

When performance problems develop, refer to the operator's manual and the following check list for the symptoms and solutions. Simple adjustments or changes can often solve what appear to be major problems.

Where possible remedies to problems are obvious, based on the possible cause, a blank space is left in the POSSIBLE REMEDY column.

TROUBLESHOOTING CHART		
PROBLEM	**POSSIBLE CAUSE**	**POSSIBLE REMEDY**
PLOW WON'T PENETRATE	Sweeps not level.	Adjust hitch to level plow fore-and-aft.
		Level plow laterally—see manual.
		Tilt points down as much as a half-inch (13 mm) below blade ends for hard ground.
	Inadequate weight.	Firmly attach additional weight to plow frame.
	Sweeps too dull.	
TRASH PLUGGING	Coulters not turning.	Lubricate or replace bearings if needed.
		Set coulters lower so they cut deeper into the soil.
	Coulters set too deep.	Set coulters just deep enough to keep turning. If set too deep, trash is pushed instead of being cut.
	Trash is bunched.	Use straw spreader on combine.
		Scatter bunches or shred residue.
	Trash too wet.	
	Coulters too dull.	
PLOW RAISES TOO SLOWLY, OR INCOMPLETELY	Inadequate hydraulic pressure.	Check tractor hydraulic system for malfunction.
		Install larger cylinder on plow.
	Lift linkage binding or damaged.	Check linkage—lubricate, repair, or replace parts as needed.
	Tractor low on hydraulic fluid.	

SUMMARY

Wide-sweep plows are used in areas where lack of moisture is a problem. Actually, they're not plows at all—they're more closely related in function to rod weeders, for they kill weeds effectively with minimum disturbance of erosion-controlling surface trash and minimum moisture loss. They also loosen the upper 3 to 6 inches (76 to 152 mm) of soil (depending on depth of operation) for better rainfall infiltration.

Wide-sweep plows are used immediately after harvest, just prior to planting, or for weed control in summer-fallowed fields. The V-shaped blades may be from 4 to 8 feet (1.2 to 2.4 m) long, and implement capacity varies from a single sweep for small tractors to six or more if power permits. Larger units have folding outer sections for easier transport.

CHAPTER QUIZ

1. What are the primary objectives of using wide-sweep plows?

2. What is the normal operating depth of wide-sweep plows?

3. Why are rolling coulters used on wide-sweep plows?

4. (True or false) Wide-sweep plows perform best in normal conditions when the ends of the sweeps are about a half-inch lower than the point.

5. Hot or cool (choose one) days are best for using wide-sweep plows?

6. Name two ways of increasing penetration of wide-sweep plows in hard soil.

9
Disk Tillers

Fig. 1—Disk Tiller in Operation

INTRODUCTION

Disk tillers fall between the disk harrow and moldboard and disk plows in function (Fig. 1). They consist of spherical blades, which normally throw the soil to the right, mounted on a common axial shaft (or gang shafts for flexible types).

Disk tillers also are known as wheatland disk plows, one-ways, one-way disks, diskers, seeding tillers, vertical disk plows, and by other names. They were developed in the Great Plains area about 1927. They quickly were blamed for the dust storms in those parts in the 1930's, and even were accused of "poisoning the ground." But today they are considered as invaluable tillage tools when properly used.

Disk tillers should never be permitted to produce a smooth, bare, powdery surface condition which encourages wind erosion and runoff of surface water. Good management practices, and use of supplementary implements, such as the field cultivator, chisel plow, and rod weeder, help to meet particular farming conditions and maintain good soil texture.

The disk tiller is normally considered to be a dryland implement for seedbed preparation and planting small grain, but has been successfully used in many soil types and moisture conditions, including riceland. It is being used on limited scale in the Southeastern United States to double-crop soybeans in stubble immediately after small-grain harvest.

Because disk tillers are somewhat difficult to adjust, they are experiencing competition from heavy tandem and offset disk harrows.

Tillers are efficient tools in terms of horsepower-hours per acre (KW-h/L) for working large acreages. Average disk-tiller draft at a depth of 3 to 5 inches (76 to 152 mm) varies from 180 to 400 pounds per foot (2.6 to 5.8 kN/m) of cut, depending on type and condition of soil. Therefore, to operate a 16-foot tiller would require a tractor with 2,880 to 6,400 pounds (12.8 to 28.5 kN) of drawbar pull—31 to 68 drawbar horsepower (23 to 51 drawbar kW).

The normal function of a disk tiller is to cut and mix soil and plant residues with a minimum of soil pulverization. A considerable amount of residue is normally left protruding from the tilled soil to control wind and water erosion. Depth may be varied from 2 to 8 inches (50 to 203 mm), depending upon blade size and spacing.

Disk tillers are normally used for primary tillage, and in subsequent operations for summer fallowing. Their weed control is excellent. By adding seeding and fertilizing attachments, they may be used to prepare a seedbed, seed, and fertilize in a single, efficient operation (Fig. 2).

Single frame tillers are designed to operate with the right tractor wheels in the furrow. However, crawler tractors and tractors with large tires or dual wheels can be operated on land by installing a wide hitch attachment (Fig. 3). Multiple frame tillers also use a wide hitch attachment so that tractors can operate on land.

TILLER TYPES AND SIZES

Disk tiller types include:

- *Drawn, which is attached to the tractor drawbar.*

- *Integral, which is attached to the tractor 3-point hitch.*

- *Reversible, which moves soil to right and left on alternate passes.*

Let's look at each of these more closely.

DRAWN TILLERS

The largest tillers are the drawn type. These tillers are available with either *rigid* or *flexible* frames, either single or multiple frames (Fig. 4).

Rigid-Frame Tiller

Gangs on rigid-frame tillers are attached to the frame and cannot flex with the contour of the land. The frame weight aids penetration (Fig. 4). A typical tiller may have 9-inch (230 mm) disk spacing with 22-, 24-, or 26-inch (559, 610, or 660 mm) - diameter disks. A typical 12-foot (3.7 m) tiller may use 18 disks with a maximum cut of 11 feet, 8 inches (3.6 m). The 16-foot (4.9 m) size may use 23

Fig. 2—Tilling, Seeding and Fertilizing in Once-Over Operation

Fig. 3—Wide Hitch Attachment Permits On-Land Tractor Operation

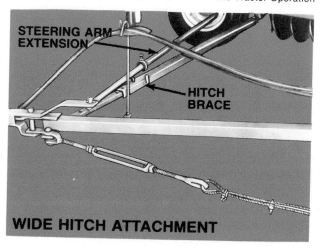

STEERING ARM EXTENSION

HITCH BRACE

WIDE HITCH ATTACHMENT

Fig. 4—Rigid-Frame and Flexible-Frame Tillers

Fig. 6—Hydraulic Cylinders at Each End Control Depth

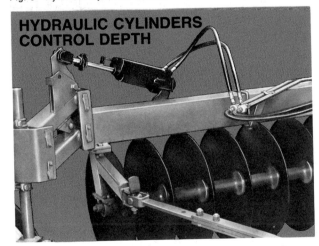

Fig. 5—Rear Coulter for Better Tiller Control

Fig. 7—Hand Jacks are Available if Preferred

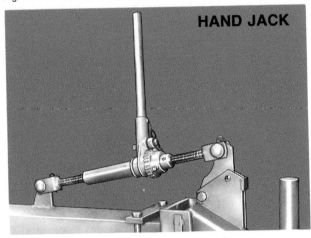

Fig. 8—Gangs Can Flex Over Obstructions

disks to cut 14 feet, 10 inches (4.6 m) in the widest set-ting, while the 20-foot (6 m) model may cut a maximum of 18 feet, 3 inches (5.6 m) with 28 disks. Multiple hitching permits two or three of these tillers to be pulled by one large tractor for greater daily acreage.

A rear coulter, which helps keep the tiller running straight even in difficult conditions, is standard equipment on this type (Fig. 5).

Working depth is controlled by a remote hydraulic cylinder at each end of the frame (Fig. 6) or by handscrew lift jacks (Fig. 7).

Flexible-Frame Disk Tiller

Flexible-frame tillers are designed to allow the gangs to follow ground contours and flex over obstructions (Fig. 8). These tillers ordinarily are available in 8-, 12-, 15-, 16-, 18- or 20-foot (2.4, 3.6, 4.5, 4.9, 5.4 or 6 m) sizes. Two or three machines may be combined for double or triple widths up to 45 feet. Normal operating depth is 2 to 5 inches (50 to 127 mm).

Mutliple hitching and use of seeding and fertilizer attachments on flexible tillers permit faster seeding with fewer man hours and reduced equipment investment (Fig. 9).

Heavy-Duty Disk Tiller

Heavy-duty disk tillers may have up to 10-inch (254 mm) blade spacing with 24- or 26-inch (610 or 660 mm) diameter disks for tilling as much as 9 inches (229 mm) deep (Fig. 10). Flexible, 5-disk gangs permit this tiller to flex over ground contours, stones, and other obstructions.

HEAVY-DUTY TILLER

Fig. 10—Heavy-Duty Tillers can Work to 9 Inches (229 mm) Deep

Fig. 9—Multiple Hitching Increases Field Capacity

MULTIPLE HITCHING INCREASES CAPACITY

Fig. 11—Cast-Iron Wheel for Tilling—Rubber Tire for Transport

A heavy cast-iron furrow wheel helps keep the tiller in line. If additional ballast is required, rocks may be carried in the weight box. A rubber-tired transport wheel is available to raise the cast rear wheel off the ground for transport (Fig. 11).

Integral Disk Tiller

Integral tillers are attached to the tractor 3-point hitch, and are limited in size by tractor lift capacity and front-end stability. Operating depth of the front disk and lateral leveling are controlled by the tractor right-hand lift link. The tractor top link levels the tiller fore-and-aft. A sharp rear wheel cuts into the furrow bottom behind the rear disk to hold the machine in the proper operating position. The right tractor wheel is operated in the furrow. Common sizes have four to seven disks with 10-inch (250 mm) spacing and 24- to 26-inch (610 to 660 mm) blades.

Reversible Disk Tiller

Reversible tillers turn soil to the right and left on alternate passes, thus eliminating a dead furrow in the middle of the field and the need to plow out the corners.

Hydraulic cylinders are used to reverse the frame, position the disks, and set the rear wheel to its operating position on each reversal. Because of the reversing mechanism and other extra components, this is a more expensive machine and has limited usage.

PRINCIPLES OF DISK-TILLER OPERATION

Disk tillers may be adjusted for maximum cutting width consistent with desired results and minimum draft. They are operated around the field, or worked in lands to fill the previous furrow so the field is left level with no furrows, wheel tracks, or uncut ridges. When operating around the field, an extra round is required at each corner to plow out the spots left uncut on each turn (Fig. 12).

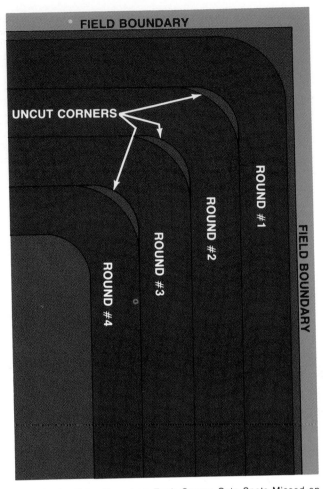

Fig. 12—An Extra Round on Each Corner Cuts Spots Missed on Turns

Some of the factors involved in tiller operations are:

- **Disk diameter**
- **Cutting angle**
- **Soil forces**
- **Weight**
- **Line of draft**
- **Furrow wheels**

Let's look at each factor in more detail.

DISK DIAMETERS

Disk diameters are commonly 18 to 26 inches (457 to 660 mm) with blade spacings of 7 to 10 inches (178 to 254 mm), depending upon soil type and work to be done. If the tiller will be used primarily for summer fallowing and seeding, smaller disks and closer spacings are commonly recommended. Larger disks and wider spacings are used for deeper and primary tillage.

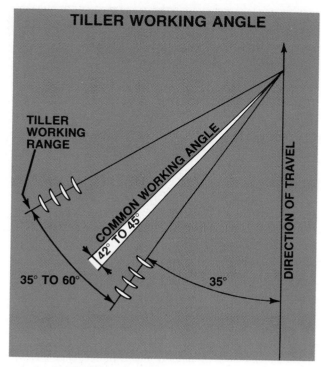

Fig. 13—Disk Tiller Working Angle

CUTTING ANGLE

There are two methods of measuring the cutting angle of disk tillers.

First is to measure the angle between disk gangs and the direction of travel. The second is to measure the angle between disk gangs and a line perpendicular to the line of travel.

Tiller WORKING ANGLE is commonly 42 to 45 degrees (Fig. 13), but it varies from about 35 to 60 degrees from the line of travel, depending on machine, soil conditions, and desired results. This is less than disk-harrow working angle, which is usually measured from a line *perpendicular* to the line of travel.

SOIL FORCES

Soil forces on an individual disk are very complex. The net forces of cutting soil and trash and pulverizing and turning the soil, plus friction losses, may be expressed in several ways. However, the operator is primarily interested in total forces on the entire machine.

When a tiller is operated at maximum width (LARGE ANGLE) the blades tend to roll on the surface like wheels. Thus, considerable downward force (frame weight plus additional weight) is necessary to obtain penetration.

Other than pulling the extra weight, draft load is fairly low because very little energy is applied to the soil and speed can be increased with relatively low pul-

verization of soil. Such a setting is normally used for light work, such as seeding or summer fallowing.

When disk blades are operated at a SMALL ANGLE to the direction of travel (reduced width of cut), disk concavity aids penetration and the downward vertical force required is less than when operating at a wider width of cut. Speed of disk rotation is markedly reduced and the width of cut per disk is decreased. More trash is covered with disks in this setting, and soil pulverization is increased, especially at higher speeds.

However, total machine draft may be increased because this setting is normally used in hard soils, heavy trash, or for deep work. In the same soil and operating conditions, draft is generally reduced by operating at a disk angle of approximately 45 degrees compared to both lesser and greater angles from direction of travel. Keep in mind that reducing total width of cut improves penetration in hard ground.

Disk tillers are usually operated between these extreme disk angles, using the widest possible machine angle (smallest angle of cut per disk) which will perform the quality of work desired (Fig. 14).

WEIGHT

Because the soil resistance is applied to disk blades below the ground surface, the resulting force causes the front disk of each gang to tend to penetrate deeper and the rear disk to come out of the ground.

Built-in tiller weight provides sufficient penetration for many operations. However, additional weight (Fig. 15) may be required to force blades into the ground and retain sufficient weight on the wheels for holding the tiller in its correct operating position. Extra weight added to or near the rear furrow wheel normally provides the greatest benefit. The weight may be added to the frame, to the wheel hub, or as calcium chloride solution in rubber-tired furrow wheels.

A properly weighted tractor should handle the largest tillers and even multiple units without steering problems. When a tractor does not have a swinging drawbar, the hitch should be attached to the tractor as nearly as possible on the line of draft from the tiller center of load to the tractor center of pull, as will be discussed next.

LINE OF DRAFT

The center of load of a simple disk tiller is located approximately at the middle disk and below the surface of the soil. To determine the true line of draft, stop the tiller with the right front wheel in the furrow so the front disk will cut about one inch less width than the other disks. Position the tractor where it would normally be operating — right wheel in the furrow, or near the furrow for on-land operation. Then stretch a string from the midpoint of the disk gangs (center of load) to the front pivot point of the tractor's swinging drawbar (center of pull), which is usually on the tractor centerline and slightly ahead of the rear axle.

155

Fig. 14—Operate Tiller at Widest Possible Angle to Perform Quality of Work Desired

Most tiller hitches are free to flex at the front end, which lets them adjust to the line of pull. A tractor swinging drawbar does the same thing. Hitch members may not actually be attached to the tiller frame at the center of load, but the forces and line of draft will always pass through that point.

On larger tillers this line of pull must angle toward the right when the tractor wheel is in the furrow, which means there will be some side draft on the tractor. Adjusting tractor wheel-tread, so the tractor drawbar runs as closely as possible to the center of the tractor,

will help offset objectionable side draft. When operating with tractor wheels in the furrow, always set front and rear tread equidistant, center-to-center.

FURROW WHEELS

To help offset soil forces, set the tiller front furrow wheel with about 5-degree lead to the right, toward plowed ground (Fig. 16). When the tiller is operating, weight on the land wheel is quite small and has very little holding effect on the tiller. Therefore, the land wheel is set to run straight forward, and on many tillers is designed to caster for easier steering and turning (Fig. 17).

The rear furrow wheel provides the main holding force on a disk tiller. This wheel may be mounted on rubber for easy transport, but a rubber tire may not provide sufficient holding capacity to overcome high-thrust loads, especially in hard ground or deep plowing. One solution is the use of a coulter wheel. This is a sharp wheel or disk behind the rubber tire. The wheel is controlled from a toggle lever. The wheel runs 1 to 1½ inches (25 to 38 mm) deep in the hard furrow bottom with a small lead to the right (Fig. 18). This wheel can easily be raised from the ground for transport or when not required for tiller stabilization (Fig. 5).

Other tillers, particularly heavy-duty models, have a heavy cast rear wheel with a "V" face, which either is replaced with a rubber-tired wheel for transport and summer fallow, or lifted off the ground by a rubber transport wheel for travel (Fig. 19).

Fig. 15—Frame Weights Aid Penetration and Control

FRAME WEIGHTS

156

Fig. 16—Lead Front Furrow Wheel Toward Plowed Ground

Fig. 18—Rear Coulter Cuts into Hard Furrow Bottom

TRACTOR PREPARATION

On some farms disk-tiller operation consumes more power than any other task. Therefore, adequate tractor preparation not only will save fuel and operating time, but can help prolong tractor life, and even increase yields by timely completion of field work.

Before going to the field:

1. Get a complete engine tune-up, change oil and filter and air filters.

2. Thoroughly check hydraulic system—oil level, new filters, etc. Check hoses, couplings, and cylinders for leaks or damage and repair as necessary.

3. Adjust wheel tread so line of draft falls as near center of tractor as possible.

4. Check tires for recommended inflation; consider adding calcium chloride solution for added ballast if required.

Fig. 17—Land Wheel Runs Straight Forward, and May Be Free to Caster

Fig. 19—Cast-Iron Furrow Wheel for Heavy-Duty Tiller

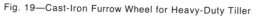

ADD BALLAST TO REAR WHEELS

Fig. 20—Add Ballast to Prevent Excessive Slippage

5. Add sufficient ballast to prevent excessive wheel slippage—about 15 percent slip is the recommended upper limit (Fig. 20). When pulling rated load, tread marks should be slightly broken and shifted. Too little weight will cause tread marks to be completely wiped out. Too much weight causes tracks to be clear and distinct, and could cause drive-train failure.

6. If using integral disk tiller, add front ballast to maintain tractor stability.

7. Set tractor drawbar at height recommended in tiller manual and free to swing in the field.

TILLER PREPARATION AND MAINTENANCE

Spending a little extra time to prepare the tiller thoroughly before going to the field or putting it into storage may save hours of field time, reduce tractor fuel consumption, and lower repair costs. Field performance will also be improved, resulting in more uniform weed control and better seedbed preparation and seeding (Fig. 21).

At The Start Of Each Season:

1. Use kerosene or diesel fuel to clean heavy oil from fertilizer attachment and other moving parts. Remove paint or other protective coating from disks.

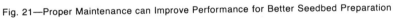

Fig. 21—Proper Maintenance can Improve Performance for Better Seedbed Preparation

PROPER MAINTENANCE CAN IMPROVE PERFORMANCE

Fig. 22—Loosen Seeder Feed Shafts

Fig. 23—Wash Seed and Fertilizer Hoppers before Storage

2. Make sure all moving parts work properly and are free of paint, rust, or dirt.

3. Inspect wheel- and disk-gang bearings—lubricate, tighten, or replace as needed.

4. Sharpen disks if necessary and replace any which are broken or cracked. After replacing disks, or when putting a new tiller into operation, retighten gang bolts every four or five hours of operation for approximately 15 or 20 hours of use, or until all slack is taken up. Follow torque recommendations in operator's manual—they may vary from 500 to 600 foot-pounds (700 to 840 N-m) of torque. Recheck tightness once or twice each season thereafter.

5. Check all bolts, set screws for tightness, and be certain all cotter pins are spread.

Daily Before Going To The Field

1. Lubricate all moving parts according to operator's manual instructions. Clean grease fittings first to avoid forcing dirt into bearings. While lubricating, watch for loose bolts, nuts, screws, or cotter pins.

2. During the planting season, turn feed shafts with a wrench in the direction they normally turn each day before filling seed hoppers (Fig. 22). If using chemically treated seed, turn feed shafts with a wrench any time the machine has been stopped an hour or more. If seed runs stick, remove foreign objects and loosen stuck parts with kerosene or diesel fuel—do not oil.

3. When seeding stops for the day, remove any chemically treated seed remaining in boxes.

4. At the end of each day's work, remove any fertilizer left in the hopper and open hopper bottom. Clean fertilizer from feed wheels and hopper bottom. If machine won't be reused for two days or more, use a water hose to wash out the entire fertilizer hopper and all moving parts. Coat all moving and unpainted parts with diesel oil to prevent rust and corrosion.

5. Before restarting fertilizer attachment, turn feed shaft with a wrench to loosen any stuck parts.

6. During operation, watch for and immediately tighten or replace any loose, worn, or broken bolts or parts. If uncorrected, such problems could cause machine damage, loss of field time, or even personal injury.

Getting Ready For Storage:

1. When the season's work is completed, examine the tiller for any broken or badly worn parts. Order any parts or accessories needed for the next season.

2. Clean and wash seed and fertilizer hoppers, metering devices, and tubes (Fig. 23). Use kerosene or diesel fuel to coat metal seed runs and portions of seed box not protected by paint. Shift seed-rate lever-control several times to work oil into joints.

3. Remove any accumulations of soil, trash, or dirty grease from the tiller which may draw moisture and cause rust or corrosion.

4. Repaint any spots where paint has been scratched or worn off.

5. Coat metal fertilizer feed wheels and unpainted portions of the hopper with oil. Shift rate-control lever several times to work oil into joints.

6. Completely lubricate the machine with recommended oils and greases to keep moisture out of bearings. Coat disks with heavy grease or special disk- and plow-bottom paint to prevent rust.

7. Remove all drive chains. Clean them with kerosene or diesel fuel and either store them in oil, or oil them well and replace them on the machine.

8. Store tiller inside. If such storage is unavailable, at least cover vital working parts and seed fertilizer boxes with a tarpaulin.

9. Place blocks under frame to take weight off tires.

10. Place boards under disks and lower to operating position to permit seed tubes to straighten out. Storing with disks raised may cause permanent bending of tubes.

FIELD OPERATION

Factors that hold a disk tiller stable in operation are:

- *Weight of tiller.*
- *Point from which it is pulled (hitch adjustment).*
- *Lead of front and rear furrow wheels.*
- *Working angle.*

Speed may also be a factor (Fig. 24). Excessive speed can cause over-pulverization of the soil and make it more difficult to maintain stable operation, particularly in some soil conditions. Top recommended operating speed for many units is 4 miles per hour (6.4 Km/h).

Each tiller is designed with means of adjustment, as explained in the operator's manual for that particular model. Consult the manual for instructions, adjustments, and operating tips before starting to use the machine.

Remember: a well-adjusted tiller pulls lighter, performs better, and requires less maintenance and repair than one that has not been adjusted properly.

Fig. 24—Speed is a Factor in Disk Tiller Operation

SPEED IS A FACTOR

Adjust the tiller hitch as outlined under *Line Of Draft* and set the front disk to run slightly shallower than the others. Lower the tiller, pull forward about 100 yards (91 m), and stop. While operating, the tiller frame should be level and the rear wheel should stay in the furrow. If the furrow wheel is equipped with a rubber tire, set it to run against the furrow wall to help absorb side thrust of the tiller. A small lead to the right on front and rear wheels helps absorb additional thrust.

Instructions for preliminary adjustments on some tillers specify setting the machine with no lead in front or rear wheels before it is lowered to the ground. As the machine reacts to soil forces, wheels assume a position with some lead to the right, which may be increased or decreased as necessary after the field is opened and normal operation begun.

Until disks are polished and final adjustments made, do not attempt to run too deep. However, final adjustments cannot be made until a proper furrow has been established and the tractor correctly located with respect to the furrow wall.

Always keep disks sharp. Dull blades will not penetrate well, especially in hard soil, and cannot cut heavy, tough, or damp trash. Dull blades also increase tiller draft.

If the front wheel has a tendency to jump out of the furrow toward unplowed land, lengthen the hitch brace (Fig. 25). Don't allow the front wheel to crowd the furrow wall. After the hitch brace is adjusted, readjust the front-wheel steering bar to maintain a small lead to the right. If the front wheel rides up on plowed land, decrease lead slightly. If a problem remains, shorten rear-wheel linkage to reduce width of cut and realign hitch and steering linkage.

Before final adjustments can be made, the rear wheel must also stay in the furrow and maintain a small lead toward plowed ground. Maintaining excessive lead on furrow wheels increases wear on tires and wheel bearings. Therefore, use only as much lead as is re-

Fig. 25—Adjust Hitch Brace and Steering Linkage to Control Front Wheel

ADJUST HITCH BRACE

ADJUST WIDTH OF CUT

Fig. 26—Adjust Cut by Changing Width-Adjusting Rod

FRAME LEVELING SCREW

LEVEL MACHINE

Fig. 27—Level Machine for Uniform Depth of Cut

quired to maintain satisfactory tiller operation—don't try to correct all performance problems by increasing furrow-wheel lead.

If the rear wheel will not stay in the furrow and tends to ride on unplowed ground, either the machine is cutting too wide or too deep for field conditions, or more weight is needed. Reduce width of cut by shortening the width-adjustment rod (Fig. 26). Then readjust hitch brace and front-wheel steering linkage as necessary to keep the front wheel operating properly. If all adjustments fail to correct the problem, gradually reduce penetration until the tiller will operate in a stable condition.

Also check depth of cut of the front disk. It should cut no deeper than the lower edge of the front wheel (except when opening a field or land). Relevel machine if front disk is cutting too deep (Fig. 27). Depending on machine design, extra weight may be added to the frame near the rear wheel, weights may be attached to each side of the wheel itself, or, if the furrow wheel has a rubber tire, the tire may be filled 75 percent full with calcium-chloride solution.

If the rear wheel tends to ride on plowed land, reduce lead of front and rear wheels slightly. Shift the rear furrow wheel to the right by readjusting the steering linkage. Then operate the tiller a short distance and see if the front wheel remains in the furrow.

Keep in mind: to operate properly, the tiller must be a balanced machine. Balancing requires a series of adjustments, each one of which may require compensating changes elsewhere on the machine. Adjustments of wheel lead and operating position may require adjustment of the hitch brace, which must be lengthened when width of cut is increased, or shortened when width of cut is reduced. Changes in depth or speed may also affect various settings.

The pay-off for "fine-tuning" a tiller is:

● *Uniformly worked soil with no ridges or depressions.*

● *Minimum draft requirement and reduced fuel consumption.*

● *Pride in knowledge that the job was done right.*

The disk tiller is normally turned to the left in field operation and may be raised either for turning or allowed to keep working around corners. Right turns can only be made with the tiller in the raised position.

On many tillers there is a possibility of striking the front tiller wheel or hitch with the rear tractor wheel on sharp right turns. If right turns must be made, raise the tiller out of the ground before turning and watch that the tractor tire doesn't turn into the tiller. Keep steering-arm stops adjusted to prevent over-centering of furrow wheels on sharp turns (Fig. 28). Right-hand turns can be made by turning 270 degrees to the left.

Operation of flexible tillers is as outlined above, plus individual gang adjustments which are described in the operator's manual. Because of the tendency of the front disk of each gang to dig in, a spring is used to push the rear end of each gang downward. Adjustments are provided to hold the front disk of each gang from running deeper than the rear. These adjustments keep the tiller from leaving ridges between gangs due to uneven operating depth. However, excessive pressure on gangs, to force penetration, may result in enough weight reduction on wheels to cause loss of tiller control.

If penetration becomes a problem, add extra weight, reduce width of cut, or sharpen disks if needed. A

STEERING
ARM STOP

Fig. 28—Set Steering-Arm Stops to Avoid Over-Centering Furrow Wheels when Turning

combination of these corrections may be required for optimum tiller performance.

SEEDING AND FERTILIZING

Disk-tiller seed and fertilizer attachments are normally driven from the tiller land wheel (Fig. 29) and are started and stopped automatically when the disks are lowered and raised. See-hopper capacity on different tillers is from 1½ to 2 bushels per foot (175 to 230 L/m) of machine cutting width. A typical fertilizer hopper holds 70 pounds (105 Kg/m) of granular fertilizer per foot.

Seeding rate on some tillers is adjusted by shifting a single lever which exposes more or less of the rotating fluted feed rollers (Fig. 30) to the seed in the individual seed cups. One seed cup is provided for each disk.

For uniform seeding, each fluted roller must move in and out of the seed cups the same distance. To check,

shift the seed-rate control lever to the fully closed position and note the position of each roller. If any are not completely closed, loosen the bolts holding the seed cup and shift it to one side or the other.

When changing seeding rate, always shift lever first to a higher-than-desired setting, then bring it back to the desired location. This takes up slack in the control linkage and helps assure correct seeding rate.

On some machines a high and low seeding rate for each roller setting is obtained by changing drive gears to alter speed of the feed shaft. Most machines are designed with a series of steps in seeding rate, while others offer infinite adjustment of seeding and fertilizing rates.

An adjustable feed gate (Fig. 30) compensates for variations in seed size and can be hinged open for cleanout of seed cup and hopper. All gates must be set at the same opening to provide uniform seeding.

Fertilizer metering and rate adjustments vary, but in each case adjustments are designed to deliver a prescribed quantity for each gate setting. As with seeding attachments, check fertilizer gate settings regularly for uniformity and readjust as necessary.

Charts showing recommended initial settings for various seeding and fertilizing rates are normally attached inside the hoppers and are included in operator's manuals. Due to variations in seed size and weight per bushel (m³), fertilizer density, and flow characteristics, these settings should only be used as starting points. Seed a small measured area and measure the quantity of seed and fertilizer applied; then compare it to the desired rate and adjust settings if necessary.

From seed cups and fertilizer metering devices, seed and fertilizer drop together through flexible tubes to be deposited close to the disk at the bottom of the furrow. Tube ends must be kept close to the disks. If not, seed and fertilizer will be mixed throughout the depth of tilled soil, resulting in poor germination, uneven

Fig. 29—Land Wheel Drives Seed and Fertilizer Attachments

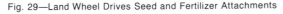

LAND WHEEL DRIVES ATTACHMENTS

162

Fig. 30—Seed Cup with Fluted Feed Roller

Fig. 31—Acreage Counter for Measuring Area Seeded

stands, and poor utilization of fertilizer. When seed and fertilizer are placed in the bottom of the furrow they are evenly covered with soil, generally within reach of adequate moisture, and seeds are more likely to grow and mature evenly.

Certain chemical seed treatments may create enough friction to retard seed flow, reduce delivery rates, and increase power requirement to operate the seeder. In such cases, add a small amount of powdered graphite to the seed to counteract the problem.

Never allow sticks, straw, clods, or large, hard lumps of fertilizer to enter seed and fertilizer hoppers. Such foreign objects may plug metering devices and cause uneven distribution of seed and fertilizer, or result in equipment damage.

Shear pins are on most machines to protect them from damage due to sticking, foreign objects, or accidental breakage. If a pin shears, find and correct the problem before replacing the pin and resuming seeding. Use only recommended shear pins as replacements—never use wire or nails which may be the wrong size, too soft, or too hard to provide adequate protection.

An acreage counter (Fig. 31) is available for most seeders to tell how much land has been seeded. Some counters read directly in acres covered, while others require a little arithmetic, based on machine width and seeding rate, to convert the counter reading to acres covered. Most counters read down to tenths of an acre and are easily reset to zero when starting on a new field.

When seeding is finished, remove grain left in hoppers and cups to keep it from drawing moisture, sprouting, molding, or causing rust and deterioration of metal parts. Never leave fertilizer or chemically treated grain in hoppers for extended periods, because fertilizer and many chemicals used are corrosive.

Trash Bars

Use trash bars if trash and soil tend to bridge between disks. They are used only on those disk spacings which do not have bearing hangers (Fig. 32).

Scrapers

Use scrapers in wet, sticky soils to keep disks clean (Fig. 33). Many are spring-loaded and can be released with a master lever when not needed. Individual spring and scraper adjustments are provided by set-screw collars or similar devices. Scrapers should lightly touch the disk at the inner end but should not rub hard enough to cause scraper or disk wear or slow disk turning.

Fig. 32—Trash Bars Prevent Bridging of Soil and Trash Between Disks

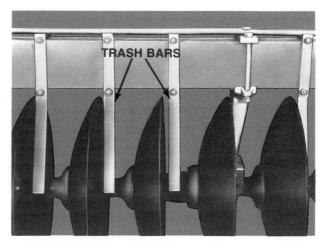

Fig. 33—Disk Scrapers Clean Wet, Sticky Soil

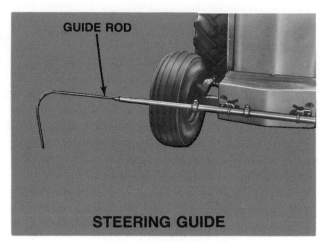

Fig. 35—Steering Guide for On-Land Tractor Operation

Fig. 34—Multiple Hitch for Wider Tilling

TRIPLEX HITCH

Two or more tillers may be hitched together, to form duplex and triplex units, to save time (Fig. 34). To obtain a better line of draft, the tractor is usually operated on land near the furrow wall. A marker or guide may be used to assist in steering accurately (Fig. 35). The front tiller is adjusted as outlined earlier for single-tiller operation, and the second machine is hitched to uniformly fill the furrow left by the front tiller.

Tillers used in a multiple hitch need not be of equal size or design, but must operate at the same angle and depth of cut. The larger tiller is hitched in front. Normally the lead machine should be 12 feet wide or larger to maintain line of draft within hitch limitations.

Some multiple hitches use a ball-and-socket coupling between tillers. The two units operate end-to-end in a straight line (Fig. 36).

The multiple-hitch attachment in Fig. 37 has a pivot arm at the rear of the front tiller. The rear tiller is attached to the pivot arm. A cable from the front end of the front tiller hitch extends to this pivot so that the pulling forces are not applied to the rear tiller directly through the frame of the front machine.

A short chain at the rear of the cable permits quick length changes as necessary when adjusting the front tiller hitch. Do not shorten this chain too much when assembling the machine. At least one extra chain link should be reserved for transport setting. A cable strut (Fig. 38) is used to keep the cable out of the caster wheel on turns. On some machines adjust the hitch so the cable doesn't touch disk blades while the tiller is operated. If the cable contacts the disk blades, excessive cable and disk wear can result. Such interference can also restrict free turning of the disks and throw off hitch adjustments.

When correctly adjusted, the rear tiller tends to keep the front tiller in its proper working position. The rear tiller usually has screw-jacks for depth control to avoid use of excessively long oil lines from the tractor.

END-TO-END HITCHING

Fig. 36—End-to-End Hitching with Ball-and-Socket Coupling

Left-hand turning with tandem units is excellent, but right turns should be avoided as much as possible.

TRANSPORT AND SAFETY

Disk tillers can be moved off the highway for short distances in regular field-operating position or in semi-transport position. However, for road movement put the tiller into endwise (full) transport position (Fig. 40).

Tillers should not be transported faster than normal tractor road speed of 20 miles per hour (32 Km/h), and never at speeds which do not permit full control of steering and stopping. Nor should tillers be transported at high speeds or for long distances with seed and fertilizer hoppers filled. Seed and fertilizer will pack down in the boxes and could severely damage the machine when restarting unless shafts are loosened with a wrench before the drive is engaged.

Fig. 37—Multiple-Hitch Pivot Arm

Fig. 38—Cable Struts Keep Cable Clear of Caster Wheel

Fig. 39—Full Transport Position for Road Travel

Most disk tillers can be placed in semi-transport position (Fig. 40) to reduce width for traveling short distances.

TYPICAL TRANSPORT PREPARATIONS

First, lower the tiller to working position to remove weight from the wheels. Shift the front wheel into vertical position (Fig. 41).

Raise the tiller keel (Fig. 42), or set rear wheel into vertical position, depending on tiller design. Put rear-wheel adjusting rod in transport position (Fig. 43).

Adjust rear-wheel lead so that rear wheel runs about 3 feet (1 m) to the right of the front furrow wheel as the tiller is transported.

Raise the tiller and install safety locks or hold-up pins before removing load from hydraulic cylinders. Cylin-

Fig. 40—Semi-Transport Position

Fig. 41—Place Front Wheel in Vertical Position for Transport

ders do not have to be removed from tiller. Then, detach hydraulic lines from the tractor and unhitch the tiller from the tractor. On some machines, remove and fold the hitch and place it on transport brackets on tiller frame (Fig. 44). On other machines, shorten the hitch brace steering arm hitch to set up the transport position.

Lock steering bar over front furrow-wheel axle lug and attach one end to tractor drawbar. Lock tractor drawbar to provide minimum transport width.

The tiller is now ready for road transport and is fully maneuverable.

Transport arrangements for other tiller makes and models are explained in their respective operator's manuals.

SAFETY

If each operator follows these and other basic rules of safety spelled out in tractor and tiller manuals, accidents are less likely to happen:

• *Keep SMV emblem clean and prominently displayed. Do the same with reflectors and warning lights as required by state and local regulations.*

• *Never allow anyone but the operator to ride on the tractor.*

• *Never ride or permit others to ride on the tiller.*

• *Lower the tiller to the ground when not in use.*

Fig. 42—Raise Rear Coulter and Set Rear Wheel Vertical

Fig. 43—Move Rear-Wheel Adjusting Rod to Transport

- *Secure the machine in the raised position by installing safety locks or hold-up pins when servicing or cleaning it.*

- *Disk blades are extremely sharp; be very careful when working or making adjustments in the disk area.*

- *Never walk close beside the rear wheel when the tiller is in operation. A sudden imbalance of forces could cause this wheel suddenly to jump to the left.*

- *Never grease, oil, or adjust the tiller while it is in operation.*

- *Escaping hydraulic oil under pressure can cause serious personal injury and infection. Therefore, be sure all connections are tight and that oil lines are undamaged. Always relieve hydraulic pressure in lines before disconnecting hoses. See a doctor immediately if escaping hydraulic oil has penetrated the skin.*

TROUBLESHOOTING

The majority of tiller operating problems can be traced to improper adjustment. The service chart that follows is designed to suggest a probable cause and solution. Special equipment may be available to help overcome some problems.

NOTE: If the machine is not working properly, attempt to correct the problem by making only one adjustment at a time. If the problem still exists after several adjustments it is usually best to return all adjustments to the suggested initial settings and begin over again. If the machine still functions improperly, contact the dealer or company representative for assistance.

Where possible remedies to problems are obvious, based on the possible cause, a blank space is left in the possible remedy column.

Fig. 44—Set Hitch for Transport

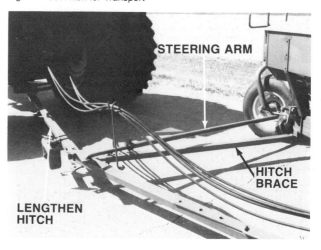

TROUBLESHOOTING CHART

PROBLEM	POSSIBLE CAUSE	POSSIBLE REMEDY
FRONT DISK CUTTING TOO NARROW	Tractor driven in furrow with duplex unit.	Operate with the tractor on the land. Use a wide-angle hitch if there is not sufficient adjustment in the standard hitch.
	Tractor tire compaction of soil adjacent to front furrow.	Increase penetration. An optional long hitch brace may be used, placing the tractor on land.
		NOTE: Under no circumstances should the front end of the tiller penetrate deeper than the rear end.
FRONT FURROW WHEEL SKIDS SIDEWAYS DURING LEFT TURNS	Too much lead to the right.	Adjust wheel to have only slight lead to the right (worked land).
DURING LEFT-HAND TURNS, FRONT FURROW WHEEL NOT STEERING ENOUGH OR DRAWBAR SWINGING TO THE RIGHT	Steering stop adjustment too short.	
TRACTOR TIRE OPERATING ON THE WORKED LAND	Tractor too far to the right of the tiller.	Use wide-angle hitch to operate on the land (with duplex units). Lengthen hitch brace and adjust steering bar on single units.
FIELD RIDGING BETWEEN GANGS	Front of gangs too low.	Raise tiller, be sure frame is level, adjust front of each gang so it will be approximately ½ to ¾-inch (13 to 19 mm) higher than the rear of the next gang.
		NOTE: Harder ground requires greater height.
	If ridging occurs directly behind left tractor tire, it is the result of compaction.	On flexible tillers adjust gang directly behind left tractor tire to be less than ½-inch (13 mm) higher than the rear of the front gang.
TILLER NOT PENETRATING DEEP ENOUGH	Too wide a cut.	Readjust hitch brace and steering arm.
	Disk diameter too small.	Replace worn disks or use optional larger disks.
	Ground speed too fast.	

PROBLEM	POSSIBLE CAUSE	POSSIBLE REMEDY
	Dull disks.	Reconditioning of disks is best achieved by using a disk roller.
	Not enough weight on rear furrow wheel or front frame.	Install special equipment weight stack adapter and weights and/or weight box.
	Drawbar pinned at the center.	Use a wide swinging drawbar and allow the hitch to swing to the left. Do not pin.
	Rear furrow wheel in hard-ground setting.	Adjust rear furrow wheel to soft-ground setting.
	Hitch brace too long.	Shorten brace and readjust front furrow wheel lead by steering bar. Width of cut may change only slightly. Move hitch brace only one or two holes for first adjustment.

Important: The front disk must cut slightly less than other disks; then the front furrow wheel does not have to run tight to the furrow wall.

PROBLEM	POSSIBLE CAUSE	POSSIBLE REMEDY
FRONT DISK CUTTING TOO WIDE	Tractor operated too far on land when pulling duplex unit or using on-land hitch.	Drive tractor closer to furrow. Use steering guide if necessary.
	Rear furrow-wheel lead set for soft ground.	Set rear furrow wheel to hard-ground position.
	Hitch brace too short.	Lengthen brace and readjust front furrow wheel lead by steering bar. Change hitch brace adjustment only one or two holes at a time.
	Rear frame to main frame adjustment too long.	Shorten turnbuckle between tiller frame and rear-axle support arm.

PROBLEM	POSSIBLE CAUSE	POSSIBLE REMEDY
IMPROPER FILLING OF THE FRONT FURROW AND CENTER FURROW ON DUPLEX UNITS, OR LEAVING RIDGES IN THE FIELD	Front disk cutting too narrow: (a) Front axle arm improperly adjusted. (b) Drawbar pinned at the center.	Increase width of cut. Use a wide swinging drawbar and allow hitch to swing to the left. Do not pin.
	Front disk cutting too wide: (a) Incorrectly adjusted front axle arm.	
	Front gang not cutting deep enough.	Lower front of tiller with leveling screw.

PROBLEM	POSSIBLE CAUSE	POSSIBLE REMEDY
FRONT GANG CUTTING TOO DEEP OR TOO SHALLOW	Incorrect frame leveling.	Turn the leveling screw to raise or lower the front end of the tiller as required. The front gang should cut a little less than the rear gang. When leveling frame, the front furrow wheel should be in the furrow.
TILLER PENETRATING TOO DEEP	Excessive spring compression on flexible gangs.	Loosen spring compression. Tighten set screw on spring collar when the gangs are on the ground.
	Cylinder stop not properly adjusted.	
	Tires underinflated.	
REAR FURROW WHEEL JUMPS OUT OF FURROW	Operating too shallow.	Increase depth and, if necessary, reduce width of cut.
	Width-of-cut setting for soft ground.	Reduce width of cut as outlined in operator's manual.
	Not enough weight on rear furrow wheel.	Add stack weights and/or install special equipment weight box.
	Wheel leading away from worked land.	Adjust wheel to lead slightly towards the worked land.
	Cutting too wide.	
	Excessive field speed.	
	Land-wheel axle stop set to lead land wheel too far to the left.	
LAND WHEEL SKIDDING	Land-wheel stop improperly adjusted.	Reduce lead on land wheel.
EXCESSIVE SIDE-DRAFT ON TRACTOR	No wide swinging drawbar, or drawbar is pinned. Tractor operating too far to right or too far back with respect to line of pull.	Lengthen hitch brace to move tractor to left. Lengthen main drawbar and hitch brace to move tractor ahead. Set drawbar so it is free to swing.

PROBLEM	POSSIBLE CAUSE	POSSIBLE REMEDY
TRACTOR HORSEPOWER REQUIREMENTS EXCESSIVE FOR OPERATING CONDITIONS	Rear furrow wheel leading too far into worked land.	Adjust for just a slight lead to the right.
	Front of gangs digging too deep.	Level gangs.
	Penetration too deep.	
	Front of tiller digging deeper than rear.	
VARIATION IN WIDTH OF CUT IN THE SAME FIELD	Tiller working both soft and hard spots in the field results in a changing side force on the machine.	A tiller will not automatically adjust itself for great changes in soil conditions. The tiller should be adjusted to meet the average conditions in the field while cutting as wide as possible and still keeping the rear furrow wheel in the furrow in hard-ground areas.
DUPLEX UNITS NOT ALIGNED	Width of cut not the same for both tillers.	Increase cut of one and/or reduce cut of other. Check lead of furrow wheels.
	Hitch cable (on some tillers) too short or too long causing misalignment.	
REAR OF DUPLEX UNIT FALLS BACK IN LOOSE GROUND	Rear furrow wheel not against stop.	Shorten hitch cable or adjust steering linkage.
	Excessive lead to right on rear furrow wheel.	Check lead of rear furrow wheel for proper adjustment.
DISK BREAKAGE	Working in stony ground at high speeds.	Reduce speed. Slower speed is especially important with new disks.
	Tiller set too narrow in stony land.	Cut as wide as possible so disks will roll over stones rather than dig into them. Reduce speed until sharp edge has worn off disk.
	Loose gangbolts.	Tighten gangbolts after first four or five hours of use. Check at 15 or 20 hours and periodically thereafter.

SUMMARY

Almost unknown a few decades ago, disk tillers (or "one-ways") have developed into major tools for primary tillage, summer fallowing, seedbed preparation, and once-over seedbed preparation, seeding, and fertilizing. A comparatively new use is double-cropping small-grain land after grain harvest by combining tillage with soybean planting.

Disk tillers and disk plows both use concave disks, but tillers have smaller blades mounted on a common axial shaft, while each disk-plow blade has its own standard and sealed bearing. Disk harrows also have concave blades, but tandem types have double gangs, one set of blades throwing to the right and the other to the left, while disk tillers, disk plows, and offset disk harrows throw only in one direction.

Disk blades, whether on tiller, plow, or harrow, mix trash into the tilled layer instead of inverting the furrow slice and burying trash the way a properly adjusted moldboard plow does. Disk tillers need comparatively low tractor power for their width, due to their relatively shallow penetration.

Disk tillers should never be used where they produce a smooth, powdery surface which invites wind erosion and inhibits moisture penetration.

Disk-tiller types include drawn (largest), integral, and reversible; the latter moves soil to right or left on alternate passes to eliminate dead furrows. Blade diameters range from 18 to 26 inches (457 to 660 mm), and overall width to 20 feet (6 m), although tandem operation of drawn tillers behind large tractors is quite practical. Normal operating depth is 2 to 5 inches (50 to 127 mm), though heavy-duty tillers (with large blades on wide spacing) can penetrate 9 inches (230 mm) in proper conditions.

CHAPTER QUIZ

1. (Fill in blanks) The disk tiller, or one-way disk, was developed in _____, about _____.

2. (True or false) The normal function of a disk tiller is cutting and mixing soil with minimum pulverization.

3. What is normal disk-tiller cutting angle (angle between disk gangs and direction of travel)?

4. Increasing or decreasing (choose one) disk-tiller cutting angle from direction of travel improves penetration in hard soil.

5. (True or false) All wheels on a disk tiller are set to run straight ahead on operation.

6. Name three important factors in stable disk-tiller operation.

7. (True or false) Tandem hitching of disk tillers requires two machines of equal size for successful operation.

8. (True or false) Disk-tiller seeding and fertilizing attachments are started and stopped by individual hydraulic cylinders.

10
Stubble-Mulch Tillers

Fig. 1—Stubble-mulch Tiller is "Hybrid" Disk Harrow and Chisel Plow

INTRODUCTION

Stubble-mulch tillers are hybrid tillage tools which combine the disk harrow and chisel plow into one machine (Fig. 1) for primary or secondary tillage. They are well-suited for stubble-mulch or deep-fallow tillage in small-grain, soybean, or corn fields.

The front section of the stubble-mulch tiller is a disk harrow which cuts trash and loosens cornstalks and other stubble. To the rear, a chisel-plow section rips up the packed ground, breaks through old plow sole, and helps mix trash into the soil.

It's often possible to work cornstalks immediately after harvest with no prior stalk chopping or disking. One pass leaves the soil mulched, broken, and rough—ready to absorb winter moisture and resist wind and water erosion. The reduced number of trips over the field helps minimize root-restricting compaction and improves water infiltration.

Used in soybean stubble, stubble-mulch tillers mix residue into the soil and bring larger, rougher clods to the surface to resist erosion.

Because of the cutting, mixing action of the disk gangs, the stubble-mulch tiller covers more trash than a regular chisel plow. This is particularly advantageous in heavy cornstalks where all of the residue may not be required on the surface for erosion control.

Adjusting the disk-gang angle between 7 and 14 degrees (measured from a line perpendicular to the direction of travel) and varying working depth permits coverage of from 25 to 75 percent of the crop residue.

Trash incorporated into the soil decomposes more rapidly, adding fertility and humus while improving water-holding capacity.

Twisted shovels are most commonly used on the chisel-plow portion of the stubble-mulch tiller, particularly for deep fall tillage, but a wide range of chisel-plow sweeps, shovels, and points is available for specific conditions.

LESS DRAFT

Generally, a tractor capable of pulling a 6-bottom, 16-inch (400 mm) moldboard plow can pull an 11-foot (3.4 m) or larger stubble-mulch tiller at equal or slightly higher speeds. Recommended available power ranges from about 12 to 20 drawbar horsepower (9 to 15 drawbar kW) per chisel shank, depending on soil type, moisture content, working depth and soil engaging tools used. Thus more ground can be worked in the same period of time. There are no dead furrows or back furrows, and machine investment is roughly equal.

Tiller working depth in the fall may be as much as 10 to 14 inches (250 to 360 mm) to break through plow sole or hardpan, or to increase the soil's winter moisture-holding capacity. Working depth of the machine is controlled by adjusting the depth stop on the remote hydraulic cylinder. Manual adjustment is provided to alter relative cutting depths of disk gangs and the chisel-plow section.

Use sweeps in spring and plow just deep enough to prepare a seedbed and obtain desired incorporation of remaining surface residue. Use sweeps for spring seedbed preparation to kill weeds, loosen and level

Fig. 2—Switch to Sweeps for Seedbed Preparation

the soil, and leave the ground ready to plant (Fig. 2).

Working the soil too deep in the spring may form hard clods or cause surface soil to run together and crust, making later tillage and planting difficult.

Where improved incorporation of residue is desired, such as in heavy infestation of weeds and grasses, or where the land was not worked in the fall, it may be desirable to shred trash before using the stubble-mulch tiller.

If additional tillage is desired before planting, usually a light disking or single pass with a field cultivator will prepare an adequate seedbed. For better seedbed preparation with a stubble-mulch tiller, a tine-tooth smoothing harrow may be attached to the rear to break up clods and smooth and level the soil (Fig. 3). The harrow rises automatically when the tiller is raised to transport position. Stubble-mulch tiller operating speed may range from 3½ to 6 miles per hour (5.5 to 10 Km/h), with 4 to 5 (6.5 to 8 Km/h) most commonly used.

MULCH-TILLER TYPES AND SIZES

Stubble-mulch tiller capacity depends on operating width, power available, operating depth, and the number

Fig. 3—Smoothing Harrow Aids Seedbed Preparation

of chisel-plow shanks used (Fig. 4). On most machines, 15-inch (380 mm) shank spacing is standard, while other shanks are on 12-inch (300 mm) centers, depending on design and working conditions. Different shank patterns are recommended for maximum soil ridging, for better soil leveling, and for maximum trash clearance.

Fig. 4—Field Capacity Depends on Machine Size and Type of Work

FIELD CAPACITY DEPENDS ON MACHINE SIZE

Fig. 5—Tandem Wheels on Walking Beam Provide More Stable Operation

STABLE OPERATION

Wheel mountings similar to those on many disk harrows and chisel plows are used for stubble-mulch tillers. Two wheels mounted directly on the rock shaft are used on some machines (Fig. 6). On others, tandem wheels mounted on opposite sides of a walking beam provide more uniform depth control (Fig. 5). If one wheel drops into a hole or gully, or rises over a stone or hump, the machine remains relatively level. This wheel arrangement also improves machine stability in transport.

A variation of the stubble-mulch tiller (Fig. 6) uses chisel-plow shanks on the rear, but has flat *coulters* instead of angled disk gangs in front. Widths available are approximately 6 to 26 feet (2 to 8 m). Coulters are spaced 7½ or 12 inches (190 or 300 mm) apart and slice through trash and roots to permit easier penetration of the chisel plow and reduce trash plugging. Coulter operating depth is adjustable, and should be just deep enough to slice through trash. Attempting to cut too deep with the coulters, especially in hard, dry soil, may tend to lift the machine from the ground and reduces the blade-soil angle. This makes trash cutting less effective, because coulters tend to push trash instead of cutting it.

Shanks must be correctly positioned on the frame to avoid uneven draft and poor performance. An equal number of shanks must be used on each side of the machine, and the location of individual shanks determined by measuring from the frame centerline.

Aggressive cutting of trash and roots is provided by large disk blades on the front gang of the typical stubble-mulch tiller. For improved leveling, the outer blade on each gang is slightly smaller in diameter. Disk spacing is usually 7½-, 9- or 12-inches (190, 230, or 300 mm).

This unit is basically designed for primary tillage in fall or spring, but may be used in lighter soil for seedbed preparation by adding a smoothing harrow. Both integral and drawn models are available. Heavy residue may be shredded before using this machine, but prior

Fig. 6—Stubble-Mulch Tiller Variation Has Coulters Instead of Disk Gangs in Front

COULTERS INSTEAD OF DISK GANGS ON FRONT

disking may leave soil too loose for coulters to cut trash properly.

Scrapers or trash bars are available to prevent buildup of soil and trash between disks or coulters and are particularly helpful in sticky soil.

Other mulch tiller variations include a machine with spherical disk blades individually mounted at 12-inch (300 mm) intervals across the front of a chisel plow frame. Chisel shanks are also spaced 12 inches (300 mm) apart.

Another version has widely-spaced disks set to throw soil outward from the center of the machine, followed by a second set of disks angled to move soil back toward the center thus leveling the surface ahead of the chisel shanks. The nominal disk spacing is 12-inches (300 mm).

Two similar rows of wide-spaced disks are used on another machine equipped with curved subsoiler shanks (V-ripper) rather than chisel shanks.

A single row of chisel shanks spaced 28 to 32 inches (710 to 810 mm) can be attached to the rear of some offset disk harrows to break compacted layers below disking depth.

Disk gangs set to pull soil toward the center are mounted on the rear of some V-subsoilers with parabolic shanks. The disks break clods, tear out weeds and reduce moisture loss from subsoiler slots.

There are also a number of disk- and coulter-field cultivator combinations similar to those with chisel shanks only built lighter for secondary rather than primary tillage.

Shanks

Several chisel-plow shank types are available for stub-ble-mulch tillers, and may be 1 × 2 inches (25 × 50 mm), 1¼ × 2 inches (32 × 50 mm), or 1½ × 2 inches (38 × 50 mm).

Principal shank types include:

- **Semi-rigid**
- **Spring-cushion**

Semi-Rigid Shanks

These shanks clamp directly to the frame and depend on flexing of the shank itself for protection from soil obstacles (Fig. 7). Some have shear bolts which release on impact with a solid object. As a result, semi-rigid shanks are recommended for economy in soils which are relatively free of obstructions.

Spring-Cushion Shanks

Springs on these shanks absorb and cushion shocks and allow the shank point to lift and pass over stones or other obstructions (Fig. 8). Cushioning also produces a vibrating action which aids in breaking and shattering clods in firm, dry soil.

PRINCIPLES OF STUBBLE-MULCH TILLER OPERATION

Stubble-mulch tillers provide more complete cutting and mixing of trash and deep shattering of the soil in one operation than either disk harrows or chisel plows alone. Trash coverage and soil pulverization are controlled by adjusting gang angle and relative working depth of disk gangs and chisel points, so the surface is left in the desired condition.

Fig. 7—Semi-Rigid Shank

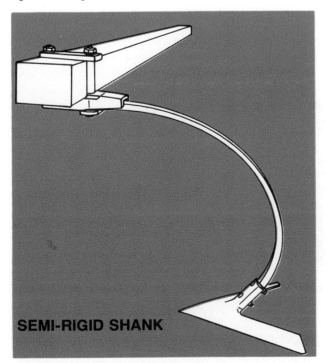

SEMI-RIGID SHANK

Fig. 8—Spring-Cushion Shank

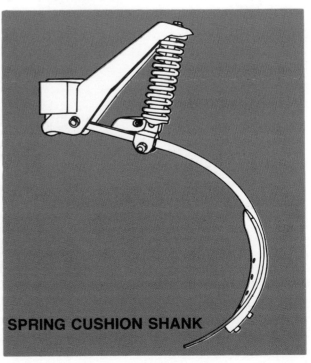

SPRING CUSHION SHANK

SCREW ADJUSTMENT FOR RELATIVE DEPTH

Fig. 9—Screw Adjustment Varies Relative Disk and Chisel Depth

Fig. 10—Disk Gang Angle is Adjustable from 10 to 25 Degrees

DISK ANGLE ADJUSTMENT

Fig. 11—Adjustment to Level Machine Fore-and-Aft

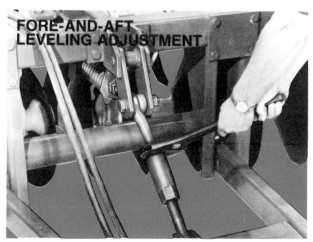

FORE-AND-AFT LEVELING ADJUSTMENT

Disk cutting depth may be adjusted (Fig. 9) as deep as 8 inches (200 mm) to meet specific working conditions. (Note: when field adjustments are made, take care to adjust both sides equally.) On some machines, a hydraulic cylinder is used to adjust disk or coulter depth in relation to chisel depth.

Angle of the disk gangs can be varied from 7 to 14 degrees on different machines (Fig. 10) to provide the desired aggressiveness. Increasing the gang angle provides better trash cutting and mixing and more complete loosening of plant roots. Decreasing the disk-gang angle and depth moves less soil and is recommended for seedbed preparation.

When changing overall working depth and the relative depth of disk gangs and chisels, maintain level fore-and-aft operation of the chisel plow by adjusting the hitch-leveling screw or turnbuckle (Fig. 11).

TRACTOR PREPARATION

The stubble-mulch tiller is a high-draft machine and the tractor must be in top condition for optimum performance. Tractor preparations are similar to those recommended for chisel-plow operation (see Tractor Preparation in the chapter on CHISEL PLOWS).

STUBBLE-MULCH TILLER PREPARATION AND MAINTENANCE

Careful preseason preparation and daily attention to service and maintenance will help assure optimum machine performance and life.

Before Each Season:

1. Check for loose nuts and bolts and worn, broken, or missing parts.

2. Check disk or coulter blades for cracks and excessive wear. Sharpen or replace as necessary.

3. Sharpen or replace dull or broken soil-engaging tools.

4. Lubricate machine as recommended in manual.

5. Check tires for proper inflation. Clean, lubricate, and tighten wheel bearings.

6. Be sure disk-gang bolts are properly tightened—see manual (Fig. 12).

Daily Before Operation:

1. Lubricate as recommended in manual.

2. Visually check for loose nuts and worn or broken parts while lubricating.

3. Check tires for proper inflation.

4. Recheck torque on coulter or disk-gang bolts daily on new machines or after replacing blades. Do the same for wheel bolts and bearings.

Before Storage At End Of Season:

1. Thoroughly clean machine of trash, soil, and grease.

Fig. 12—Tighten Disk Gang Bolts

2. Repaint spots where paint has been scratched or worn off.

3. Coat disk or coulter blades and soil-engaging tools with heavy grease or plow-bottom paint to protect them from rust.

4. Lubricate entire machine to keep moisture out of bearings.

5. Store machine inside, if possible, to prevent weathering. Block up to remove the weight from tires. Protect tires from sun.

6. Install transport safety lock and relieve pressure in hydraulic cylinder, or lower machine to ground and protect points and disks from rust by setting them on boards.

7. Fully retract hydraulic cylinders to protect cylinder rods from rust.

FIELD OPERATION

Drawn stubble-mulch tillers are attached to the tractor drawbar; they depend on remote hydraulic cylinders for depth control. **Integral** models are attached to the tractor 3-point hitch; the tractor load-and-depth control system regulates operating depth.

Operating procedures for stubble-mulch tillers are similar to those for chisel plows. Always raise the machine completely out of the ground before making sharp turns or backing up to avoid shank and disk damage. Level the frame fore-and-aft and side-to-side for uniform penetration.

As with chisel plows, stubble-mulch tillers do not require any specific pattern of field operation for satisfactory performance. Two passes at different angles may be required in extremely hard soil. Better leveling of row-crop ridges and uprooting of crop stubble is obtained if the stubble-mulch tiller crosses rows diagonally at a 20- to 30-degree angle (Fig. 13). Otherwise, it may follow the row pattern, be operated in lands, in

Fig. 13—For Better Leveling, Work Diagonally across Rows

adjacent passes across the field. Working on the contour or across slopes improves erosion control.

Headlands are worked last, and if work has progressed around the field, an extra round will be required on each corner where the machine has been lifted for turning.

Liquid or granular fertilizer may be broadcast before using the tiller and then mixed into the soil with crop residues if desired. To prevent loss of nutrients, do not apply nitrogen fertilizers in the fall while soil temperatures still exceed 55 degrees F (13°C). Some stubble-mulch tillers may be equipped with anhydrous-ammonia applicators and the tank mounted on the frame. This permits fertilizing, residue disposal, and tillage in one operation.

Fall Tillage

The stubble-mulch tiller is effective in heavy trash (Fig. 1), such as cornstalks, where it is desired to chop and mix a portion of the residue with the soil and leave enough exposed to control winter erosion. For such work, set disk gangs at maximum angle and fairly deep to provide maximum trash cutting. Such a setting will also help loosen plant roots, break up surface soil, and mix trash for faster decomposition.

Set coulters just deep enough to slice through residue. Coulters set too deep waste power and may push trash in front of the blades instead of cutting it. If soil is very hard, coulters set too deep may hold the front of the machine out of the ground.

Set chisels to penetrate the root zone and break soil open for maximum entry of air and water. The breaking, stirring action of the chisels also helps anchor

179

Fig. 14—Smoothing Harrow Breaks Clods and Levels Surface

Fig. 15—Hydraulic-Cylinder Depth Stop

residue and roughens the surface to improve its erosion resistance.

If too much residue is being covered for satisfactory erosion control, reduce disk-gang angle and cutting depth if necessary. If too much residue remains exposed, it may be necessary to shred or disk stalks prior to using the stubble-mulch tiller. If residue is disked prior to using the coulter-type stubble-mulch tiller, allow the soil to settle (hopefully rained on) between operations to permit better trash cutting by the coulters.

Spring Tillage

For fast, efficient seedbed preparation on land worked in the fall with the stubble-mulch tiller, set disk gangs at minimum angle and shallow depth. Set coulters just deep enough to cut through remaining residue. If residue is very light, coulters may be raised clear of the soil to reduce power requirements. Replace twisted shovels with sweeps and set them for shallow operation. Attach a smoothing harrow (Fig. 14) and start working. If this does not produce a satisfactory seedbed, increase disk angle and depth slightly, or go over the field with a light disk or field cultivator.

If bad weather or winter grazing prevent fall operation of the stubble-mulch tiller, it may be used in the spring with some changes in settings. Use a less aggressive angle and depth for disk gangs and reduce the chisel depth. Use the smoothing harrow to break up clods immediately. Do not work the soil when it is too wet.

If heavy residue is present in the spring, it may be advisable to shred or disk the trash before using the stubble-mulch tiller.

Fluted coulters or disk openers on the planter will help cut through remaining residue, making planting easier and providing more uniform stands.

Operating Tips

1. Control basic depth of operation by adjusting remote hydraulic-cylinder depth stop (Fig. 15).

2. Adjust relative depths of disks and chisels as outlined in operator's manual. Be sure both sides of frame are adjusted equally on manually-adjusted models.

3. Adjust disk-blade scrapers so they are just flush with disk—never tight enough to restrict disk rotation.

TRANSPORT AND SAFETY

There is no substitute for a careful operator, in the field or on the road. Following these simple safety precautions and those in the operator's manual and using common sense can help prevent accidents.

1. Never transport the stubble-mulch tiller at excessive speeds; reduce speed when traveling over rough or uneven terrain.

2. When transporting equipment, always use lights, reflectors, and the SMV emblem as prescribed by law.

3. Keep the SMV emblem clean and in place at all times. Replace emblem if reflective material becomes faded.

4. Provide adequate tractor ballast for tractor front-end stability and to prevent excessive wheel slippage.

5. Never permit riders on tractor drawbar or stubble-mulch tiller in operation or transport.

6. Allow only the driver to ride on the tractor.

7. Always install transport safety lock and relieve the pressure in the hydraulic cylinder before transporting the stubble-mulch tiller. Never trust the tractor hydraulic system to support the machine in transport.

8. Never allow anyone to stand or walk close to the stubble-mulch tiller while operating, raising, or lowering the machine.

9. Never allow children to play on or near the machine while it is in operation, transport, or storage.

10. Disk blades, coulters and soil-engaging tools are

TROUBLESHOOTING

Understanding the function and proper operation of the stubble-mulch tiller will permit quicker detection and easier correction of improper performance.

TROUBLESHOOTING CHART		
PROBLEM	**POSSIBLE CAUSE**	**POSSIBLE REMEDY**
EXCESSIVE TRACTOR-WHEEL SLIPPAGE	Insufficient ballast.	Add wheel weights or put fluid in tires. Do not exceed recommended limits. See tractor manual.
	Tractor overloaded.	Reduce penetration by lowering wheels or reducing disk angle.
DISKS WOBBLE IN OPERATION	Gang-bolt nut loose.	
INADEQUATE PENETRATION	Wheels holding machine up.	Adjust depth stop on cylinder.
	Chisel shanks not set right.	Level machine fore-and-aft. Set chisel plow deeper in relation to disk gangs.
	Worn disk or coulter blades.	
	Improper disk-gang angle.	
EXCESSIVE PENETRATION	Wheels carried too high.	Raise depth stop on lift cylinder.
	Too much disk angle.	
	Disks too deep in relation to chisel plow.	
	Machine not working level.	
DISKS OR COULTERS NOT CUTTING TRASH	Disks or coulters not cutting deep enough.	Lower blades in relation to chisels, or lower hydraulic cylinder depth stop.
	Wrong direction of travel.	Cut diagonally across rows.
	Insufficient disk angle.	
	Worn disk blades.	
DISK OR COULTER GANG WON'T REVOLVE	Bearing frozen.	Lubricate or replace.
	Scrapers improperly adjusted.	Set scrapers to barely touch disks.
	Obstruction between disks and frame or scrapers.	Check for stones, roots, and mud.
TRASH PLUGS AROUND WHEELS AND SHANKS	Excessive trash.	Shred or disk trash before using stubble-mulch tiller. Rearrange shanks for maximum clearance.
	Disk cutting angle too severe, trash moved to outside.	
	Shanks improperly located on frame soil and trash too wet.	
	Disk or coulters not deep enough to cut through trash.	

PROBLEMS	POSSIBLE CAUSE	POSSIBLE REMEDY
INADEQUATE TRASH INCORPORATION	Improper tools on chisel plow shanks.	Twisted shovels recommended.
	Operating speed too slow.	Increase speed to 5 or 6 miles per hour (8 to 10 Km/h).
	Chisel plow tools not operating deep enough.	Increase chisel depth relative to disks.
	Insufficient cutting by disks.	Increase disk angle and depth.
	Excessive trash present.	Shred or disk trash first.
EXCESSIVE RIDGING	Excessive disk angle.	Decrease angle to reduce volume of soil moved.
	Excessive tractor speed.	Decrease.
	Excessive overlapping.	Drive further away and don't overlap at all, or overlap more to level ridge.

sharp—be careful when working close to such objects.

11. Always lower the stubble-mulch tiller or install safety lock and shut off the tractor before adjusting, repairing, or lubricating the machine.

12. Always lower the machine or install the safety lock when storing the stubble-mulch tiller.

SUMMARY

A cross between disk harrows and chisel plows, effectively combining many good features of each, integral and drawn stubble-mulch tillers work well for fall or spring primary and sometimes secondary tillage in grain and row-crop farming.

Briefly, large disk blades (sometimes plain or fluted coulters) on the front gang cut trash and stubble; then deep-penetrating (as much as 10 or 12 inches; 250 to 300 mm) chisel points rip the rootbed, break hardpan and plow sole, and mix trash into the soil.

Adaptability is increased by the variety of disk blades, shanks, and soil-engaging tools for the shanks. Another plus factor is the wide range of independent adjustment of both disk and chisel-plow halves of the total implement. Among useful attachments are trailing tooth-type harrows for seedbed preparation and fertilizer applicators.

Stubble-mulch tillers usually have somewhat less draft than moldboard plows, permitting greater working width and increasing acres covered per day with the same tractor power. However, as with moldboard plows, prior chopping or shredding may be needed for severe trash conditions.

CHAPTER QUIZ

1. (Fill in blanks) The stubble-mulch tiller is a combination of _____ and _____.

2. Describe normal soil condition after use of the stubble-mulch tiller.

3. How is trash coverage and soil pulverization controlled when using a stubble-mulch tiller?

4. (Fill in blanks) For fall tillage in heavy trash, such as cornstalks, set disk gangs for _____ angle and depth, and set the chisels _____ (deep or shallow).

5. (True or false) Disk gangs should be set for minimum angle and chisels replaced with sweeps for spring tillage.

11
Rotary Tillers

TILLAGE OPERATIONS VARY

Fig. 1—Rotary Tillers are Used for many Tillage Operations

INTRODUCTION

Once-over seedbed preparation and reduced draft are among the most frequently mentioned reasons for using rotary tillage. By applying engine power to the soil through the PTO rather than tractive force through tires, less power is lost, and tractor weight and soil compaction are reduced.

As a result, rotary tillers are being used to:

- *Shred stalks and mix them with soil*
- *Replace the plow, disk, and harrow* (Fig. 1)
- *Cultivate row crops*
- *Renovate pastures*
- *Reclaim wasteland*
- *Till orchards and vineyards*
- *Landscape*
- *Strip-till while planting*
- *And other tasks*

Several tiller configurations have been developed over the years—vertical axis, longitudinal axis, and the horizontal axis, which operates at right angles to the direction of travel (Fig. 2).

We will discuss only the horizontal axis, which is the most common. Four types of horizontal-axis tillers are available:

- *Self-propelled, heavy-duty models (commonly used in construction and road building; little current agricultural application).*

- *Tractor-drawn or mounted, with auxiliary engine for tiller power.*

- *Tractor-drawn or mounted, powered by tractor PTO.*

- *Self-propelled, walk-behind garden tillers.*

Rotary tillers powered by the tractor PTO currently are the most common type on farms in North America and will be featured in this discussion, although many of the factors apply equally to the other machines. Models are available for tractors from 15 to 160 PTO horsepower (11 to 120 PTO kW) and in a wide variety of widths and operating depths.

ROTARY VERSUS CONVENTIONAL TILLAGE

Interest in rotary tillage has been spurred because of soil compaction and the extreme amounts of weight required to develop sufficient traction with large tractors to pull big plows and disk harrows. Compaction is reduced by making fewer trips over the field and by reducing tractor weight. Traction is no problem with rotary tillers because of negative draft (see discussion below).

MINIMUM TILLAGE

Some proponents of rotary tillage refer to it as minimum tillage because of the limited number of operations required to prepare a seedbed. But, if not properly used, rotary tillage may actually be excessive tillage that damages soil structure and results in surface puddling and crusting.

NEGATIVE DRAFT

In the most cases rotary tillers actually provide negative draft. That is, the rotating blades tend to push the tractor forward as the soil is tilled (Fig. 3). Therefore, the tractor used must be operated in a direct-drive transmission gear, as opposed to the free-wheeling gear ranges provided by some planetary, shift-on-the-go transmissions. Most of these tractors do have direct gear drive in either high or low range so they can be used for tiller operation.

Fig. 2—Horizontal Axis Rotary Tiller

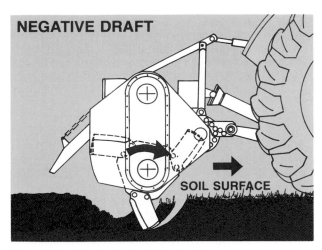

NEGATIVE DRAFT

SOIL SURFACE

Fig.3—Blades Tend to Push the Tractor Forward

POWER REQUIREMENTS

Specific energy required for rotary tillage—energy used per cubic foot (m³) of soil moved—is about three times as much as that required for moldboard plowing. Energy used may even exceed the total required for plowing, disking, and harrowing.

A rotary tiller working in medium soil has an average **draft equivalent** of 30 psi (21 N/cm²) of tilled-soil cross-section, compared to 5 to 9 psi (3.4 to 6.2 N/cm²) for moldboard plowing and 3 to 5 psi (2 to 3.4 N/cm²) for a disk in similar soil.

Drawbar horsepower required to pull a tillage implement is calculated by the following formula:

$$\text{Drawbar horsepower} = \frac{\text{Draft in lbs. x speed in mph}}{375}$$

Therefore, an 80-inch tiller operated 5 inches deep in 30-psi soil at 3 miles per hour would require a tractor with 96 drawbar horsepower (equivalent to approximately 113 PTO horsepower):

Draft equals 5 inches x 80 inches x 30 psi, or 12,000 lbs.

$$\frac{\text{Drawbar}}{\text{horsepower}} = \frac{12,000 \text{ lbs. x 3 mph}}{375} = \frac{96 \text{ drawbar}}{\text{horsepower}}$$

Drawbar horsepower = about 85 percent of drawbar horsepower

(This is only an example. The percentage can vary considerably because of tractor weight, soil conditions, and draft of implement.)

Approximate PTO horsepower = 96/0.85 = 113.

Using the metric system:

$$\text{Drawbar kilowatts} = \frac{\text{Draft, kN} \times \text{speed, km/h}}{3.6}$$

A 2000 mm tiller operated 130 mm deep in 21 N/cm² soil at 5 km/h would require a tractor with approximately 76 drawbar kilowatts, or about 89 PTO kW.

Convert 2000 mm to 200 cm and, 130 mm to 13 cm, for convenience.

200 × 13 × 21 = 54,600 N; or 54.6 kN

$$\frac{54.6 \times 5}{3.6} = 76 \text{ drawbar kW}$$

Approximate PTO kW = 76/0.85 = 89 PTO kW

Heavier soils, a faster operating speed, or greater operating depth would require even more horsepower for the same-size machine.

Thus adequate rotary-tiller capacity for extremely large acreages may require tractor power greatly exceeding that required for other farm operations or for conventional tillage. Even when soil is rotary tilled once in the fall and once in the spring, tiller capacity may delay finishing the planting unless equipment is carefully matched to the acreage to be covered.

Although a rotary tiller with a given capacity will cost more than an individual implement required for conventional tillage—plow, disk, harrow—the total investment is generally less, particularly if a stalk shredder is also eliminated.

COMMON ROTARY TILLAGE METHODS

Typical rotary tillage methods are described in the following discussion. These are general practices, and may vary in different geographical areas.

CHOPPING AND MIXING

Rotary tillers are good for chopping and mixing trash with the soil, but they do not cover trash as completely as moldboard plows. Many farmers make one pass with a rotary tiller in the fall to chop and mix cornstalks or other crop residue with the soil (Fig. 4). The tiller is operated to provide minimum soil pulverization. This eliminates separate stalk-chopping and disking or chiseling operations in the fall and provides more-complete mixing of residues into the soil for faster decomposition.

Fig. 4—Rotary Tillers Chop and Mix Residue into the Soil

After large amounts of trash have been incorporated, additional applications of nitrogen fertilizer usually are required to maintain fertility and speed decomposition. This nitrogen is used by the microorganisms, which decompose the residue, and prevents competition between them and the crop seedlings for available nitrogen.

A single pass in the spring with the rotary tiller is usually the only additional tillage required for seedbed preparation. The spring operation finishes the shredding and mixing of remaining residue.

If spring weeds are a problem, the field may be tilled early, 10 to 14 days prior to planting, and weed seeds allowed to germinate. At planting time, a second pass with the tiller at a reduced depth and higher rotor speed kills early weeds and prepares a final seedbed.

TILLING AND PLANTING

To reduce operations even further, many rotary tillers may be equipped with toolbar-mounted unit planters, or a hitch for towing a planter, for tillage and planting in one operation. By rearranging blade flanges on the rotor shaft, most tillers may be used for strip tillage in the seedbed only, although shorter blades are often used between the rows for weed control.

Fertilizer and chemicals may be incorporated by spreading in bands or broadcast from tanks or hoppers on the tractor or the tiller or by broadcasting ahead of the tiller. Sprayer booms may be mounted on the tractor or the tiller. Many tillers are equipped with a stub powershaft on the gearbox for driving a chemical

Fig. 5—Rotary Tiller Set Up for Row-Crop Cultivation

ROW-CROP CULTIVATION

or fertilizer pump. On some models with extra lift wheels on the rear, spray tanks may be mounted directly on the tiller frame. The stirring, mixing action of rotary tillers provides excellent incorporation of pre-emergence chemicals.

CULTIVATING

By rearranging blade flanges on the rotor shaft, most tillers may be used for cultivation. Plant shields, attached to the front and rear frame members, may be used to protect small plants from thrown soil. Such shields may be used during cultivation to keep from disturbing bands of preemergence herbicides in the row, or the shields may be used to protect crops when applying herbicides to row middles during cultivation (Fig. 5). Rotor diameter and operating depth are limiting factors in the size of plants that may safely be cultivated in this manner.

POOR WEATHER

If poor weather has prevented fall tillage and stalk chopping, and weather again delays spring work until planting time, once-over seedbed preparation with a rotary tiller may be a solution. Even if two passes are required with the tiller, the second pass will usually be shallow and fast. The total time will still beat the time required for conventional operations.

MOISTURE CONSERVATION

Because soil is stirred only once or perhaps twice, compared to several times for plowing, disking, and harrowing, more moisture is retained in the soil. This can be particularly important in dry areas and seasons and when working small-grain stubble for double-crop soybeans.

SPECIAL APPLICATIONS OR CONDITIONS

ORCHARD TILLING

Rotary tillers are used in vineyards, citrus groves, and orchards because it is easy to control operating depth to prevent damage to tree or vine roots (Fig. 6). These tillers also provide a surface mulch, which conserves moisture during dry seasons in non-irrigated areas. They also help reduce damage from mice and moles by destroying surface cover and mulching the soil to destroy burrows. Some tiller models are built for offset operations; the offset permits working under trees without injuring low branches (Fig. 7). Offsetting the tiller on small tractors where wheel-tread exceeds tiller width permits operating the tiller in lands, similar to plowing, to eliminate tractor tracks from tilled soil.

RICE FIELDS

Rotary tillers are adapted for work in wet rice fields. The rotating blades help push the tractor forward and reduce tire slippage where it would be nearly

CONTROLLED OPERATING DEPTH

Fig. 6—Tillers may be Used in Vineyards, Citrus Groves, and Orchards

impossible to gain sufficient traction for pulling heavy-draft implements such as plows and disks. Heavy rice straw may be chopped and mixed with the soil to improve fertility and soil structure, rather than burned. Special blades are available for some machines to aid in disposal of rice straw.

ROTARY TILLER TYPES AND SIZES

Rotary tiller sizes vary from 16-inch (400 mm) garden models on up. Maximum size continues to increase with the climb in tractor horsepower, with 10-, 12- and even

Fig. 7—Some Smaller Tillers may be Offset from Tractor Centerline

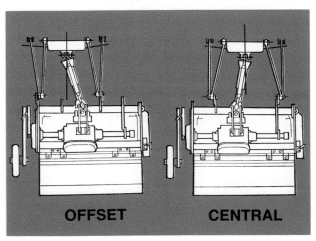

OFFSET   CENTRAL

SEEDBED PREPARATION

Fig. 8—Rotary Tiller Working in Rice Field

15-foot (3, 3.7 and 4.6 m) models available for tractors with 160 or more PTO horsepower (120 PTO kW).

Although most rotary tillers are used for both primary and secondary tillage, some are built for secondary tillage only. These lighter-duty units may be used for primary tillage in lighter soils without heavy trash, but are basically designed for seedbed preparation and planting (Fig. 8). Rotor flanges and blades may be rearranged to permit broadcast or strip tillage, or changed after planting to allow cultivation of small plants.

Most rotary tillers built for open-field tillage are mounted on the tractor 3-point hitch in the conventional manner and driven by the tractor PTO. However, semi-integral mounting of larger tillers is normally recommended, especially those equipped with planters or spray tanks. Semi-integral mounting reduces the required lift capacity of the 3-point hitch and also improves tractor stability by maintaining more weight on tractor front wheels when transporting. Actually, a tractor may be able to handle a semi-integral machine with as much as twice the weight of the recommended hitch capacity. Caster mounting of transport wheels provides excellent maneuverability and easy turning.

NOTE: To permit flexing between the semi-integral tiller and tractor as they pass over uneven ground, the tractor top link must not be used.

To loosen soil to a greater depth, chisels may be attached to the rear of some rotary tillers and set to work from 3 to 9 inches (75 to 230 mm) below the rotor blades. These chisels can help break up plowpan, loosen tight soils for better root penetration, and counteract the forward thrust of the rotor. Deeper soil is not pulverized, and total specific energy input is much less than for operation of the rotor to the same total depth. Chisels, spaced about 20 to 25 inches (500 to 650 mm) apart, must be used in relatively dry soil to accomplish their purpose.

Toolbars on the rear of some tillers permit mounting of middlebreakers or bed-shapers to till and form the soil for planting in one operation.

BLADES

The L-shaped blade is most common because the L shape is usually superior to others in heavy trash. They are better for killing weeds and generally cause less soil pulverization.

Penetrating hard ground may be easier with C-shaped blades. They are curved more than L-shaped blades and are recommended for heavy, wet soil because they have less tendency to clog.

When strip-tilling and planting in one operation, long blades should be used in the row area to make a seedbed. Shorter blades are then used between rows to kill weeds and mulch the surface with less power requirement.

Tillers designed for special purposes may be equipped with soil-engaging tools such as straight spike points that penetrate the soil to fracture clods and mix soil into a uniform seedbed. The straight spike points are generally used in previously worked soil or for shallow weed control. Operating speeds up to 8 miles an hour (13 Km/h) are possible—double that of ordinary blade tillers.

Another tiller adaptation features straight knife blades which simply slice through the soil to aerate and loosen turf, eliminate surface compaction, and improve moisture penetration. Primary application for these blades is for lawns, parks, and golf courses.

Some machines designed primarily for secondary tillage and cultivation have thin, flat, straight, or nearly-straight blades with narrow, rectangular edges that flail the soil to break clods and tear out weeds.

LOCATION OF THE DRIVE

Depending on machine design and operating width, the rotor may be driven from one end, both ends, or from the center. Each design has advantages and disadvantages.

Driving from the **center** reduces the total number of drive-system components. Power is transmitted directly from the PTO shaft, through the gear-box, to the rotor. However, driving from the center usually requires special close-fitting blades next to the gearbox. This means that a small center strip will generally remain uncut. A small chisel may be installed in the center to remove the unworked strip for seedbed preparation, but it should be removed for primary tillage.

Driving smaller tillers from **one end** eliminates the problem of an uncut center strip but requires more drive components. Power may be transmitted to the rotor from the gearbox or upper drive shaft by a series of gears or through heavy-duty roller chains.

Driving from **both ends** permits more-uniform application of power for longer rotors on large machines and again eliminates an unworked center strip. It requires a more complex drive train, but may allow use of lighter components in some places because power is split. The use of end drives increases overall machine width; this may not pose problems in field operation, but could be a factor in orchard or vineyard work.

PRINCIPLES OF ROTARY TILLER OPERATION

Rotary tillers apply tractor-engine power directly to the soil, through the PTO, without wheel slippage or excessive tractor weight (Fig. 9). But, rotary tillers cannot cure poor tillage management, and they may even make poor management look worse. However, with good management and proper operation, they can provide a valuable tillage alternative to plowing, disking, and harrowing.

ROTOR OPERATION

The tiller rotor normally turns in the same direction as tractor wheels, although some research has been done on reverse rotation. On most newer tillers, blades are bolted to flanges which in turn are clamped to the rotor shaft and may be shifted or removed for varying tillage patterns. Normally, two or three pairs of right-hand and left-hand blades are used on each flange, depending on soil conditions and the degree of pulverization desired.

A left-handed blade is one where, when viewed from the rear of the tiller, the cutting end points to the left. A right-hand blade points to the right (Fig. 10).

Wet Conditions

Three pairs of blades per flange are usually recommended for general operation on most tillers. In wet, sticky soil it may be desirable to remove one pair of blades per flange so that soil is more easily thrown from the rotor. Rotor speed must be increased when

Fig. 9—Tractor Engine Power Applied Directly to Soil through the PTO

POWER FROM TRACTOR PTO

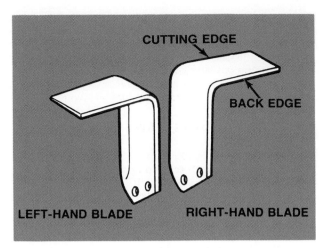

Fig. 10—Right-and Left-Hand Blades

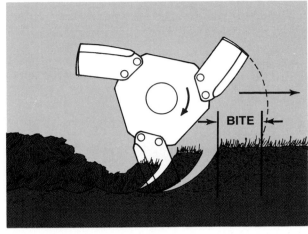

Fig. 12—Blade Bite is Determined by Rotor and Tractor Speeds

using only two pairs of blades, and the remaining blades relocated on the flanges to obtain the proper cut. The increased speed moves the soil faster and helps prevent plugging.

Normal Conditions

Under normal conditions most tiller blades are self-sharpening. They are arranged in a scroll or spiral pattern on the rotor (Fig. 11) so that no more than one blade strikes the soil surface at a time. Failure to maintain the scroll pattern will cause uneven work and excessive vibration and thumping when more than one blade hits the ground at the same time.

SIZE OF CUT OR BITE

The size of the cut or bite of rotary-tiller blades (Fig. 12) is determined by tractor forward speed, number of blades per flange, and rotor speed.

Slower rotor speeds:

- *Require much less power*
- *Help maintain soil structure*
- *Reduce blade wear*
- *Reduce fuel consumption*
- *Save time*

ROTOR SPEED

Rotor speed should be increased as soil moisture increases to keep soil moving through the rotor. However, avoid working soil that is too wet. Most soils, if worked too wet, will dry so hard that preparation of a good seedbed is next to impossible, and plant germination and growth are severely retarded. On the other hand, working soil too dry causes excessive dust and increases blade wear.

Fig. 11—Blades Arranged in Scroll Pattern for Smooth Operation

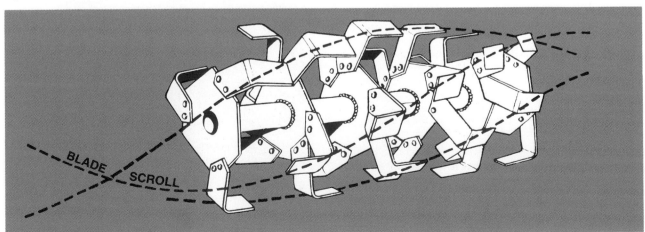

Larger blade bites mean less soil pulverization and reduced power requirements. It is usually best to use the slowest rotor speed and largest bite which will provide an acceptable tilth or degree of pulverization.

However, excessive blade wear and soil friction may result if forward speed is too fast for the rotor speed being used. As the rotor turns, this pushes the outer surface of the blade against uncut soil, causing troweling of the surface and wasting power.

ROTOR SPEED AND FORWARD SPEED

When rotor speed and forward speed are properly matched, a gap will be present between uncut soil in front of the rotor and the outer surface of the blades as the rotor revolves. As operating depth increases, this clearance will gradually decrease, and blades will be in continuous contact with undisturbed soil.

Important: all soil cut by the blades is carried inside the rotor until it reaches the back side. Then it must be discharged between the blades before they again reach their cutting position and pull in more soil. If soil is not discharged, the rotor will soon plug. Obviously, rotor diameter, speed, blade arrangement, soil moisture, and volume of trash all have a bearing on soil movement within the rotor.

CHANGING ROTOR SPEED

Rotor speed on most tillers may quickly be changed in the field by shifting a lever on the gearbox or by in-

stalling different gears in an easily accessible gear case. Rotor speeds may vary from about 140 rpm to nearly 300 rpm for special applications where extremely fine tilth is desired. Medium speed is generally adequate for most field conditions.

SPEED AFFECTS TILTH

If rotor speeds remain constant, blade bite varies by changes in travel speed. Slow travel produces fine tilth, while faster speeds produce progressively rougher conditions. For this reason, adequate tractor power must be available to maintain sufficient travel speed to prevent over-pulverization. Normal working speeds for most tillers range from 2½ to 5½ miles per hour (4 to 9 Km/h), depending on soil conditions and desired results.

HOOD ADJUSTMENTS AFFECT TILTH

Tilth also depends on the shape of the hood, soil shield, and shield adjustment. Raising or lowering the soil shield controls the amount of soil shattered as clods leave the rotor (Fig. 13). When the shield is raised, soil cut by the blades is not broken by impact with the shield; larger clods, trash, and weeds remain on the surface. This method is recommended for fall-working soil that is to be reworked and planted in the spring. The rough surface and residue help control soil blowing and water erosion. Raising the shield also reduces power requirements and lowers the tendency for damp soil to stick to the shield or plug the rotor. Lowering the shield increases soil pulverization and

Fig. 13—Till with Shield Raised or Lowered
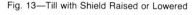

SHIELD RAISED AND LOWERED

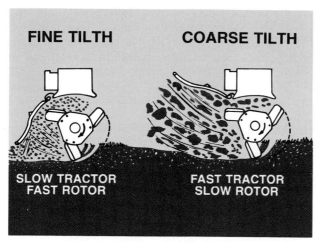

Fig. 14—Provide Desired Tilth by Adjusting Soil Shield, Rotor Speed and Tractor Speed

Fig. 15—Gauge Wheels Control Operating Depth

provides a finer seedbed, incorporates trash better, and helps level the surface.

For fine tilth: lower the soil shield, select a fast rotor speed, or slow tractor speed, or both. **For coarse tilth:** raise the soil shield, select a slow rotor speed, or a fast tractor speed, or both (Fig. 14).

SOIL TILTH FACTORS

Tilth produced by a rotary tiller depends on these factors:

• **Soil type**—heavy or light—cannot be controlled by the operator.

• **Soil moisture** can be controlled only by delaying work or irrigating.

• **Rotor speed** is easily controlled on most tillers.

• **Forward travel** speed is easily controlled if sufficient power is available.

• **Position of soil shield** is easily controlled.

• **Number of blades per flange** is controllable by removing or adding blades.

• **Amount of residue present** is controllable by removal, burning, or adding more. Incorporate all the residue into the soil whenever possible for the best soil condition.

OPERATING DEPTH

Regulate the operating depth of the rotary tiller by adjusting the gauge shoes or wheels on the machine (Fig. 15). Tractor hitch links must be free to float when operating either integral or semi-integral tillers. Don't attempt to use load-and-depth control to regulate tiller working depth because the **negative** draft created by the rotor will cause the load-and-depth control to operate improperly.

Depending on machine design and options available, gauge wheels may be controlled by adjusting clamps or cranks, changing pins, or by actuating remote hydraulic cylinders for on-the-go variations.

TRACTOR PREPARATION

Rotary tillers require much power. The tractor must be in top mechanical condition and tuned for maximum power output and optimum fuel consumption.

To provide the desired degree of pulverization, the tractor must have sufficient power to maintain standard PTO speed at the desired tiller operating depth and forward travel speed. Adequate hydraulic-lift capacity must also be available, or the tiller must be equipped with lift wheels for semi-integral operation.

Weighting rear tractor wheels is unnecessary for normal tiller operation, and it may be desirable to remove weights required for other operations to reduce soil compaction.

Use adequate front-end weights for stability when turning and transporting integral tillers (Fig. 16). Tractor load-and-depth control is not used, so front-end stability during operation usually poses no problems.

If tiller width permits, adjust the tractor tread so tires run on untilled soil and the tread is wide enough for good tractor stability. To eliminate wheel tracks in tilled soil with smaller tillers, it may be necessary to offset the tiller hitch and operate in lands.

To prevent tiller damage when lowering the machine, set the hitch-drop rate to operate as slowly as possible, or install restrictors in the hydraulic lines if the drop rate is too fast.

Install hitch stabilizers, sway blocks, or sway chains to restrict side-sway in operation and transport. The hitch must be free to float in operation with no down-

Fig. 16—Adequate Front-end Weight is Necessary for Integral Tillers

pressure applied and no load-and-depth control used. Adjust stops to limit raising and lowering to prevent PTO shaft damage from excessive angle or over-extension.

Install PTO shields, if not already in place, and be certain that shaft-mounted shields function properly.

PTO Shaft

To prolong equipment life and reduce problems, the PTO shaft must operate in as straight a line as possible (Fig. 17). A range of vertical hitch-attaching points on most tillers permits operation with a variety of tractors. but may require hitch relocation for some tractor-tiller combinations.

Match shaft length to the tractor being used and re-check length any time the tiller is switched to a different tractor model.

To prevent universal-joint chatter and possible damage, it is usually recommended that maximum PTO angle not exceed approximately 40 degrees when the tiller is fully raised or lowered (Fig. 18). This may require limiting the lift height of the tiller or rearranging the hitch points. Some manufacturers recommend limiting the maximum distance the tiller may drop below normal ground level to about 10 inches (250 mm).

Because of the high power requirements of rotary tillers, take care to assure adequate lubrication of the PTO shaft for easy telescoping as the tiller moves up and down in relation to the tractor. Special PTO shafts which telescope freely under load have been developed and may be desirable if problems arise. If the shaft cannot telescope freely while operating, vertical flexing of the hitch could result in damage to bearings and seals in the tractor PTO and the tiller gearbox.

A slip clutch is usually provided in the tiller drive train to protect the machine and tractor in case blades strike a solid object. Depending on the shape of the obstacle, tiller blades may simply lift the tiller and "walk" over some objects.

MACHINE PREPARATION AND MAINTENANCE

As with any implement, the performance and machine life of rotary tillers depend on the care provided in operation, servicing, and maintenance. Time spent to prepare and care for the machine will be rewarded by better field performance, longer machine life, reduced repair costs, and safer operation.

Fig. 17—Operate PTO in as Straight a Line as Possible

Fig. 18—PTO Angle must not Exceed 40 Degrees on most Tillers

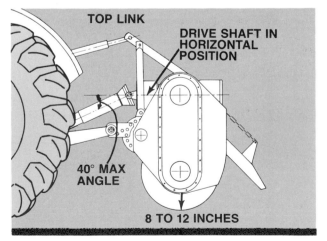

TOP LINK

DRIVE SHAFT IN HORIZONTAL POSITION

40° MAX ANGLE

8 TO 12 INCHES

192

Before Each Season:

1. See that hitch brackets are properly installed for the tractor to be used.

2. Check lubricant level in gearboxes—drain, flush, and refill them at recommend intervals.

3. Check all bolts, nuts, and set screws for tightness. Be sure all cotter pins are spread.

4. Provide proper PTO shaft for the tractor model used.

5. Check position of rotor flanges for work to be done—broadcast tillage, strip tillage, cultivation, etc.

6. Check rotor for proper scroll pattern of blades.

7. Check condition of blades. Straighten or replace bent or damaged blades. Use only bolts provided or specified by the tiller manufacturer. Tighten blade bolts to recommended torque, and recheck after a few hours of operation.

Before Daily Operation:

1. Lubricate the entire machine as recommended; don't overlook telescoping portions of the PTO shaft.

2. Check rotor for loose, bent, or broken blades. Check for wear on the outer surface of blades—wear here means travel speed is too fast for rotor speed.

3. Visually check entire machine for loose bolts and loose or broken parts.

4. Check rotor, especially near bearings, for wire, twine, or trash wrapped around the shaft. If not removed, such material may ruin seals and bearings.

On some machines, weed-cutting blades at each end of the rotor prevent weeds and heavy trash from wrapping on the rotor next to bearings. Check these blades. Adjust them regularly.

Before Storage At End Of Season:

1. Thoroughly clean machine of soil, crop residue, and greasy dirt.

2. Repaint spots where paint has been scratched or worn away.

3. Coat rotor, hood, and soil shield with rust-preventive paint or heavy oil to prevent rust and corrosion.

4. Lubricate entire machine as recommended.

5. Store machine inside a dry building, if possible with rotor off ground. Take weight off tires. Fully retract hydraulic cylinders to protect cylinder rods from rust.

FIELD OPERATION AND ADJUSTMENTS

Few adjustments are possible or necessary for satisfactory rotary-tiller performance. Make the following adjustments when starting to operate, when changing fields, when soil conditions vary, or when changing operations.

1. Level machine fore and aft; adjust top link on integral tillers; adjust hitch and wheel settings on semi-integral and drawn units.

2. Level tiller side-to-side (Fig. 19) with 3-point-hitch leveling crank or wheel adjustments, according to machine design and type.

3. Adjust gauge shoes or wheels for desired working depth.

4. Change rotor speed to obtain desired "bite."

5. Raise or lower soil shield to obtain desired tilth.

6. Change number of blades per flange according to soil conditions.

7. Arrange flanges on rotor according to work to be done—broadcast tillage, strip-tilling, etc.

If large amounts of crop residue (such as cornstalks, rice straw, etc.) are to be incorporated into the soil, do such work in the fall to speed decomposition of the material. Till when trash is as dry as possible to permit easier cutting and prevent plugging and wrapping of the rotor.

When tilling cornstalks, set the rotor speed and soil shield to obtain desired tilth at the intended operating speed. Then work at approximately a 20-degree angle to the corn rows to assure even distribution of trash, better leveling of ridges, and more-complete cutting and mixing of trash. Usually, slow rotor speed and high shield setting provide best results on the first pass, especially in the fall.

Spread straw from small grains, rice, and soybeans during combining. Operate the tiller diagonally across rows, as with cornstalks.

Rotor Care During Operation

Always straighten or replace bent, badly worn, or damaged blades immediately to avoid power waste and possible machine damage. When replacing or relocating blades, follow instructions in the operator's

Fig. 19—Tiller Operates Level Side-to-Side

LEVEL SIDE-TO-SIDE

manual. Most blade bolts are specially shaped and hardened to match blade and flange design. Therefore, use only those bolts recommended by the tiller manufacturer. Use of standard bolts to attach blades may result in premature bolt failure, equipment damage, and voiding of the machine warranty.

Fall Tillage

If a crop is not to be planted until spring, one pass of the tiller is usually enough to:

• *Chop and mix crop residue*

• *Kill weeds*

• *Leave soil rough to absorb moisture and avoid erosion (Fig. 20)*

Follow suggestions listed below to do the best job of fall tillage.

1. Use highest practical forward speed, but don't overrun the rotor.

2. Use slowest possible rotor speed compatible with travel speed for maximum bite.

3. Raise soil shield to provide desired soil tilth and trash mixing.

4. Leave surface rough, with residue exposed, to reduce soil blowing and water runoff.

Spring Tillage

On light soil, or land worked with a rotary tiller in the fall, one pass with the rotary tiller in the spring is usually enough for seedbed preparation. However, it may be desirable to till early to kill weeds already started and to encourage sprouting of weed seeds in the soil. To kill the newly-sprouted weeds, follow up 10 to 14 days later with a fast, shallow tilling, and then plant.

For row crops, two passes of the rotary tiller are usually required to handle heavy trash in the spring if the trash has not been previously chopped or tilled. The first pass, at approximately 4 inches (100 mm) deep, cuts trash and loosens the soil for fast drying. The second tilling at full depth will complete the mixing of trash with the soil and leave the ground ready for planting. If possible, make the second pass 10 to 14 days after the first to allow weeds to sprout and the soil to settle and dry.

1. Use a fairly fast rotor speed in relation to forward travel, but don't over-pulverize the soil.

2. Lower the soil shield to break up clods.

3. To avoid bringing more weed seeds to the surface for germination, do not work the soil deeper than necessary to provide a seedbed.

Pattern Of Operation

With smaller tillers it is usually advisable to operate in lands, as with moldboard plows, to:

• *Provide proper leveling of the surface*

• *Maintain uniform depth of operation*

• *Prevent running tractor wheels on worked soil*

When tiller width exceeds overall tractor width, operation may be in lands or in adjacent passes from one side of the field to the other. Headlands are tilled after the rest of the field is worked.

Weed Control

Some critics say rotary tillers simply chop weed roots into little pieces that will each produce a new plant. However, even troublesome perennial weeds like quack and Johnson grass, which spread from roots, have been controlled satisfactorily by these machines.

As many as five passes at 2- to 3-week intervals may be required for complete weed eradication (Fig. 21), so the land may have to lie fallow for a portion of the year. Repeated tillings after the rhizomes have again started to take root and throw off new green leaves will keep the weeds from regaining their strength. Researchers say it is better to wait a little too long between tillings than to till too quickly. By waiting the maximum, depletion of food reserves stored in the roots will occur as the weeds attempt to become reestablished.

Operate the tiller with a fast rotor speed and the tiller hood up so that as many roots as possible are thrown out on the surface to die. Make successive passes with the tiller progressively deeper in 2-inch (50 mm) increments to assure cutting all of the roots. Till when soil is relatively dry so that roots are shaken loose from the soil and have some time to dry and start dying instead of immediately rerooting.

Fig. 20—Fall Tillage Should Leave Soil Rough

ROUGH SOIL ABSORBS MOISTURE

Fig. 21—Several Passes Required for Complete Weed Eradication

Pasture Renovation

Pasture renovation with a rotary tiller involves a shallow first pass to break up and kill the sod. After two or three weeks, lime, fertilizer, and herbicides are applied and a second pass is made at greater depth to produce the seedbed and mix residue deeper into the soil. Seeding may then immediately follow the second pass.

If the goal is simply improvement of an existing stand of grass, one pass is generally sufficient to:

- **Incorporate fertilizer**
- **Kill weeds**
- **Level the surface**
- **Prepare the seedbed**

Operating Tips

1. Do not leave tractor seat while engine is operating; shut off engine before lubricating, inspecting, or adjusting the tiller.

2. Do not permit anyone near tiller while rotor is operating.

3. Lift the tiller clear of the surface before starting the rotor; engage the PTO before lowering the machine to the ground.

4. Raise the tiller before making turns.

5. If engine speed drops while tilling (PTO speed falls below standard 540 or 1,000 rpm), shift tractor to next lower gear or reduce working depth.

6. Use slowest rotor speed which will provide desired tilth; this uses the minimum of power input and results in less wear.

7. Remember there is less danger of underworking soil with a rotary tiller than of overworking it.

TRANSPORT AND SAFETY

When an operator ignores or violates the rules of safe operation of equipment, accidents are a likely result. However, most serious problems can be avoided by being aware of hazards and potential dangers inherent in the use of any equipment. Follow the safety suggestions above and below and the specific precautions in the operator's manual for safer operation of any rotary tiller.

1. Don't leave the tractor seat while the engine is running or the rotor is turning; allow the rotor to stop and shut off the engine before lubricating, making adjustments, or cleaning the tiller.

2. Never permit anyone to walk beside, stand behind, or ride on the tiller while rotor is turning.

3. Allow only the operator to ride on the tractor in operation or transport.

4. Do not transport at high speeds over rough terrain or where steering and stopping capabilities are impaired.

5. Don't attempt to block the rotor by hand to test the slip clutch; use a securely positioned bar or block of wood.

6. Keep PTO shields in place; don't work near or step over the PTO shaft while it is running.

7. Maintain adequate tractor front-end weight for stability when transporting integral tillers. (It is recommended that at least 55 percent of the static tractor front-end weight be retained on front wheels when the tiller is raised for transport.)

8. To prevent possible damage or breakage of PTO shaft, avoid sharp turns with drawn rotary tillers or lifting integral tillers beyond recommended limits (PTO shaft angled more than 40 degrees).

9. Use only recommended PTO shaft; if shaft is too short, it may become disconnected when raised; if too long, sections may "bottom out" and damage the gearbox or tractor drive train.

TROUBLESHOOTING

Improper operation and adjustment are the usual causes of unsatisfactory rotary-tiller performance. If the tiller is not functioning correctly, check the symptoms below, and make the recommended corrections according to instructions in the tiller manual. Where possible remedies to problems are obvious, based on the possible cause, a blank space is left in the possible remedy column.

TROUBLESHOOTING CHART

PROBLEM	POSSIBLE CAUSE	POSSIBLE REMEDY
ROTOR WON'T TURN	Seizure or failure of PTO shaft or gearbox.	Recondition or replace as needed.
	Slip clutch worn or out of adjustment.	Adjust or recondition— see manual.
	Obstruction in rotor.	Check rotor for stones, wire, or trash.
ROUGH, UNEVEN OPERATION	Machine too high or too low for PTO joints.	
	Blades not arranged in scroll pattern.	
	Universal joints worn.	
EXCESSIVE NOISE FROM TILLER	Gear failure in gearbox or drive train.	Stop tiller immediately. Determine cause if possible. Replace damaged parts.
	Bearing Failure.	Same as above.
	Gearbox oil level too low.	Add oil to recommended level, and check for leaks.
	Stone or obstruction wedged between blades.	Stop tiller, and remove object. Straighten or replace damaged blades.
TILLER SWAYS TO ONE SIDE	Stabilizers or sway blocks missing or improperly placed.	Install or adjust to prevent sidesway.
	Tiller not working level.	Adjust tractor hitch-lift link or gauge wheels or shoes or all of these as necessary.
BLADES WON'T PENETRATE OR HOLD CORRECT DEPTH	Tractor hydraulics improperly set.	Allow links to float— do not use load-and-depth control.
	Tractor travel too fast.	Shift to lower gear or shift to direct-drive gear (as opposed to free-wheeling gear).
	Blades bent or not arranged in scroll pattern.	
	Rotor speed too slow.	
SEVERE BLADE BREAKAGE	Blade bolts not tight.	Tighten bolts to recommended torque. Use only recommended blade bolts. Check flanges for elongation of bolt holes from running with loose bolts. Replace damaged flanges.

PROBLEM	POSSIBLE CAUSE	POSSIBLE REMEDY
	Working in soil with too many obstructions.	Use slower rotor speed to reduce blade impact on obstacles. Make first pass at shallower depth. Gradually increase depth on later passes.
EXCESSIVE POWER REQUIREMENT	Rotor speed too high.	Slow rotor by changing gears in gearbox.
	Blades bent, broken or incorrectly arranged.	Straighten or replace damaged blades; be sure blades follow scroll pattern.
	Trying to work too deep in one pass.	
	Tractor speed too fast.	
ROTOR PLUGS WITH SOIL	Soil too wet to work.	Wait till soil dries. Make very shallow first pass to speed drying.
	Rotor speed too slow for travel speed.	Speed up rotor by changing gears.
	Blades bent or incorrectly arranged.	Straighten or replace damaged blades, and check scroll pattern.
	Wet soil sticks to soil shield.	Raise shleld, or wait till soil dries.

SUMMARY

Potential purchasers of farm-size rotary tillers must consider all advantages and disadvantages of their use in comparison to conventional plowing, disking, and harrowing—even some types of cultivation.

Rotary tillers are versatile and useful implements, but the principles of their operation are unique, and must be thoroughly understood for satisfactory results. For example, one danger is over-pulverization of the soil (especially if too wet), not difficult to prevent, but harmful if it occurs.

Rotary tillers require little tractor draft power and can produce "negative" draft by tending to propel themselves forward. On the other hand, they are heavy consumers of energy, and may use as much per cubic foot (m³) of soil moved as all three operations of plowing, disking, and harrowing. Time- and man-power saving is possible, implement investment may be less than for traditional tillage methods, tractor-wheel soil compaction is reduced, and less soil moisture is lost by one operation than would be by three.

As with so many agricultural alternatives, each farm operator must determine for himself whether he should include rotary tillage in his soil management, maybe even substitute it entirely for other methods. Frank discussion with a nearby farmer who already has adopted rotary tillage would be an excellent preliminary to so important a decision.

CHAPTER QUIZ

1. Name four rotary-tiller applications.

2. (True or false) More tractor weight per horsepower is required for rotary tillers than moldboard plows.

3. What operating changes are required to reduce soil pulverization and power requirements with rotary tillers?

4. How can soil pulverization be increased with a rotary tiller?

5. (True or false) Rotary tillers are always centered behind the tractor.

6. (True or false) If a rotary tiller is used for two passes in the spring, the second pass should immediately follow the first.

7. Some people describe rotary tillage as minimum tillage. Others say it is maximum tillage which damages soil structure. Which is right (explain why)?

8. Why are rotary tillers well adapted for use in flooded rice paddies?

9. How can soil be loosened below the normal rotary-tiller working depth without making a second pass with another implement?

10. (True or false) Rotor blades are normally welded to the shaft to prevent loosening during operation.

11. (True or false) Lowering the rotor soil shield reduces soil pulverization.

12. (Fill blanks) Rotary-tiller operating depth is normally controlled by _____ or _____.

12
Listers and Bedders

Fig. 1—LIsters Shape Ridges but do not have Planting Attachments

Fig. 2—Lister-Planters Ridge the Soil and Plant in the Furrows

INTRODUCTION

Lister, bedder, lister-planter, lister-bedder, disk bedder, middlebuster, and middlebreaker are terms used in different areas to describe the same equipment. The local name usually pertains more to the operation being performed than to the implement, and much confusion in names has come about from different applications of similar tools. Reshaping of existing beds is sometimes called "hipping."

Listers (Fig. 1) are normally classified as primary-tillage tools which open furrows or make beds. Lister-planters (Fig. 2) are primary-tillage tools which not only open furrows but plant simultaneously in the bottoms of the furrows.

These implements may have solid moldboard-type bottoms or small disk gangs. Choice depends on soil and moisture conditions and operator preference. Disks may work better in soils where lister bottoms do not scour well or if there are many soil obstructions. Lister bottoms develop suction to aid penetration. Disks have little or no suction and often are used after previous primary tillage has loosened the soil. Lister shares may need to be sharpened or replaced more often than the disks.

Lister planting is usually done in dryland area to get seed down to moist soil for fast germination and quick growth. It is used in parts of the South to plant double-crop soybeans or sorghum following small-grain harvest. Such planting reduces the time between grain harvest and planting to provide maximum growing season for the second crop and conserve available soil moisture by reducing tillage to a minimum.

However, soil in the furrow may be cold enough in wet weather to delay germination and early crop growth. Hard rains and minor flooding of the furrows may cover seeds and small plants with silt or cause soil crusting, reducing stands and yields.

If the crop is to be irrigated, additional cultivation is used to furrow the soil between rows for water movement. By harvest, the furrow has been shifted from the row area to between the rows by several cultivations.

Ridges between rows protect some young crops from wind and blowing soil which can cut small plants off at the surface. Lister planting on the contour provides maximum conservation of soil and water because each ridge acts as a small terrace to slow runoff.

Providing once-over tillage and planting, the lister-planter can speed spring work, especially if already delayed by bad weather. It actually is a form of minimum tillage. When lister planting is done in previously worked soil it is called "loose-ground," as opposed to "hard-ground" done with no previous tillage.

In most areas, listed land is cultivated flat during the growing season. This provides effective weed control by tearing weeds out of the space between rows and covering those in the row area. The additional soil provides root support for the growing plants.

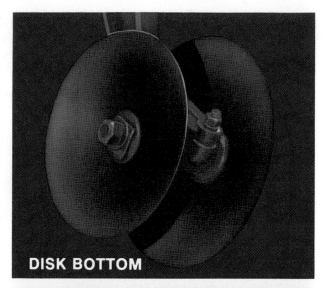

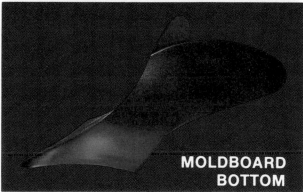

Fig. 3—Listers and Betters may have Disks or Moldboards

PROFILE PLANTING

A variation of lister planting, sometimes called profile planting, puts the crop row on a small ridge or raised bed in the bottom of the lister furrow (Fig. 4). This places seed in moist soil, protects young plants from wind and blowing soil, and reduces the chances of water standing on the row or washing over the seed or seedlings during a heavy rain.

In some areas beds are shaped soon after fall harvest and disposal of crop residue. If the soil was previously listed or bedded, the old ridges or beds are usually split and new ones formed over the old furrows. Before planting, the new ridges may be reshaped, or split again (Fig. 5), and reformed in the same area as ones of the previous year. Fertilizer may be applied under the row area during the listing operation.

If the land has been listed for some time and soil has eroded into the furrows, ridges may be reshaped and furrows cleaned with the lister. Crops to be planted, soil conditions, and local custom usually dictate the practices and number of operations performed in preparing land for planting.

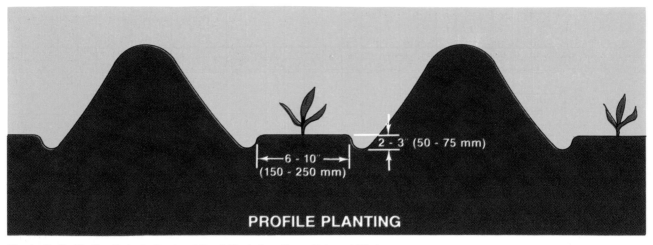

Fig. 4—Profile Planting Protects Seed and Small Plants from Heavy Rain and Wind

SEVERAL PLANTING METHODS

Beds may be planted in various ways. A unit planter with a large, specially shaped sweep (Fig. 6) may be used to cut away dry soil on top of the bed and plant a single row of corn, sorghum, cotton, or other crop on each bed. In drier areas the furrows can be used for later irrigation.

A bed-shaper (Fig. 7) may be used to firm and shape beds. Then, depending on crop and bed width, from one to four rows of vegetables, sugar beets, sorghum, or other crop is planted on each bed. Such units are sometimes called sled tool carriers, because of the runners which operate in the furrows to guide and support the implement.

MINIMIZE HEIGHT VARIATIONS

The runners or skids on sled carriers minimize height variations in the beds and provide excellent uniformity in placement of chemicals, fertilizer, and seed for optimum plant growth and weed control. On some units,

Fig. 5—Lister Ridges are often Split and Reshaped during Planting

the bed-shaping plates and unit planters may be replaced with cultivating tools for very precise, high-speed cultivation. Operating in the same furrows followed when planting, the runners provide extremely accurate guidance with a minimum of operator control.

Disk or cone-shaped guide wheels (Fig. 8) can be used instead of the runner. They roll along bed edges for precision guidance of cultivating tools close to small plants. Vertical skid adjustment permits shaving weeds from the beds without disturbing the crop.

Bed-shapers may be used on previously bedded soil, in loose soil prepared with plow and disk, or be attached to a rotary tiller.

Some soils develop impervious layers, just below normal tillage depth, which restrict root growth and water infiltration. In such cases subsoiling may be used to break up the hard layer and provide a channel

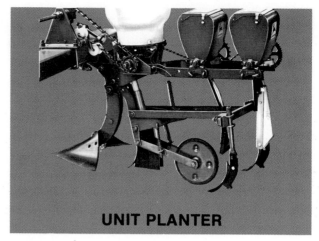

UNIT PLANTER

Fig. 6—Planting Attachments can Place one Row on Top of each Bed

Fig. 7—Bed-Shapers Provide Very Accurate Planter Control

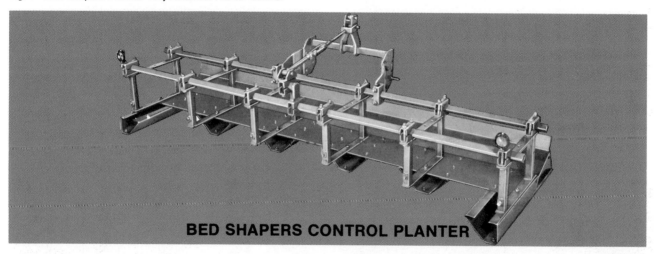

BED SHAPERS CONTROL PLANTER

Fig. 8—Cone or Sled Guides Provide Precision Guidance on Beds

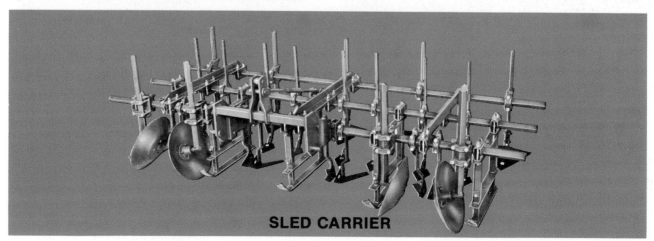

SLED CARRIER

BEDS OVER SUBSOILER SLOTS

Fig. 9—Placing Beds over Subsoiled Slots can Improve Plant Growth and Yield

for roots and water into the subsoil. By placing beds directly over the subsoiled slot with a single operation, plant roots are provided a direct path to more soil volume, moisture, and nutrients. Once-over preparation also conserves time and fuel. Parallelogram linkage between subsoiler bar and bedder bar (Fig. 9) lets the rear bar flex independently for uniform bed-building in varying soils or rolling land. Adjustable gauge wheels control operating depth of the bedder, and may also be used on the subsoiler bar.

Fig. 10—Integral Bedders and Listers Provide Maximum Convenience

INTEGRAL BEDDERS AND LISTERS

Ridge Or Bed Planting

Planting on top of beds is common in high-rainfall areas where surface drainage is a problem. Soil in beds dries and warms faster than in furrows or on flat land, permitting faster seed germination and better early crop growth. This prevents water from standing on rows after heavy rain and possibly killing small plants. Depending on the bed-shaping and cultivation methods used, weed control may be more of a problem. Higher herbicide application rates may be required.

In a system developed at Iowa State University, corn is planted on the same ridges year after year. Advantages are said to be that soil warms faster on the ridges in the spring, residue is left in the furrows by spring chopping of old stalks, planting is no problem through the old corn stubs, and some weeds are smothered by the mulch between rows. Ridges are reshaped with a disk cultivator during the growing season, but there is no overall tillage at any time.

LISTER AND BEDDER TYPES AND SIZES

Formerly, many listers and bedders were front-mounted on tractors. This provided good visibility of machine performance, and improved traction for tractor tires in wet areas where listing or bedding were the first primary-tillage operation.

However, front-mounted units are somewhat limited in size and require more time for attaching and removal. Because most land now is disked or chisel plowed before listing or bedding, the trend is toward rear-mounted or integral toolbars with their fast, convenient hitching and unhitching (Fig. 10).

An added advantage of toolbar mounting is the wide range of row spacings available, from about 28 inches (711 mm) on up. The number of rows is limited only by tractor power and lift capacity and toolbar length. However, 4-, 6-, and 8-row sizes are most common.

Until recent years, most listing and bedding was done with "conventional" lister bottoms (Fig. 11) which

Fig. 11—Lister Bottoms Penetrate Well in Most Soils

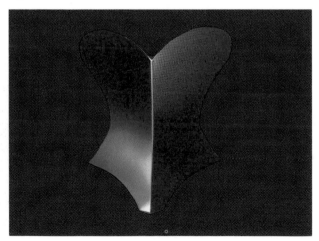

Fig. 12—Disk Bedders have Light Draft

appear to be made from right and left hand moldboard plow bottoms joined together. Many farmers now prefer disk bedders (Fig. 12) because of lighter draft, fewer scouring problems, and less maintenance.

Various bottoms have been developed to meet particular soil conditions. The general-purpose bottom (Fig. 13) works well at fairly high speeds in most soil conditions. Hard-ground bottoms (Fig. 13) have very small share and moldboard and are designed for use with disk openers (Fig. 14) which serves as "rolling moldboards" to open hard-baked soils. Blackland bottoms

(Fig. 13) with smaller moldboards are designed for better scouring in sticky soils.

Disk bottoms (Fig. 3) may have two or three blades, depending on blade size and desired bed dimensions. Disk angle may be preset, with no field adjustment required or even possible, or gangs may be designed to be tilted and turned to match soil conditions.

The working angle of fixed gangs usually offers the best performance over the widest range of conditions and avoids improper adjustment in the field. If adjust-

Fig. 13—General-Purpose, Hard Ground, and Blackland Bottoms

GENERAL-PURPOSE

HARD-GROUND

BLACKLAND

Fig. 14—Hard-Ground Bottoms with Disk Openers

Fig. 16—High Pointed and Rounded Beds

able gangs are not correctly positioned they may increase draft, penetrate unevenly, and make non-uniform beds.

The typical disk gang (Fig. 12) has blades 16 and 18 inches (406 and 457 mm) in diameter, spaced 8 inches (203 mm) apart, on sealed and permanently-lubricated ball bearings.

Disk gangs may be arranged for either staggered or opposed operation on the toolbar. For opposed operation (Fig. 15) gangs forming each bed work directly opposite each other and roll soil together into the bed. This forms a high pointed bed (Fig. 16) and is recommended for light, sandy, or loam soils.

Staggering disk gangs (Fig. 17) feathers the soil together for a flatter, rounder bed (Fig. 16). This arrangement performs best in slabby or gumbo soils.

Disk bottoms will roll up and over many obstructions which would trip or break moldboard-type bottoms.

Some disk breakage may occur in rocky or stumpy conditions, especially with large toolbar units, because of the high total implement weight involved.

PROTECTING LISTER BOTTOMS

Lister bottoms may be protected from soil obstructions by safety-trip beams (Fig. 18) or friction-trip beams (Fig. 19). Required tripping force for safety-trip beams is adjusted by tightening or loosening the upper bolt (arrow, Fig. 18), but it should never be set so tight that tripping is prevented altogether. Friction trips, used where few soil obstructions are encountered, can be adjusted to some extent for tripping resistance by varying tightness of the trip bolt. Overtightening the pivot bolts on either safety-trip or friction-trip bottoms can also interfere with proper tripping action and reduce the protection provided.

Fig. 15—Opposed Disk Gangs Roll Soil Together

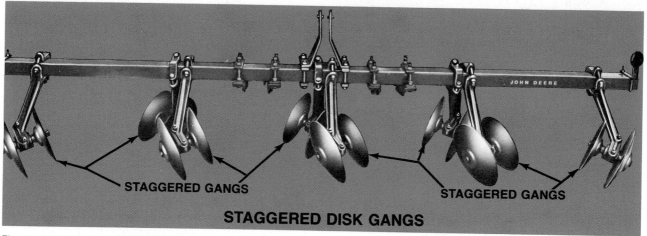

STAGGERED DISK GANGS

Fig. 17—Staggered Disk Gangs Work best in Gumbo

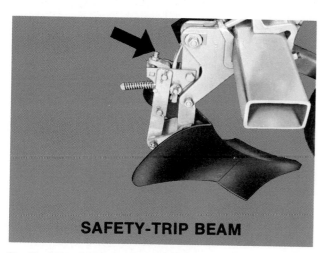

SAFETY-TRIP BEAM

Fig. 18—Safety-Trip Beam Protects Bottoms from Obstructions

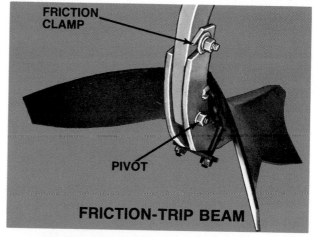

FRICTION-TRIP BEAM

Fig. 19—Friction-Trip Beam Provides Economical Protection

Fig. 20—High-Clearance Beams (Left) have more Trash Clearance than Regular Beams (Right)

Fig. 21—Main Bar is 4 × 7 Inch (102 × 457 mm) Box-Steel with 2¼-Inch (57 mm) Square Solid-Steel Second Bar

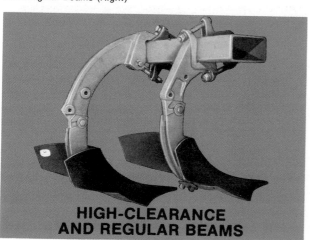

HIGH-CLEARANCE AND REGULAR BEAMS

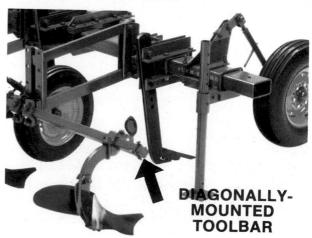

DIAGONALLY-MOUNTED TOOLBAR

207

Some high-clearance beams (left, Fig. 20) provide as much as 9 inches (230 mm) more space to the rear and 3 inches (76 mm) of additional space under the toolbar, compared to regular beams (right, Fig. 20). This improves trash clearance and permits deeper penetration.

Lister and bedder toolbar size varies according to the number of rows and variety of equipment. Most single-bar setups now use a box-steel hollow toolbar measuring from 4×7 (100×180 mm) up to 6×8 inches (150×200 mm). Others use square bars, such as 5×5 or 7×7 inches (130×130 or 180×180 mm). Similar bars are used for the front bar of subsoiler-bedder combinations (Fig. 21). The rear bar is usually a diagonally-mounted 2¼- or 2½-inch (57 or 64 mm) square solid-steel bar.

Fig. 22—Hydraulic Folding Eliminates Unhooking for Transport

HYDRAULIC
FOLDING
TOOLBAR

Some long toolbars are hinged and folded hydraulically (Fig. 22) for transport, while others may be equipped with transport attachments and pulled endways (Fig. 23).

PRINCIPLES OF OPERATION

Lister planting puts each crop row in a furrow where moisture is more readily available and the new plant is protected from wind and blowing soil. Listing may be done in previously tilled soil or in unworked ground. Commonly listed crops are corn, sorghum, and cotton.

During cultivation, listed land may be cultivated flat, or furrowed between the rows to permit irrigation. Cold, wet weather may delay germination and early growth of listed crops, but they usually catch up later because of more moisture availability in the listed rows.

Bedding may be done with lister bottoms or small disk gangs, with the crop planted later in the furrow or on top of the bed.

Bed shapers are used to firm bedded soil and make uniformly shaped beds for precision placement of seed, chemicals, and fertilizer. Furrows between beds are used to guide bed shapers and cultivators, and as paths for tractor wheels. Planting on beds places crops away from standing water in poorly-drained fields, and up where soil dries and warms fastest in the spring.

MACHINE PREPARATION AND MAINTENANCE

Satisfactory performance is directly related to the care and servicing provided for equipment. Well-maintained equipment performs better, increases operator satisfaction, and generally encourages continued careful attention because of the better performance achieved.

Before Each Season:

1. Lubricate entire machine as recommended.

2. Check bolts and nuts for tightness; replace worn, broken or missing parts.

3. Inflate gauge-wheel tires to recommended pressure.

4. Be sure all units are properly located on the toolbar for the desired row spacing.

5. Sharpen or replace dull, broken, or badly-worn shares or disks.

6. Hand-release all friction-trip and safety-trip beams and remove any rust which might restrict tripping. Avoid overtightening pivot and friction-trip bolts.

7. Refer to FMO Manual, Planting, for planter preparation.

ENDWAYS TRANSPORT ATTACHMENT

Fig. 23—Endways Transport Reduces Width of Large Units

Daily After Operation:

1. Check for loose or missing bolts, nuts, and parts, and underinflated tires.

2. Lubricate entire machine as recommended.

3. Coat all soil-engaging parts with oil to prevent rusting.

4. Empty fertilizer and chemical hoppers to prevent corrosion.

Before Storage At End Of Season:

1. Thoroughly clean the machine and repaint spots where paint is scratched or worn.

2. Coat soil-engaging parts with heavy grease or plow-bottom paint to prevent rust.

3. Wash fertilizer and chemical hoppers inside and out immediately after use and coat insides with diesel oil or kerosene to prevent rust.

4. Store inside to prevent weathering.

5. Park on boards to prevent soil-engaging parts from sinking into the soil and rusting.

FIELD OPERATION

Because of the great variety of equipment combinations possible when assembling toolbar units

each operator should familiarize himself with specific operating instructions furnished with his equipment.

HITCHING

Toolbar-mounted listers and bedders are attached to the tractor 3-point hitch with or without quick-couplers (Fig. 24). The toolbar is leveled laterally by setting hitch lift links for equal length, and the top link is adjusted to level the machine fore-and-aft. Correct adjustment of the top link is particularly important on disk bedders, so right-hand gangs will balance their side draft against left-hand gangs. Lack of this balance will cause steering problems, and lack of side draft is the best indication that the bedder is level fore-and-aft.

IMPORTANT: be sure the correct hitch pins are used before hooking the quick-coupler to the toolbar.

If the toolbar is equipped with a support stand, put it in storage position before starting operation. If markers are used, connect hoses or chain and slowly raise and lower the markers through one complete cycle to be sure everything is working correctly.

When first hitching to a new lister or bedder, when attaching it to a different tractor, or after changing row spacings, always raise the toolbar very slowly the first time to be sure there is adequate clearance between tires and implement.

Lower the support stand before detaching the toolbar from the tractor.

PICK-UP-AND-GO
WITH QUICK COUPLER

Fig. 24—Pick-Up-and-Go with Quick Coupler

DEPTH CONTROL

Working depth of bedders and listers may be controlled by the tractor load-and-depth or draft-control system, or by gauge wheels (Fig. 25). Many large bedders and listers have gauge wheels as standard equipment for more uniform operation over uneven ground.

Do not use the tractor load-and-depth control with gauge wheels except for subsoiler bedders; with these, set the load-and-depth control in mid-range to provide some weight transfer to the tractor.

Set the tractor lift links in "float" position so the toolbar will raise and lower automatically as the gauge wheels follow ground contours. Gauge-wheel height may be adjusted by changing height of a turnbuckle (Fig. 26).

Operating ranges are provided on this particular wheel to permit operation in the shallow setting for bedding flat land, or running the gauge wheel on top of the bed; or the deep setting to place the wheel in the furrow while hipping beds or lister planting. The turnbuckle provides final positioning.

Fig. 25—Guage Wheel

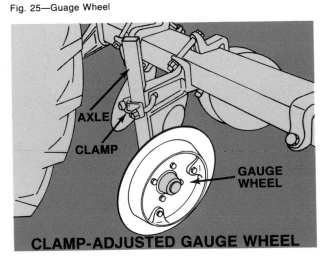

AXLE

CLAMP

GAUGE
WHEEL

CLAMP-ADJUSTED GAUGE WHEEL

Fig. 26—Turnbuckle Adjustment is Fast and Simple

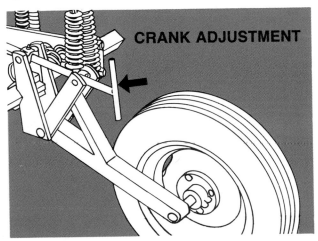

CRANK ADJUSTMENT

For maximum convenience and optimum implement control, use hydraulically-controlled gauge wheels. A decal on the cylinder indicates working depth. Both turnbuckle- and hydraulically-controlled gauge wheels are available with single or dual wheels. Duals flex over uneven ground to maintain the bedder bar at more uniform height.

TROUBLESHOOTING

Some unique operating problems may develop with listers and bedders, depending on the combination of equipment used and individual design features. So, when trouble arises, look for the problem and possible causes and remedies here, and then in the operator's manual for specific instructions.

Where possible remedies to problems are obvious, based on the possible cause, a blank space is left in the possible remedy column.

TROUBLESHOOTING CHART		
PROBLEM	**POSSIBLE CAUSE**	**POSSIBLE REMEDY**
POOR PENETRATION	Toolbar not leveled fore-and-aft.	Shorten top link to tip points down.
	Extremely hard ground.	Disk or chisel ground first.
	Excessive trash.	Shred or disk trash.
	Using too much load-and-depth control with gauge wheels.	Allow hitch to float with gauge wheels controlling depth; set load-and-depth control closer to "zero."
	Gauge wheels set wrong.	Raise gauge wheels to allow deeper penetration.
	Dull shares or disks.	
DISKS WON'T TURN	Bearing seized.	Replace or repair.
	Disks plugged with mud.	Install scrapers
	Stone, stick, or trash between disk and hanger.	
DISKS OR BOTTOMS WON'T SCOUR	No land polish.	Stop frequently to clean soil from disks or bottoms until they scour.
	Soil too wet.	Disk lightly or chisel plow to speed drying.
	Paint or rust on bottom of disk.	
	Not using disk scrapers	
FURROWS AND BEDS UNEVEN	Toolbar not level side-to-side	Adjust hitch lift links to level toolbar.
	Gauge wheels improperly adjusted.	Set gauge wheels to same level.
	Improper spacing of standards.	Readjust gangs or standards on toolbar.

PROBLEM	POSSIBLE CAUSE	POSSIBLE REMEDY
EXCESSIVE TRACTOR SIDE-DRAFT	Disk-bedder toolbar not level fore-and-aft.	Adjust top link to eliminate side draft.
	Toolbar not level side-to-side.	Adjust hitch lift links, or adjust gauge wheels to same level.
	Subsoilers not set at same depth.	
	Lister bottoms not set with equal "suck".	
	Tractor lift links not set to equal length.	

SUMMARY

Cropping methods in North America fall into several general categories, depending on crop, rainfall, soil, and other factors. For instance, Corn Belt farmers' methods may range from moldboard plowing, disking, harrowing, and planting to stubble-mulch or minimum-tillage practices.

But in some areas, particularly southern states and drier western areas, farmers use listers and bedders for primary tillage and often combine tillage and planting in one operation. These implements form ridges or beds, with furrows in between. Seed is planted either in the furrow, usually if moisture is deficient, or on top of the beds if excess moisture or cool climate are problems. Cornbelt farmers in drier areas of the Missouri valley also use this method.

Though the terms bedder, lister, and middlebreaker often are used interchangeably, bedders only make or reshape beds. With disk bedders, planting is a separate operation. Listers make or reshape beds, and—when equipped with planting attachments (lister-planters)—plant one row to each furrow.

Toolbars have broadened the range of equipment, from soil-shaping tools to unit planters, available for farmers who raise listed crops. Often the same carrier may be used for several types of operations by changing its equipment, thus reducing overall implement costs.

CHAPTER QUIZ

1. What is the commonly accepted definition of bedding? Of listing?

2. In what conditions are disk bedders superior to listers with moldboard-type bottoms?

3. What is one advantage of lister bottoms?

4. What are major advantages of lister planting?

5. List four advantages of planting on the ridges.

6. What is profile planting and what are its advantages?

7. Name several advantages of sled-type bed-shapers.

13
Subsoilers

Fig. 1—Subsoilers Break Hardpan

INTRODUCTION

Subsoiling usually is done to break up impervious soil layers below the normal tillage depth to improve water infiltration, drainage and root penetration (Fig. 1). To be effective in improving crop yields, subsoiling must meet these conditions:

1. It should be done when soil is relatively dry to permit shattering of the hard layer. If soil is wet, only a thin slot is sliced through the soil, which will likely reseal very quickly, and down-pressure of tractor weight and subsoilers can cause compaction.

2. Soil below the impervious layer must have excess water-holding capacity, or there will be no place for surface water to go, and no air in the deeper layers for plant-root growth.

3. Deeper soil must not be so acid or alkaline as to discourage root growth.

4. Tractors and heavy implements must run at least one foot away from subsoil slots during subsequent operations, to prevent resealing of the slot by tire compaction.

5. Subsoiling should not penetrate into a deep layer of sand if the water table in the sand drops rapidly during dry weather. Otherwise, available moisture may infiltrate beyond reach of roots when they need it most.

Some outstanding results have been achieved from subsoiling. Yield increases of 50 to 400 percent have been reported from subsoiling under the right soil and moisture conditions and in the right areas.

CHISEL PLOWS

Chisel plows, running a few inches deeper than the soil has been plowed for many years (Fig. 2), are used in some instances to break up the plow sole. However, chisel plows normally run 8 to 10 inches (200 to 250 mm) deep, and moldboard plows may run as deep as 12 to 14 inches (300 to 360 mm), so in some fields it is necessary to use subsoilers which can penetrate 16 to 22 inches (410 to 560 mm).

Where soil and moisture conditions are right, subsoiling can break up compacted layers caused from equipment operation, heavy wheel traffic, salts in irrigation water, or natural soil formations. Water from rain and snow soak in, instead of running off. Moisture soaks down faster so soil dries more quickly in spring, and crops can be planted earlier. Subsoiling also aerates the subsoil to encourage plant roots to seek available moisture and nutrients at the deeper levels.

In some conditions a mole or mole-ball (Fig. 3), attached to a subsoiler, will improve subsurface drainage. The heavy steel ball is attached by chain links to the rear of the subsoiler and drawn through the soil below the normal tillage level, deep enough so wheel traffic will not crush the tunnel immediately. Water penetration from the surface is improved by the subsoiler slot to the tunnel, and the force of the ball pressing soil outward and sliding through the ground forms a firm lining.

Slope of the drain must be enough to keep water flowing, but never steep enough to start erosion in the tunnel.

Fig. 2—Toolbar Chisel Standard

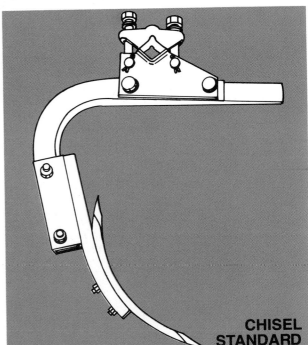

CHISEL STANDARD

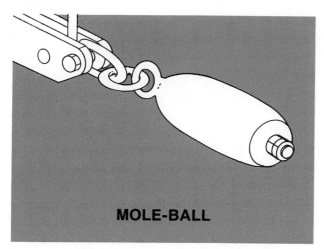

MOLE-BALL

Fig. 3—Mole-Ball on Subsoiler Can Form Subsurface Drain

Slope is usually between 6- and 24-inches (150 and 600 mm) in a thousand feet (300 m). A piece of pipe or drain tile should be placed in the outlet to prevent erosion and outlet stoppage.

MOLE-DRAIN REQUISITES

Mole drains work in damp, plastic soils, but are ineffective in soft loose soils which cave into the tunnel. An undulating surface or stones may prevent attainment of satisfactory grade for adequate drainage.

Mole drains may not work at all, may plug up in a week or two, or may satisfactorily drain the soil for some time. No ditching or tiling is required, so cost is low, and the drainage system is made as the tractor travels.

Toolbar-mounted subsoilers may be used with bedders to fracture hardpan for better root and water penetration and build beds directly over the slot (Fig. 4). (See Listers and Bedders). This centers crop rows over the soil opening for direct root access to increased moisture and growing space.

Depending on subsoiler design, power available, soil conditions, and depth of hardpan, subsoiler penetration may be as deep as 24 inches (600 mm) (Fig. 5).

SUBSOILER TYPES AND SIZES

Most subsoilers in current farm use are integral models with up to 13 standards for various tractor sizes and depths of penetration. Wheeled subsoilers have been common in some areas in the past and are still used for severe conditions. Increased tractor size has made it practical to operate larger integral subsoilers, which are more convenient and maneuverable. Single-standard subsoilers are available for Category 1 or 2 integral hitches for small to medium-size tractors.

When two or more subsoiler standards are used, they are usually attached to a box-bar or tubular toolbar which permits wide variations in spacing (Fig. 6).

Fig. 4—Beds Placed Over Subsoiler Slots May Provide Better Root Growth

SUBSOILERS

BEDS PLACED OVER SUBSOILERS

Fig. 5—Many Subsoilers Operate as Deep as 24 Inches (600 mm)

Many of these standards are equipped with a shear bolt or even better, a safety-trip (Fig. 7) to protect subsoiler and tractor from damage if the subsoiler strikes an obstacle.

A reversible, replaceable, tapered shin (Fig. 8) on some standards reduces draft and provides longer service and better soil cutting.

A subsoiler variation is the "V"-shape subsoiling chisel plow (Fig. 9), or V-ripper or V-chisel, which has from 5 to 13 standards for ripping hardpan as deep as

Fig. 6—Toolbar Mounting Permits Varying Space and Number of Standards

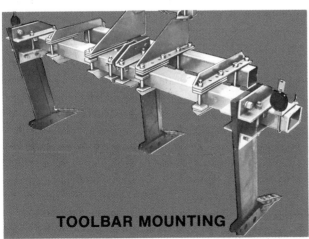

Fig. 7—Safety-Trips Protect Standards from Soil Obstacles

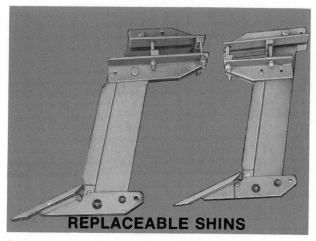

REPLACEABLE SHINS

Fig. 8—Tapered Replaceable Shins for Reduced Draft

16 inches (400 mm). Standards may be located from about 18 to 40 inches (457 to 1016 mm) apart. Standard or optional gauge wheels control depth of penetration.

PRINCIPLES OF SUBSOILER OPERATION

Subsoilers work best in firm soil where a hard layer prevents adequate root and moisture penetration. If soil is uniformly textured to the depth of subsoiling, or is too wet, subsoiling is usually not as productive.

Fig. 9—"V" Subsoiling Chisel Rips Hardpan Down to 16 Inches

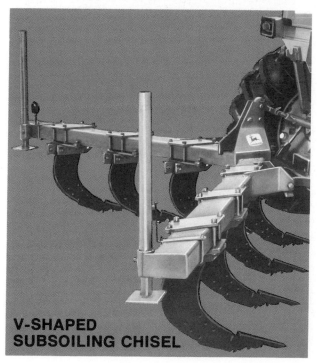

V-SHAPED
SUBSOILING CHISEL

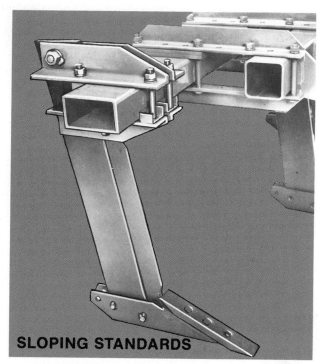

SLOPING STANDARDS

Fig. 10—Sloping Standards Lift and Break Soil

Slope of subsoiler standards and points affects draft and soil shattering. When standards are inclined forward (Fig. 10) they lift and break the soil much better than if they are vertical, or nearly so. Curved standards (Fig. 11) work under hardpan, lifting and shattering the soil ahead of and between standards.

Staggering subsoiler standards on the toolbar (Fig. 12) provides better trash clearance and permits easier operation because the front row of standards helps break the soil for the second row.

Fig. 11—Curved Standards Penetrate under Soil for better Shattering

CURVED STANDARDS

217

STAGGERING SUBSOILER STANDARDS

Fig. 12—Staggering Subsoiler Standards Reduces Draft, Improves Trash Clearance

Subsoiling chisels leave the soil surface rough, open, and loose (Fig. 13) to absorb water and reduce erosion. Deep penetration permits deeper water infiltration and increase storage for crop production.

SUBSOILER PREPARATION AND MAINTENANCE

Length of subsoiler life and maximum operating efficiency depend largely on good care and proper adjustments. Subsoilers are high-draft implements and working deeper than other tillage tools, so special care must be taken to assure proper maintenance.

Before Each Season:

1. Check for badly worn or broken points. Reverse or replace as needed.

2. Reverse (or install new) replaceable shins where needed. Points and shins often may be hard-faced locally if desired.

3. Check for loose, broken, or missing bolts, nuts, and parts. Replace as needed. When replacing bolts, always use bolts of recommended strength to prevent premature failure.

4. Check for rust and over-tightened pivots by removing shear bolts and tripping standards by hand.

5. Clean, repack, and tighten wheel bearings.

6. Be sure hitch pins are correct size if using quick coupler.

Daily Before Operation:

1. Check for worn points and loose, broken, or missing parts and bolts.

2. Check pressure in gauge-wheel tires.

3. Always replace broken shear bolts with bolts recommended by manufacturer. Softer bolts will shear too easily and delay field work. Bolts that are too hard could restrict shearing and cause equipment damage.

Before Storage At End Of Season:

1. Lower support stands before detaching toolbar from tractor.

2. Thoroughly clean the subsoiler to prevent rust; repaint scratched or worn spots.

3. Cover soil-engaging tools with heavy grease or plow-bottom paint to prevent rust.

4. Store inside to prevent weathering.

FIELD OPERATION

Subsoiler operating patterns depend on the objectives. If subsoiling and bedding are a single operation, the pattern must follow the desired future row pattern. Where soil is irrigated, the same row pattern is usually followed year after year. It may be desirable to subsoil and place new beds over old row middles. This will break down the old bed, uproot crop stubble, and help cover trash.

For mole drainage, subsoiling must follow ground contours from the intended outlet through the area to be drained. Adequate slope must be maintained in the channel for proper water flow.

Toolbar subsoiler spacing is adjustable and is usually 3 feet (1 m) or more. "V" subsoiling chisels may be arranged from 18 to 40 inches (457 to 1016 mm) apart on the bar (Fig. 14). Single-standard subsoilers are usually operated at 3- to 8-foot (1 to 2.5 m) intervals, depending on soil conditions and operator preference.

Operating the subsoiling chisel on the contour or across the path of prevailing winds will help reduce erosion and increase water-holding capacity.

Level the subsoiler laterally by setting hitch lift links to equal length for uniform, vertical penetration of all standards. Leveling fore-and-aft by adjusting the top link assures proper angle of penetration.

When using gauge wheels to control depth, place tractor load-and-depth control in a mid-setting to allow the load-and-depth system to provide weight transfer. Place lift links in "float".

V-SHAPED CHISELS LOOSEN SOIL

Fig. 13—Soil is Left Loose and Rough by Subsoiling Chisels

TRACTOR PREPARATION

Subsoilers are relatively high-draft tools and require top tractor performance for optimum efficiency. Performance will be improved by proper tractor operation.

1. Tune engine for maximum power and optimum fuel consumption.

2. Clean or replace oil, air, and hydraulic filters.

3. Check hydraulic system for adequate pressure and oil level. Repair leaks in hoses, couplings, or cylinders (Fig. 15).

4. Adjust wheel tread to match row spacing, center-to-center, or use widest practical tread for operation on steep slopes.

5. Inflate tires to recommended pressure.

6. Provide front-end weight according to recommendations in tractor and implement operator's manuals. Use maximum allowable front-end weight when operating integral machines in lower gears.

7. Weight rear wheels to avoid excessive slippage. When pulling rated load, tracks should be slightly broken and shifted. Inadequate weight causes rapid

Fig. 14—Subsoiling Chisels are Spaced 18 to 40 Inches (457 to 1016 mm)

SUBSOILING CHISELS

tire wear and wastes time and fuel. Excess weight also wastes fuel, and increases soil compaction.

8. For drawn implements, set drawbar at recommended height and pinned in the center or free to swing as recommended in implement manual.

9. Most integral implements are leveled laterally by adjusting hitch lift links to equal length. Fore-and-aft loading is adjusted by the center link.

10. When using toolbars with gauge wheels, set lift links to float so that each end of the toolbar can move up or down independently.

TRANSPORT AND SAFETY

Most accidents are caused by someone's carelessness, neglect, or oversight. Following basic safety precautions at all times—in the field, on the road, or when storing equipment—can greatly reduce the number and severity of equipment-related accidents.

1. Always use lights, reflectors, and SMV emblem as required by law when transporting equipment—day or night.

2. Avoid peak traffic periods and heavily-traveled roads if at all possible.

3. Reduce implement width as much as possible for transport.

4. If transport width exceeds normal vehicle width, be especially cautious to avoid collisions with other vehicles and to miss holes, drains, or ditches along the road edge.

5. Install transport lock on drawn equipment and relieve pressure in hydraulic cylinders.

6. Never exceed recommended transport speed, or tractor road speed if maximum is not stated. Reduce speed when turning or crossing rough areas and slopes.

7. Always lower parking stands on integral equipment before detaching from tractor.

8. Always lower equipment or install transport lock before servicing, lubricating, or repairing equipment, and when the machine will be left unattended.

Fig. 15—Check and Repair Leaks in Hydraulic Hoses

9. If integral tools must be serviced while in the raised position, block frame securely before starting work.

10. Never lubricate, repair, or adjust any implement while it is in operation.

11. Never allow children to play on or near equipment during operation, transport, or storage.

12. Allow only the operator to ride on the tractor, and never allow anyone to ride on the implement.

13. Provide adequate tractor front-end weight for stable operation and transport. Use enough weight to hold front wheels on the ground at all times, but do not exceed manufacturer's recommended limit. If enough weight cannot be added for safe operation without exceeding the limit, do not use that tractor-implement combination.

SUBSURFACE KNIVES

If you depend on a ripper for weeding as well as deep tillage, add subsurface knives. These knives cut weeds deep under the soil.

TROUBLESHOOTING

Improper subsoiler adjustment results in inefficient operation, poor performance, rapid wear, and possible breakage. Prompt correction of operating problems permits maximum equipment performance with minimum expense. Where possible remedies to problems are obvious, based on the possible cause, a blank space is left in the possible remedy column.

TROUBLESHOOTING CHART		
PROBLEM	POSSIBLE CAUSE	POSSIBLE REMEDY
POOR PENETRATION	Tractor hydraulic system improperly adjusted.	Set load-and-depth control more toward maximum.
	Gauge wheels improperly adjusted.	Set gauge wheels the same on both sides for desired depth.

	Frame not level—rear end low.	Shorten top link.
	Dull or broken points.	
EXCESSIVE BREAKAGE OF SHEAR BOLTS	Numerous large stones in soil.	Reduce operating depth.
	Excessive side-play in clamp.	Tighten pivot bolt and shear bolt to specified torque.
	Using improper bolts.	
	Working too deep in very hard soil.	
UNEQUAL PENETRATION OF STANDARDS	Toolbar not level laterally.	Adjust hitch lift link to level toolbar.
	Some points worn and others sharp.	Sharpen dull points.
	Gauge wheels not set the same.	
	Adjustable standards not set at same height.	

SUMMARY

Subsoiling is intended to break up impervious layers of soil below normal tillage depth and improve root and water penetration. The value of this high-draft operation ranges from excellent to poor, depending on soil conditions.

Results usually are good if the soil is hard (not sandy), if subsoil has excess water-holding capacity and is not acid or alkaline, and if winter freezing and thawing do not normally break up hard pan, plow sole, and other layers.

In proper conditions a mole-ball, pulled behind the subsoiler point, may be used to make an underground drainage conduit; this method is not recommended for sandy soils where the conduit will quickly close.

CHAPTER QUIZ

1. What is the purpose of subsoiling?

2. In what conditions is subsoiling most advantageous?

3. (True or false?) Mole drains made with a subsoiler are effective in all conditions.

4. Which has least draft, vertical or inclined subsoiler standards?

5. Should fertilizer be applied in subsoiler slots? Explain.

6. How deep should subsoilers penetrate?

14
Disk Harrows

INTRODUCTION

Using disk blades for tillage is believed to have originated in Japan. Shortly after the Civil War the Nishwitz harrow, with slightly concave disks on an A-shaped frame to throw soil outward in both directions, aroused considerable interest. Not until the late 1870's, when the LaDow and Randall disk harrows appeared, were such tools widely accepted.

Most early disk harrows were built in local blacksmith shops about 1880, when factory production began in Sterling, Illinois. Notched disk blades were introduced in the early 1890's.

From those crude beginnings came modern disk harrows, now used around the world—machines with sealed bearings, rubber-tired transport wheels, high-strength alloy-steel blades, remote hydraulic controls, and many other improvements.

WIDE VERSATILITY

Disk harrows are employed in nearly every kind of soil condition (Fig. 1). Heavy-duty harrows are used for primary tillage—breaking unworked ground, cutting and mixing heavy crop residue, and mulching stubble.

Disking cornstalks and other heavy crop residue before plowing loosens the surface, cuts up trash, and mixes it into the soil (Fig. 2). This provides better trash coverage when the land is plowed later, and more loose dirt for better soil-trash contact and faster decomposition of residue.

SECONDARY TILLAGE

Secondary tillage—seedbed preparation (Fig. 3), summer fallowing, chemical incorporation, and weed

Fig. 1—Disk Harrows Work in Many Soil Conditions

DISK HARROWS WORK IN VARIED CONDITIONS

Fig. 2—Disking Chops Residue and Mixes it into the Soil

Fig. 3—Disk Harrows Prepare Uniform Seedbeds

control—can be done with light to medium harrows, or with a properly adjusted and equipped heavy-duty model. Disk harrows generally provide better incorporation of chemicals than chisel plows or field cultivators, because of the more complete mixing action of the disks.

Disk harrows can be used to prepare seedbeds in fall- or spring-plowed land and to cover broadcast seed, such as oats, although this use is not as common as in the past.

Disking after plowing pulverizes lumps, closes air spaces, mulches the surface, and firms the soil underneath to provide a smooth, uniform seedbed.

A strip-tillage effect, with only the crop-row space finely tilled, may be obtained with a disk harrow by leaving soil open and rough after disking and using fluted coulters or disk openers on the planter to cut through residue and clods.

Fig. 4—Integral Harrows are very Maneuverable

INTEGRAL HARROW

Disk harrows, if they have sufficient strength and weight, will penetrate soils where other implements cannot function, and with the help of disk scrapers will cut sticky soils. They are also good for working land with many stones or stumps because disk blades can roll over many obstructions.

DISK-HARROW TYPES AND SIZES

Variations in disk-harrow construction are common as manufacturers seek to meet specific needs of farmers in various areas by modifying basic designs, or as they apply different solutions to similar problems. However, most disk harrows fall into two distinct classes:

● **Integral or tractor-mounted.**

● **Drawn—with or without transport wheels.**

Within each of these two classes are four primary types:

● **Single-action.**

● **Double-action, tandem, or double-offset.**

● **Offset.**

● **Plowing disks.**

Integral disk harrows are attached to the tractor 3-point hitch and are very maneuverable for turning and transport (Fig. 4). Tractor load-and-depth control is normally used to regulate working depth, which also is controlled by disk angle. Integral disk-harrow size is limited by tractor front-end stability and hitch-lift capacity. Single-action, tandem, and offset disks are all available in integral models.

Drawn disk harrows are attached to the tractor drawbar and are available in all four primary types. Most current tandem and medium-sized offset models are wheel-mounted (Fig. 5), which helps in preservation of grass waterways, crossing soft spots, easy turning, and fast, simple transport. This eliminates the

225

WHEEL-MOUNTED HARROW

Fig. 5—Wheel-Mounted Harrows are Easy to Turn and Transport

need for loading the harrow for movement between fields or on highways. Wheels are raised and lowered by a remote hydraulic cylinder.

Single wheels are standard on most smaller harrows, but dual wheels for better flotation are available on many models. The main section of most larger harrows has duals as standard equipment. Wings may have single or dual wheels (Fig. 6), depending on machine size and design.

Single-action disk harrows have two gangs of disks placed end-to-end which throw soil in opposite directions (Fig. 7). These harrows were quite common when power was limited to horses and small tractors, but now are primarily used for splitting beds, ridges, or irrigation borders, and similar specialized tasks. Single-action harrows, with gangs spread and turned in instead of out, are also used for building crop beds

Fig. 6—Wings of Large Harrows may have Single or Dual Wheels

Fig. 7—Single-action Disks have Two Opposed Gangs

SINGLE OR DUAL WHEEL WINGS

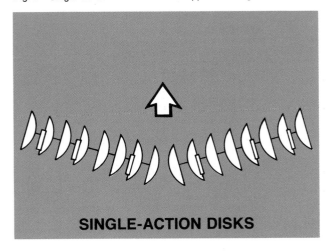

SINGLE-ACTION DISKS

Fig. 8—Tandem Harrows have Two Pairs of Opposed Gangs

Fig. 9—Middlebreaker Attachment Removes Ridge Left between Front Gangs

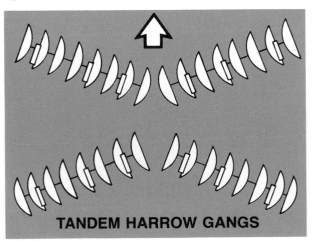

TANDEM HARROW GANGS

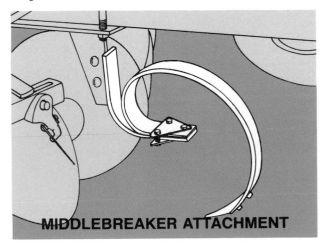

MIDDLEBREAKER ATTACHMENT

226

and irrigation borders, and in wet-soil conditions to dry out fields.

Single-action disks can be overlapped halfway on each pass to provide complete cutting, mixing and leveling of the soil. This provides essentially the same results as a double-action disk. End sections of wider single-action disks can be folded over to reduce width for transport and passing through gates, or to increase weight of center sections for better penetration. Sizes range from small border disks with two or three blades per gang to 18- or 20-foot (5.5 or 6 m) open-field models.

Double-action, tandem, or double-offset disk harrows have two opposed front gangs, like single-action harrows, plus two opposed rear gangs which pull soil back toward the center of the implement (Fig. 8). Thus the soil is tilled twice with each pass and is left more nearly level, compared to single-action disking. The small furrow left on each side by the outside rear blade may be reduced or leveled somewhat by various attachments.

A small ridge of uncut soil is left between the front gangs of single-action and most tandem disks. A chisel plow or springtooth type of shank or middle-breaker can be attached to most harrows to remove this ridge (Fig. 9). The middlebreaker may be equipped with a sweep, chisel-point, or twisted shovel. It can be removed when operating in hard ground, cutting cornstalks and similar stubble, or in other severe conditions to avoid breakage. Some are spring-loaded and do not need to be removed.

By offsetting and overlapping the front gangs (Fig. 10) the center ridge is eliminated without requiring a middlebreaker attachment. Offsetting rear gangs in the same manner allows them each to pull in as much soil as was moved out by the front gangs. Ridging between rear gangs is also reduced, because soil can spread out from the end of each gang instead of rolling together and piling up in the center.

This design essentially combines two offset disks for more uniform operation and improved leveling, compared to conventional tandem-harrow design.

LARGER TRACTORS, LARGER HARROWS

Tandem disk harrows are available for nearly every tractor size, with widths varying from 5 feet (1.5 m) up (Fig. 11). Maximum widths are closely tied to tractor horsepower—whenever a larger tractor is introduced, a correspondingly larger or heavier disk harrow usually becomes available.

Greater versatility is provided in some cases by hitching two or more smaller tandem harrows to a dual or squadron hitch for larger tractors. This offers several advantages, compared to purchase of one large harrow, when tractor size is increased.

1. If an existing small harrow is in good condition, a second smaller harrow may be purchased for less money than a new large one.

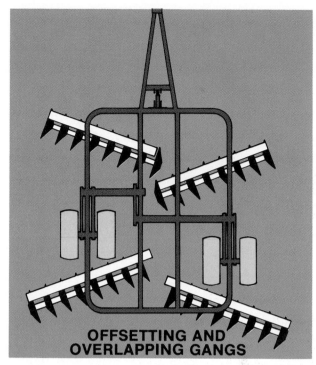

OFFSETTING AND OVERLAPPING GANGS

Fig. 10—Offsetting and Overlapping Gangs Provides Smoother Disking

2. Greater capacity is provided for easier disking jobs.

3. One machine may be used alone for extremely heavy work.

4. The dual hitch may also be used for pulling other implements—chisel plows, field cultivators, planters, and drills.

5. Individual machines may be hitched one behind the other for transport.

Fig. 11—Disk Harrows are Available to Match Nearly Every Tractor

Fig. 12—Tandem Disk Harrow with Wings Folded for Transport

Fig. 13—Hand-Folded Wing Sections

But dual hitches are not without problems. Implements extend much farther behind the tractor with such a hitch, and more time is required for hitching and preparing for transport. To assure uniform operation, care must be taken that exactly the same settings are made on each machine, and that they perform comparable work.

FOLDING THE WINGS

In most cases, tandem disk harrows more than 14 feet (4.3 m) wide are designed to permit folding outer ends

of each gang over the center section to reduce width for transport (Fig. 12). Some harrows may be operated with wings folded to improve penetration by increasing weight applied to each cutting blade. Care must be used when disking with wings folded, because gang components usually are not designed for this extra weight.

On most 14-foot-or-wider (4.3 m) machines, the outer end of each gang is hinged and folded either by hand (Fig. 13) or hydraulically (Fig. 14), although some use a crank and cable arrangement. If folded hydraulically, both front and rear gangs on each side are raised and lowered together (Fig. 15). A variation of the wing design provides

Fig. 14—Hydraulically Folded Wings

Fig. 15—Front and Rear Gangs are Folded Simultaneously

Fig. 16—Two-Section Harrow Splits for Transport

Fig. 17—Backing Tractor Swings Rear Section Around

a center-fold arrangement (Fig. 14) with only one hinge near the middle of the frame. The two sections are allowed to flex in operation so that each can follow ground contours.

Fig. 18—Guide Rods Align Sections as Tractor Keeps Backing

Another means of reducing width for transport applies to two offset harrows used side-by-side. The two are easily separated in front and trail one behind the other for transport (Fig. 16). Width can be reduced more by telescoping the hitch brace on the front harrow to place the tractor directly in front of both harrows.

Backing the tractor (Fig. 17) begins to pivot the rear disk into working position. As the tractor continues to back, the two sections swing closer to each other (Fig. 18) and a funnel-shaped tube on the left section catches an extended rod on the right section and guides the two together until they automatically latch (Fig. 19). The entire operation can be accomplished from the tractor seat.

To return to transport, a latch is released — from the tractor seat — and the tractor driven ahead. The left harrow automatically swings into transport position. The entire operation, in either direction, can be completed in a few minutes.

Offset disk harrows have a front gang moving soil in one direction and a rear gang turning soil the opposite way (Fig. 20). Due to action of soil forces on the gangs the hitch point and line-of-pull of an offset disk harrow is considerably to one side of the center of the tilled strip. Hence the name, offset.

SECTIONS LATCH AUTOMATICALLY

Fig. 19—Sections Latch Automatically when they Reach Working Position

These harrows are usually designed for right-hand offset operation (Fig. 21), but are also available in left-hand configurations (Fig. 22). The offset action of these harrows makes them particularly suited for working under low-hanging branches in orchards and groves, and in some areas they are known as orchard disks (Fig. 23).

Offset harrows may be integrally mounted on the tractor 3-point hitch, wheel-mounted for easy transport, or drawn with no wheels. These harrows generally have considerably more weight per foot of cutting width, so they are better suited for heavy primary tillage (see chapter cover illustration), and are quite limited in sizes which may be tractor-mounted.

Increases in offset-harrow operating width, greater weight, larger disk-blade size, and availability of more tractor power have resulted in substitution of offset disk harrows for disk plows in many areas. Some heavy offset harrows have blades as large as some disk plows.

WHEELED HARROWS CONVENIENT

Wheel-mounting of disk harrows (Fig. 24) provides easier turning and greater maneuverability than non-wheel models. Other advantages are faster transport, no loading or unloading between fields or farms, and easier depth control in varying soil conditions. Transport-wheel attachments, available for some trailing

Fig. 20—Front and Rear Gangs on Offset Disk move Soil Opposite Directions

GANGS MOVE SOIL IN OPPOSITE DIRECTIONS

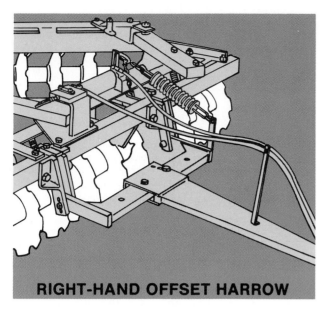

RIGHT-HAND OFFSET HARROW

Fig. 21—Right-Hand Offset Disk

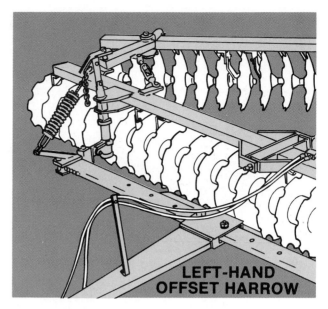

LEFT-HAND OFFSET HARROW

Fig. 22—Left-Hand Offset Disk

Fig. 23—Offset Disks are Ideal for Orchard and Vineyard Work

harrows, permit pulling the harrow endwise with gangs folded parallel.

Plowing disks (Fig. 25) may be either tandem or offset disk harrows specifically designed for tough ground and heavy trash. Built for the most adverse disking conditions, they have heavier frames, larger blades, wider blade spacing, and greater overall strength than other harrows.

Where the deep plowing and complete trash coverage provided by moldboard plows is not desired or needed, plowing disks may be used for primary tillage. These heavy-duty disks make an acceptable seedbed in one or two passes by mixing and mulching trash into the top 6 to 8 inches (150 to 200 mm) of soil, and even deeper in lighter soils.

Plowing disks can cover more acres per day than moldboard plows. For instance, a 13-foot (4 m) disk may cover as much or more ground as two 5-bottom, 16-inch (400 mm) plows, saving manpower, tractor fuel, time, and equipment cost.

By leaving a partial surface mulch, plowing disks help reduce wind and water erosion, catch precipitation, and reduce run-off of available moisture.

Plowing disks are not limited to any geographic area. Their use has increased and become more widespread in recent years, primarily as farmers attempt to decrease the number of field operations required to prepare a seedbed.

Maximum penetration of plowing disks and the best trash coverage are obtained in light to medium soil

Fig. 24—Wheel Mounting Provides Greater Maneuverability

WHEEL-MOUNTED HARROW

conditions. Heavy trash and heavier soils restrict penetration, reduce coverage to some extent, and may require an extra disking for satisfactory results.

DISK-HARROW FEATURES

Blade Types

Most disk-harrow blades are shaped like portions of a hollow sphere (Fig. 26). The spherical radius may vary, even for blades of the same diameter, resulting in flatter or deeper blades.

Blades may be sharpened on the convex or outer side (outside bevel), or on the concave side (inside bevel). Sharpening the concave side increases penetration in hard soil, while outside beveled blades perform well in general disking conditions and in rocks.

CONE-SHAPED BLADES

Cone-shaped disk blades are available for many harrows and in a variety of sizes. These blades (Fig. 27) appear to be cut from a cone of approximately 25 degrees. The distance between working surfaces of adjacent cone disks is equal, top to bottom (Fig. 28), which permits easier movement of soil between blades, reduces soil packing, and improves penetration. Spherical blades (Fig. 26) curve in slightly different arcs as soil passes between them, which tends to compress the soil.

Cone-shaped blades are generally more aggressive than spherical blades, but may have more tendency to plug in certain sticky-soil conditions.

Notched disk blades (Fig. 29) penetrate better than plain blades in hard soil because of the reduced contact area around the outer edge. Both cone and spherical blades are available with notched edges.

Some farmers prefer notched blades on front gangs for better penetration and trash cutting, and plain blades on the rear for improved pulverization and leveling — and economy, as notched blades usually cost more. Notched blades generally offer less benefit compared with plain blades as blade diameters increases.

Blade Sizes

Disk-harrow blade diameters range from 16 inches (400 mm) for small, light-duty models, to 32 inches (813 mm) for some heavy offset and plowing types. Each size (Fig. 30) has its place in today's tillage operations.

Small blades, equally weighted, will penetrate better in hard soil than larger-diameter blades because of the reduced blade-soil contact area. But larger blades cut trash better than small blades because of the angle between soil surface and cutting edge when working at equal depths.

Thickness of disk-harrow blades varies from about 1/8-inch to 3/8-inch (3 to 9.5 mm) to cover all needs from light seedbed preparation to heavy primary tillage in extremely adverse conditions. Some disk harrows have thicker blades on front gangs to compensate for faster wear of the front blades.

Fig. 25—Tandem Plowing Disk in Heavy Trash

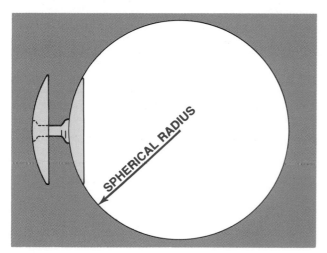

Fig. 26—Spherical Blades Appear to be Sliced from a Ball

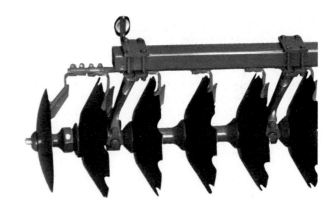

CONE-DISK BLADES

Fig. 27—Cone-Disk Blades

Fig. 28—Soil Moves more Freely Between Cone Blades (Left) than Spherical

Fig. 29—Notched Disk Blades

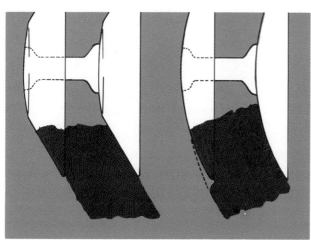

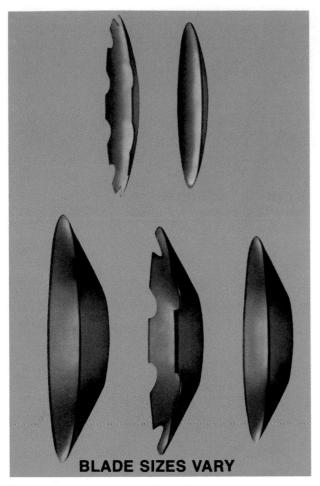

BLADE SIZES VARY

Fig. 30—Blade Sizes are Available to Match Requirements

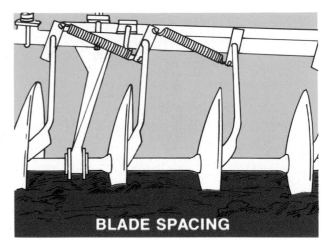

BLADE SPACING

Fig. 31—Blade Spacing Depends on Tillage Objectives

The choice of blade thickness and diameter will depend on these factors:

- *Harrow type, size, and weight*
- *Application—primary or secondary tillage*
- *Soil type and moisture—dry and hard, soft, sticky*
- *Expected operating depth*
- *Type and amount of trash to cut*
- *Stones, stumps, or other obstructions in the soil*

Blade wear-rate and impact resistance depend on type and hardness of steel used. Best wear resistance in abrasive soils is provided by hard-faced blades.

Blade Spacing

Selection of blade spacing is based on desired results, because the spacing between disks (Fig. 31) is directly related to tillage objectives. Spacings range from 7 inches (180 mm), for lighter harrows designed for fine-finish seedbed work, to almost 15 inches (380 mm) for deep-working, heavy-duty offset harrows with very large

blades. The many in between spacings permit matching of harrow design and operating goals.

Spacing selection must include consideration of soil and trash conditions, size of blades, expected working depth, and intended use (primary or secondary tillage). Some compromises may be required if the same harrow is to be used for primary tillage and seedbed preparation, or for cutting heavy trash and working plowed ground. The following guide can help in making the blade-spacing choice.

- *7-inch (180 mm) range—seedbed preparation in plowed soil, with relatively little trash cutting or extremely hard ground to work*

- *9-inch (230 mm) range—general disking, chopping cornstalks and other trash, incorporating chemicals, and preparing seedbeds*

- *11-inch (280 mm) range—deep disking in heavy trash and hard soil, seedbed preparation if soil pulverizes easily; otherwise, should be followed by lighter disk or field cultivator*

- *13-inch (330 mm) range—extra-deep work with very large blades in extreme soil and trash conditions; used only for primary tillage in very severe conditions*

COMPARING WEIGHT

Blade spacing must be considered when making comparisons between disk harrows, particularly with reference to machine weight. For instance, some machines are referred to as having a certain weight per blade, while others are compared on the basis of weight per foot of cutting width. Both are valid figures, but various harrows must be compared on the same basis. Also consider—does the given weight include hydraulic cylinders and hoses, tires and wheels, scrapers, or any other attachments? Are blades of equal size and thickness?

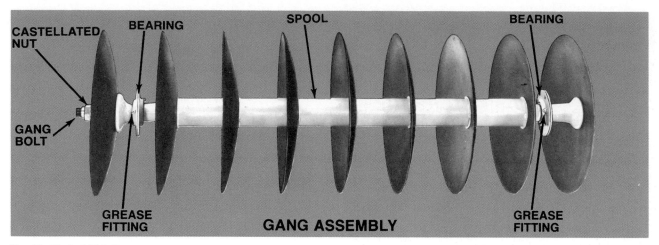

CASTELLATED NUT   BEARING   SPOOL   BEARING

GANG BOLT

GREASE FITTING   **GANG ASSEMBLY**   GREASE FITTING

Fig. 32—Typical Disk Gang

Comparing weight on the basis of pounds per blade gives a direct picture of the downward force available on each blade for penetrating hard soil and cutting trash. The weight per foot provides a quick means of comparing total weight of two machines. Harrow weights may vary from about 30 pounds (13.6 Kg) per blade, for light integral disks, to as much as 400 pounds (181 Kg) per blade for extra-heavy offset models.

The possible weight variations between the two methods of comparison may be seen by comparing the advertised weights of two disk harrows of the same model with equal width, the same size blades, and 7½- and 9-inch (190 and 230 mm) blade spacings, as shown below.

Spacing in. mm	Cutting Width ft m	Total Weight lbs Kg
7½ 190	16¼ 5	3783 1716
9 230	16¼ 5	3615 1640

Disk Gangs

Disk gangs (Fig. 32) are made up of blades; spools (or spacers) between blades; a threaded gang-bolt made of high-carbon steel and measuring 1 to 2¼ inches (25 to 57 mm), round or square; a large plate or washer at each end, and a castellated nut and locking pin. Two or more spools on each gang will include bearings for attaching the gang to hangers on the gang frame.

Many large harrows are designed with each gang in two sections for greater strength and dependability. This permits use of shorter gang bolts with less tendency to stretch and loosen, which could cause blade damage and wear. Drive couplers (optional or standard) may be used between these gangs as well as between main gangs and wings to assure positive disk rotation in difficult conditions (Fig. 33).

C-shaped springs cushion disk gangs to protect blades and frames from damage when working rocky soils. The springs absorb shocks and permit blades to roll over obstacles without lifting the entire machine.

On other disks, heavy coil spring shock absorbers permit gangs to deflect upward as blades pass over obstructions. The springs then immediately force blades back to normal working position.

Disk-gang bearings over the years have included hard wood (usually oil-impregnated maple), white iron, sealed, self-aligning ball bearings, and double-tapered roller bearings. Wood bearings are no longer used extensively because of increased machine sizes and operating speeds. White-iron bearings are very

No. of Blades	Weight Per Ft.	Weight Per M	Weight Per Blade lbs. Kg.
52	233	347	73 33
44	222	330	82 37

Fig. 33—Drive Coupler Provides Positive Rotation of Split Gangs and Wing Extensions

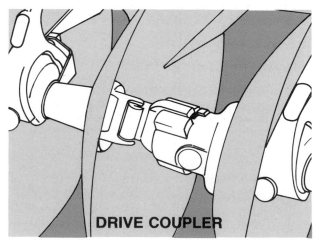

DRIVE COUPLER

SEALED GANG BEARING

Fig. 34—Sealed Gang Bearing on Steel Sleeve for Increased Strength

Fig. 35—Cast-Iron Blade Spacers

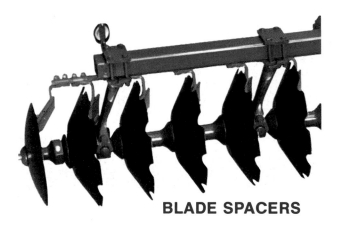

BLADE SPACERS

Fig. 36—Disk Gangs must be Properly Tightened to Prevent Possible Blade Damage, Excessive Wear, and Loss of Parts

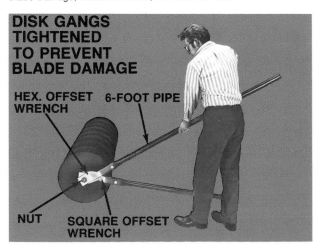

DISK GANGS TIGHTENED TO PREVENT BLADE DAMAGE

HEX. OFFSET WRENCH   6-FOOT PIPE

NUT   SQUARE OFFSET WRENCH

hard and brittle with excellent resistance to wear from the constant exposure to dirt, but they require frequent lubrication to keep dirt flushed out and are used less and less on modern harrows.

Sealed, self-aligning bearings are designed for the high stresses and dirty operating conditions found in disking. Triple lip seals help keep dirt out, and most bearings may be greased regularly to flush any dirt which has entered. On many harrows the bearing is mounted on the outside of a sleeve which fits over the gang bolt (Fig. 34). A split spool then slides over the sleeve to provide proper blade spacing. This arrangement spreads the bearing load over a wider portion of the gang bolt and strengthens the entire gang.

Replaceable wear-guards often are available for attachment around the bearings to increase bearing life in sandy, abrasive soils.

Gang spools on most disk harrows are made of cast-iron with machined ends for close fitting against disk blades (Fig. 35). Some harrows have hollow, fabricated steel spools which are lighter in weight.

Spool ends are square and fit flat against the baldes for spherical blades which are flattened in the center. Spool ends for cone and other spherical blades are curved to match blade shape and assure uniform strength, tightness, and blade support.

Proper tightening of disk-gang bolts is absolutely essential for satisfactory disk-harrow operation. Necessary torque ranges from 450 pound-feet (610 N.m), on smaller harrows, to 1,000 or more pound-feet (1356 N.m) on large offset disks. Proper tools make tightening of gang bolts considerably easier; use either special wrenches from the manufacturer (Fig. 36), large pipe wrenches, or other wrenches with adequate strength for the required torque. Always block gangs to prevent accidental rolling and possible injury from the blades. It is also advisable to wear protective gloves when working with or near blades.

Tap the gang-bolt head with a hammer as the bolt is being tightened, to help take up slack in the gang bolt and improve the fit between spools and blades.

Check gang-bolt tightness before going to the field with a new harrow or any time the gang has been disassembled for repairs or replacement of parts. Recheck tightness after each four or five hours of operation until all slack has been removed, and at least once each season thereafter.

Disk-Harrow Frames

Disk-harrow frames may be classified as flexible or rigid. Flexible designs permit individual gangs to ride over obstructions without affecting the operation of other gangs.

Rigid-frame harrows are preferred for deep penetration and work in hard soils. In this case the weight of the entire machine holds gangs in their proper operating positions, regardless of changes in contour or obstructions. Such harrows tend to level soil irregularities better than flexible models, but there is more danger of blade breakage if obstructions are present.

Greater versatility is provided in some harrows by hinging the frame between front and rear gangs. The hinge may be pinned for rigid operation or the pins removed to allow flexing between front and rear gangs as each follows ground contours; or the rear gangs may be raised and pinned so that the entire harrow weight is concentrated on front gangs for maximum cutting and penetration.

Disk-frame members range from bolted or welded angles, of many different sizes, to half-inch-thick 4 × 10-inch (100 mm × 254 mm) box-steel sections. Gang frames also vary from angles to heavy box steel.

Disk-Harrow Weight

During the 1950's there was a trend toward reduced disk-harrow weight, with greater dependency on design improvements for strength and penetration. Available tractor power and performance requirements have increased since then, so more emphasis again is placed on weight as an important factor in choosing the best disk for a particular operation.

Weight of individual harrows in a particular product line may be varied by altering the thickness and shape of frame members, by increasing the size of cast-iron spools between blades, and by modifying overall frame configuration. In many cases weight can be added for difficult conditions by attaching weight boxes, which are filled with stones or other ballast; by adding cast-iron weights, such as tractor front-end weights, or even by filling gang-frame members with gravel, sand, or liquid.

Disk Scrapers

Scrapers keep disks from plugging in sticky soil and trash and help assure steady work in severe conditions. When adjusting scrapers, place the blade point near the junction of blade and spool on spherical disks (Fig. 35), or close to the shoulder of cone blades (Fig. 36) for best cleaning.

Scrapers are usually set to touch blades, especially at the scraper point, to prevent buildup of soil and trash between scraper and disk, but must never be set tight enough to blades to restrict gang rotation. In very abrasive soils, set scrapers close to blades but do not let them touch to reduce wear.

Harrow Selection

Optimum disk-harrow performance depends on proper matching of blades, gang design, and harrow construction.

Start the matching process by looking at the types of work to be done and how harrow features fit the requirements.

PRINCIPLES OF OPERATION

Each disk gang consists of a set of concave blades mounted on a common shaft or gang bolt. When the gang is operated at or close to a right angle from the

MATCH FEATURES TO THE JOB								
Type of Work	Blade Type	Blade Size		Blade Thickness	Blade Spacing		Typical Weight Per Blade	
		Inches	mm		Inches	mm	lbs	Kg
Final seedbed preparation	Plain	16-22	400-560	Min.	7-9	180-230	Light (30-110)	13-50
General disking,	Notched	20-24	500-600	Med.	8-10	200-250	Medium (80-160)	35-75
stalk cutting,	Plain							
chemical incorporation,	Notched front	16-20	400-500	Med.	7-9	180-230	Light to Med.	
seedbed preparation	Plain rear							
Heavy trash and seedbed preparation without plowing	Notched or plain	24-26	600-660	Heavy	9-12	230-300	Heavy (150-260)	70-120
Very hard soil, very heavy trash, light brush	Notched or plain	26-32	660-810	Max.	12-14	300-360	Maximum (200 or more)	90

Fig. 37—Set Scraper Point Close to Spool on Spherical Blades

Fig. 38—Set Scrapers Close to Shoulders of Cone Blades

Fig. 39—Friction Between Soil and Back Sides of Disks Decreases as Cutting Angle Increases

direction of travel, blades tend to roll over the ground like wheels and do very little cutting. As the angle increases disk rotation slows down, penetration increases for a given amount of disk-harrow weight, and blades scoop and roll soil as they rotate. More soil is turned and trash coverage improves as angle increases. Soil pulverization is also increased, particularly at high speeds, up to 7 mph (11 Km/h).

There is quite a bit of friction between the back side of disk blades and the soil as disks are cutting. This friction decreases as cutting angle increases because there is less undisturbed soil behind each disk blade (Fig. 39). The effect of this friction can be seen in the field by noting that the back or convex side of disk blades always polishes before the concave side.

Soil immediately below blade working depth is compacted by the disks. In some cases, this compaction is severe enough to restrict root growth and development. Compaction is increased by working the soil too wet.

DISK-GANG FORCES

The lateral center of load of a disk gang is considered to be located at the center of the gang when all blades are penetrating equally. The natural forces on a disk gang cause the concave end of the gang to penetrate more than the convex end.

Various combinations of hitch settings and relative front and rear gang angles will further affect the amount of offset and side draft encountered with offset harrows. To change operating position of the harrow behind the tractor, to work under trees, keep wheels off of worked soil, etc.:

1. To offset harrow further to the right with minimum side draft—

A. Angle hitch to the left.

B. Increase cutting angle of front gang (on wheeled offsets) or increase angle between front and rear

Fig. 40—Hinge on Offset Disk must Resist Torsion between Gangs

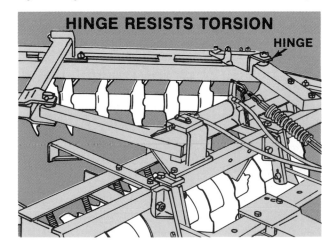

gangs.

2. To offset harrow to the left with minimum side draft

A. Move hitch to the right.

B. Reduce cutting angle of rear gang or close angle between gangs.

A stiff hinge (arrow, Fig. 40), including appropriate adjustments to level gangs with respect to each other, must be provided on offset disks to overcome the torsional effect of disk-gang operation. To further complicate offset-harrow adjustments, the rotational tendency of front gangs is generally greater than that of rear gangs because front blades cut harder soil.

This may cause the entire right side of the harrow (with right-hand offset) to run deeper than the rest of the machine. Pulling up on the right side of the harrow by lowering that side of the hitch, or using transport wheels to control depth, will help maintain more uniform penetration.

To permit wide offset harrows to work over uneven ground, and on various contours, some have been designed with squadron hitch, and others have flexible gangs with linkage between ends of front and rear gangs to maintain uniform penetration.

GANG ANGLE

Normal disk gang cutting angle ranges from 10 degrees to 25 degrees from a line perpendicular to the direction of travel on single-action and tandem disks, but may be up to 50 degrees on some offset disks. Increasing gang angle increases disk penetration, trash cutting and coverage, and power requirement. Front gangs of tandem and offset harrows perform the most tillage and cutting.

When changing gang angle on winged machines, make certain that wing gangs are properly lined up with main gangs; that blade spacing between wing and main gangs is correct, and that distance is equal between front and rear edges of the end blade on the main gang and the first wing blade.

Many factors affect the choice of disk-gang angle.

1. Desired tillage results—seedbed preparation, cutting and mixing trash, primary tillage.

2. Soil hardness—increasing gang angle provides better penetration; minimum angle may be used in seedbed preparation on previously worked soil.

3. Soil moisture—reduce gang angle to prevent plugging in wet soil; increase angle for penetration of dry soil.

4. Desired operating depth—maximum gang angle provides maximum working depth.

5. Amount of trash present—use maximum angle to cut and cover the most trash; reduce angle to expose more residue for erosion control.

Many gang-angling methods have been provided by manufacturers. In the past, hand-operated levers

Fig. 41—Front or Rear Gangs may be Angled by Removing Pins and Shifting Gangs with a Lever

changed front and rear gangs. Early tractor-drawn single-action and tandem harrows often had a rope-controlled pin-and-slide arrangement. Similar devices are currently used to control hitch angle and angle between gangs on some offset, non-wheeled harrows. In most cases, these controls may be replaced by remote hydraulic cylinders, if desired. Pulling the rope to release the pin, and then backing the tractor or driving forward, angles or straightens gangs, depending on harrow type and design.

Some harrows have a trip-rope-activated ratchet-type control on the gang axle, similar to early trip-rope lifts on drawn plows. Pulling the rope angles or straightens the gangs.

When remote hydraulic cylinders became available they were quickly adapted to angling disk gangs, which was especially convenient for turning or crossing waterways. With the advent of wheeled harrows, powered angling became less necessary and most wheeled and integral harrows are now angled manually after removing pins

Fig. 42—Gangs on some Harrows may be Angled by Loosening Bolts and Sliding Gangs Forward or Back in Slots

(Fig. 41) or loosening bolts (Fig. 42). Gang angle on some winged harrows is fixed for average conditions and cannot be altered.

Important: Right and left-hand gangs on single-action and tandem disk harrows must be angled exactly the same to provide uniform penetration and balanced draft.

FURROW-BOTTOM PROFILE

To minimize the small, uncut soil ridges left between gangs (Fig. 43), rear blades should be set to cut approximately halfway between front blades, not follow in the same tracks. The theoretical size of these ridges depends on blade size and spacing and gang angle. In practice, much of the ridge will be broken by movement of adjacent soil, particularly if ground is dry.

DISK-HARROW PENETRATION

Uniform, full-width penetration is the most important factor in disk-harrow operation. Satisfactory penetration requires ample strength for field conditions, proper weight distribution, and careful matching of features to desired results.

Some factors influencing penetration are controlled by disk-harrow design and selection while others are not.

Penetration factors dependent upon harrow design:

• *Angle of gangs–increasing angle improves penetration*

• *Total harrow weight and weight per blade–in-*

Fig. 43—Set Rear Blades Halfway between Front Blades to Reduce Ridges

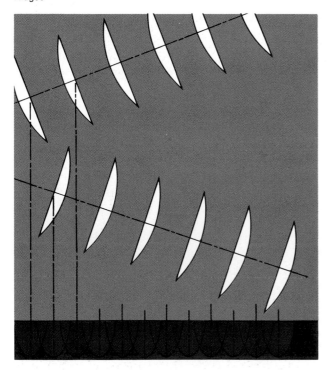

creasing weight improves penetration; increasing blade spacing increases weight per blade

• *Blade diameter–small blades penetrate better; larger blades work deeper in soft soils*

• *Blade sharpness–sharp blades cut soil and trash better; inside-beveled blades penetrate better in hard soil*

• *Hitch angle (horizontal is best)–pulling up on hitch reduces penetration; pulling down on hitch cuts tractor traction*

• *Soil type–sand, loam, gumbo, clay, peat*

• *Soil moisture–wet, moist, dry, sun-baked*

• *Soil hardness–loose, firm, hard, plowed, unplowed*

• *Amount and kind of trash–standing or shredded cornstalks, rice straw, bean stubble*

Maximum penetration is obtained in difficult conditions by using maximum gang angle, maximum weight per blade and—with wheeled disks—raising wheels off the ground for added weight.

DISK-HARROW LEVELING

The harrow must be leveled laterally and fore-and-aft for uniform penetration and satisfactory performance. Integral and wheeled harrows are easier to level than drawn models without wheels.

If integral or wheeled harrows are operated with their frames tipped lower in front, outer ends of front gangs (or the concave end of offset disks) will penetrate deeper compared to the inner end, and soil will be worked unevenly. This problem becomes worse as gang angle increases. Rear gangs also penetrate unevenly if the harrow is not leveled fore-and-aft.

Integral harrows may be leveled fore-and-aft by adjusting the tractor top-link length (Fig. 44). Adjustments in hitch or wheel linkage are provided for leveling wheeled harrows (Fig. 45) to match tractor drawbar height, work-

Fig. 44—Level Integral Disks Fore-and-Aft by Adjusting Tractor Top-Link Length

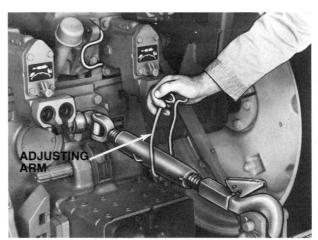

ADJUSTING ARM

Fig. 45—Adjust with Crank Handle to Level Wheeled Disks Fore-and-Aft

ing depth, and soil conditions. Some have separate leveling adjustments for transport and operation.

Adjusting the tractor 3-point-hitch lift link (Fig. 46) provides lateral leveling for integral disks. Rigid-frame wheeled disks generally require no lateral leveling. If a harrow is not level, check for unequal tire sizes, bent or damaged frame, or add shims on one side between wheel frame and main frame. Refer to operator's manual for specific instructions on lateral leveling of flexible-frame harrows when changing disking conditions. These adjustments vary by machine and may involve

changing spring tension or compression, or adjusting linkage between gang and frame.

Lateral leveling of offset harrows is discussed under, "How An Offset Harrow Works."

SOIL LEVELING

Disk harrows should leave soil level, but not necessarily smooth. That is, soil should be free of ridges or valleys after disking, but may deliberately be left rough to control erosion.

Leveling is easily done with offset disks by operating in lands and always filling the furrow left by the previous pass. Proper adjustment and careful driving can provide excellent leveling.

If a ridge is left in the center by a tandem harrow, increasing angle of front gangs and decreasing rear-gang angle will help reduce the ridge. If a valley or low spot is left in the center, reduce front-gang angle; increase rear-gang angle; relevel the harrow, or increase speed up to 6 miles per hour (10 Km/h) or the recommended limit for the harrow being used.

Ridges at outer ends of the disk can be leveled by moving rear gangs out to pull in more soil, or by reducing operating speed. Reducing angle of front gangs also reduces amount of soil thrown out for rear gangs to move back.

Furrow-filler blades (Fig. 47) may be added to outer ends of rear gangs on most disks to fill and level the furrow left by the outer blades when operating in loose soil. Depending on blade type and soil conditions, filler blades are usually from 2 to 6 inches (50-150 mm) smaller in diameter than other blades on the disk. These smaller diameter blades do not affect the amount of soil cut, or normal operation of rear gangs, but simply pull loose soil into the furrows to provide a more-level surface. Unfilled furrows not only make the field rough, but may become starting points for serious erosion of finely worked seedbeds if disking is done up-and-down slopes. The

Fig. 46—Level Integral Disks Laterally by Adjusting Length of Hitch Lift Link

Fig. 47—Furrow-Filler Blades on Rear Gangs Help Level the Surface

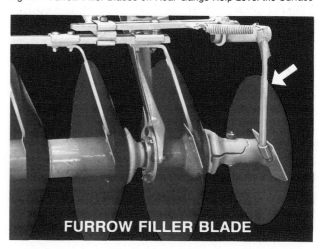

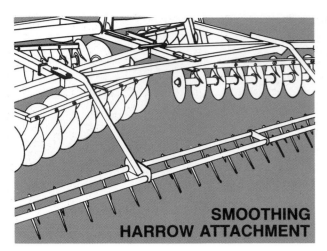

SMOOTHING HARROW ATTACHMENT

Fig. 48—Smoothing Harrow Attachment for Disk Harrow

WINGS FOLD FOR TRANSPORT

Fig. 49—Flexible Wings Fold Easily for Transport, Flex over Uneven Ground

best bet is a combination of leveling furrows and disking across slopes.

Smoothing harrow attachments are available from some manufacturers (Fig. 48). They are attached to the rear of the disk-harrow frame to break up clods and smooth and level the surface. Smoothing harrows lift automatically when the disk is raised, and are usually free to float over uneven ground.

WING FLEXING

Shorter wings on some wide disk harrows are locked solidly to the frame during operation, because the wings alone do not have adequate weight to provide satisfactory penetration.

Some manufacturers offer a choice of fixed or flexible wings in some model sizes to match terrain and type of work. Flexible wings equipped with gauge wheels to control working depth, and carry wings when the machine is raised for turning, are standard equipment on

many larger harrows (Fig. 49). Wing flexing permits the harrow to follow ground contours, even to work on three planes at once (Fig. 50).

TRACTOR PREPARATIONS

Disk harrows are usually selected by type for the desired results, then size for the tractor power available. For economical operation and maximum output, it pays to have the tractor in top condition before going to the field.

1. Be sure engine is tuned for maximum power output and optimum fuel economy.

2. Check hydraulic system for adequate oil level. Repair any leaks in lines, hoses, couplings, or cylinders. Refill reservoir after filling and working air out of new large cylinders.

3. Provide adequate rear-wheel weight to limit slippage to about 15 percent. If you can readily see

Fig. 50—Flexible Wings Follow Ground Contours—Even Three Planes at Once

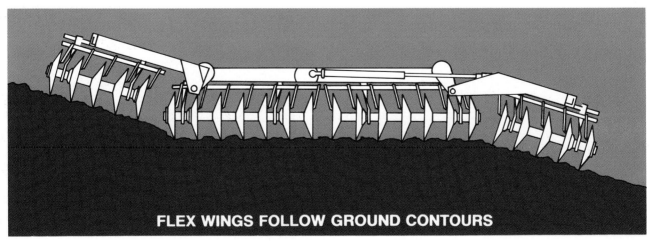

FLEX WINGS FOLLOW GROUND CONTOURS

Fig. 51—Examine Tractor Tire Tracks for Effects of too Much or too Little Weight

wheels slipping, it's too much. Tracks should be slightly broken or shifted when pulling rated load. Too much weight leaves a firm, hard track, and may damage the tractor drive train. Too little weight wastes time, fuel and tires (Fig. 51).

4. Provide adequate front-end weight to maintain tractor stability in operation and transport. Ballast front end to maximum permissible level when operating integral harrows in lower gears (see tractor manual).

5. Adjust 3-point hitch for integral harrows as recommended in tractor and harrow manuals.

6. Install sway blocks or chains to permit lateral flexibility when working but eliminate sway in transport.

7. For drawn disks, raise 3-point hitch to maximum height and set drawbar to height recommended in harrow manual.

8. Pin tractor drawbar in center, or allow to swing, according to harrow manual instructions. Recommendations vary by model and manufacturer.

9. Inflate tractor tires to recommended level (see manual).

10. Set wheel tread to match row spacing if disking with the rows. Reduce tread to keep tractor wheels on unworked soil when using small harrows.

DISK-HARROW PREPARATION AND MAINTENANCE

Careful preparation and maintenance of disk harrows pay off with more uniform tillage, lighter draft, lower fuel consumption, reduced field time, and longer machine life. In addition to recommended daily service, the proper pre-season preparation can cut repair bills and downtime in the field. Careful examination and proper preparation for storage reduces deterioration between seasons and means better performance.

Before Each Season:

1. Lubricate entire machine as recommended in manual. Avoid excessive lubrication, as surplus grease only gathers dust.

2. Inflate tires to recommended pressure.

3. Check bolts and nuts for tightness; look for worn, broken, or missing parts and repair or replace.

4. Sharpen or replace worn, cracked, or broken blades.

5. Torque gang bolts as recommended in manual.

6. Clean, repack, and tighten wheel bearings (see operator's manual).

7. Install tension springs between pairs of scrapers (Fig. 52).

Fig. 52—Install Tension Springs between Pairs of Scrapers

TENSION SPRING

INSTALL TENSION SPRINGS

Before Storage At End Of Season:

1. Thoroughly clean all trash, soil, and dirty grease from the harrow.

2. Repaint any spots where paint is missing.

3. Clean disk blades and coat with heavy grease or plow-bottom paint to protect from rust.

4. Lubricate entire machine to keep moisture out of bearings.

Daily Before Operation:

1. Lubricate as recommended in manual. Use only a hand gun to lubricate sealed bearings; pressure guns may damage seals.

2. Visually check for loose bolts, broken, or missing parts, and underinflated tires.

3. Recheck torque on gang bolts after every few hours of use on new machines, or after replacing blades, until all slack is taken up; same for wheel bolts and bearings. NOTE: Blades broken because gang bolts were loose are not covered by warranty.

4. Some breakage of scrapers and disk blades must be expected when operating over stones, stumps, or roots. Keep extras on hand for replacement if frequent breakage occurs.

5. Store inside to protect from weathering. Remove weight from tires.

6. Install safety locks and relieve hydraulic pressure in cylinders, or lower machine to the ground; protect blades from rust by placing them on boards.

7. Fully retract cylinders to protect cylinder rods from rust.

8. Order replacements for any missing or broken parts and install them prior to the next disking season.

BLADE BREAKAGE

Blade breakage must be expected when operating under severe conditions, but can be reduced or avoided in many cases by careful operation and proper equipment selection.

If soil obstructions or highly abrasive soils create blade problems, the extra cost of heavier disk blades is easily justified because of less breakage, longer life, and less downtime for replacing blades.

Breakage is reduced by lower operating speed in stony ground, particularly with new blades on which the edges have not yet worn smooth. In soil with many obstructions, reducing gang angle permits blades to roll more easily over stones or stumps without breaking.

Some blade breakage is covered by warranty because disks were obviously defective. But other damage—from striking obstructions to operating with loose gang-bolts—is not covered. How do you tell the difference? Policies differ among manufacturers, but

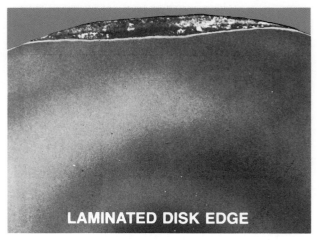

Fig. 53—Laminated or Split Disk Edges may Indicate Defective Material

whether blades are replaced under warranty or not, it pays to recognize the cause of damage.

If blade edge appears laminated or split (Fig. 53), the fault is defective material and will likely be covered by warranty. A straight directional break is also generally caused by defective material and should be checked for warranty coverage.

An irregular, non-directional break, where the blade appears to have been torn apart (Fig. 54), is caused by striking rocks or stumps and is not normally covered by warranty. Pieces torn from blades (Figs. 55

Fig. 54—Irregular Breaks Indicate Damage from Stones or Stumps

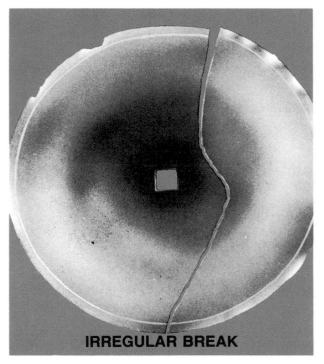

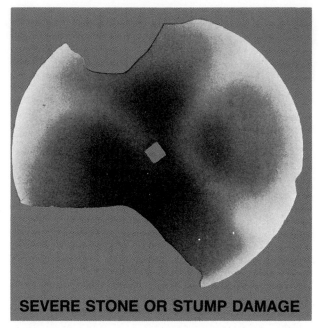

CHUNK TORN FROM BLADE

Fig. 55—Chunks Torn from Blades Indicate Damage from Soil Obstructions

SEVERE STONE OR STUMP DAMAGE

Fig. 56—Severe Stone or Stump Damage Causes Irregular Blade Breakage

and 56), or chips around the blade edge (Fig. 57) also indicate stone or stump damage.

If the blade center is cracked, torn, or broken out (Fig. 58), damage may have been caused by soil obstruc-

tions, but more likely from operating with a loose gang bolt. If gang bolt is not properly tightened, blades may wobble and be subject to breakage and wear.

Fig. 57—Chips and Dents Around Blade Edges are Usually caused by Soil Obstructions

Fig. 58—Center Torn from Blade—Caused by Loose Gang Bolt or Soil Obstructions

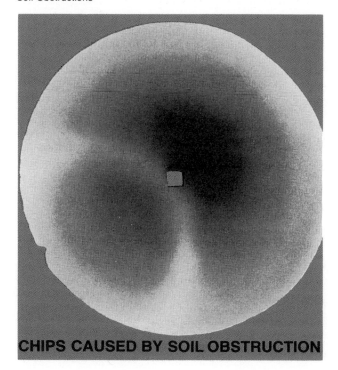

CHIPS CAUSED BY SOIL OBSTRUCTION

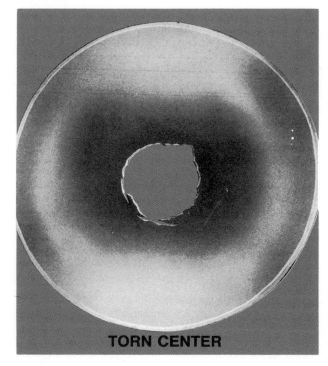

TORN CENTER

FIELD OPERATION

FIELD PATTERN

Many patterns may be followed in disking a field. If cutting row-crop stubble, running diagonally across rows (Fig. 59), at least the first time over, helps tear out old roots and level ridges. Another method involves disking back and force with each pass adjacent to the previous one, as with a 2-way plow. Or, lands wide enough for comfortable turning may be laid out and each land finished as work proceeds across the field. Offset disks are normally operated in lands to provide proper leveling of the furrow left by the rear gang. Headlands, approximately twice the harrow width for easy finishing, are worked last.

If a second disking is necessary, working at right angles or diagonally to the first pass improves results.

OFFSET DISK PROCEDURE

In open-field work, an offset disk is used so each pass fills the furrow left by the preceding pass, leaving the field level and free of furrows. Each pass with a right-hand offset harrow is made on the right side of the previous pass. Left-hand offsets operate to the left of the last pass.

The left-end blade of the front gang of right-hand offset harrows runs in the furrow to eliminate overlap and to use the full harrow width for tillage.

Furrow filling is affected by travel speed, soil conditions, disking depth, blade type, and gang angle, each of which may require different adjustments.

The right rear blade, when disking deep, may leave a furrow and small ridge (Fig. 60). These may be minimized by adding a furrow-filler blade to the end of the rear-gang axle to pull loose soil into the furrow.

Most of the loose soil which was thrown onto unworked ground at the right front blade is pulled back on the next round to help fill the furrow.

To leave most of the soil for filling the furrow, adjust the hitch as far to the left as practical. This turns the entire disk clockwise, shifts the front gang to the right and rear gang to the left. This reduces side draft and

Fig. 59—Diagonal Disking Provides better Leveling and Trash Cutting

DIAGONAL DISKING

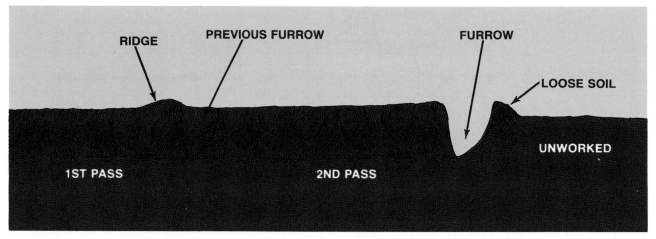

Fig. 60—Furrow from First Pass with Offset is Filled on Second Pass

increases cutting angle of the rear gang so that sufficient soil is moved to fill the previous furrow.

After adjusting the hitch, operate the disk for a short distance. If necessary, shift the rear gang laterally so the left rear blade approximately divides the space between the first two left front blades (Fig. 61). Additional lateral gang movements may be needed, depending on soil conditions, to obtain desired leveling.

Important: Whenever furrow size or amount of loose soil are changed, two passes must be made with the disk to be able to see the entire effect of the changes.

HITCHING AND UNHITCHING

Integral disks may be attached to the tractor 3-point hitch in the normal manner, or with an implement quick-coupler.

Fig. 61—Rear Blades Should Cut Halfway between Front Blades

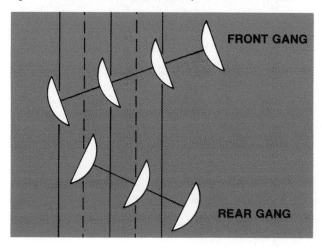

To hitch drawn disks, back the tractor into position. If the hitch is not supported by a jack, shut off the tractor. Then relieve hydraulic pressure to remote hose couplers by shifting remote valve lever back and forth. Couple the hoses and raise the tongue for easy attachment. Many harrows are designed to permit raising and lowering of the tongue for easy hitching by adjusting the remote cylinder without raising the harrow from the ground.

Before detaching from the tractor drawbar, raise the disk and install safety or transport lock, or lower the harrow to the ground. Shut off the engine and relieve pressure in hydraulic cylinders before uncoupling hoses. Lower the hitch jack (Fig. 62), if any, and remove drawbar pin.

NOTE: Always use a hitch pin with adequate strength for the tractor-harrow combination, and lock the pin in position to prevent possible loss, particularly when transporting the harrow.

OPERATION

Before starting to disk, raise the harrow to full transport height and remove safety lock (on wheeled harrows).

Determine soil and trash conditions of field and make preliminary disk settings. Final adjustments will be made in the field. When the disk is under full load at operating speed, front and rear gangs must cut equal depth.

Use maximum disk-gang angle, particularly for primary tillage, if maximum working depth is desired, if trash is extremely heavy, or if soil is extremely hard. For moderate soil hardness and trash conditions, use intermediate settings. Provide minimum gang angle for light trash and loose soil, such as seedbed preparation in plowed ground. For a combination of aggressive cutting and seedbed preparation, set rear gangs at a lesser angle than front disks.

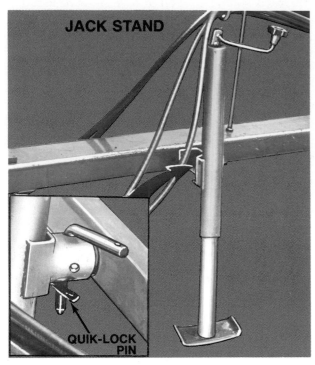

Fig. 62—Jack Makes Hitching and Unhitching Easier and Safer

Fig. 63—Set Scrapers Close to Spools and Close to Blades for Best Cleaning

Disking depth is controlled by:

• *Tractor load-and-depth control for integral disks.*

• *Adjusting depth stop on hydraulic cylinders on wheeled disks to control wheel position, and by varying gang angle.*

• *Adjusting gang angle of drawn, non-wheeled disks.*

Set disk scrapers close to spools and flush with blades (Fig. 63), but not close enough to slow disk rotation.

If more than one hydraulic cylinder is used to control transport-wheel height, make sure depth stops are set the same.

TRANSPORT AND SAFETY

Accidents cause pain and possible death, and are financial burdens as well. They cost time to recover from injuries and to make repairs. They cost money for medical treatment, repair bills, and delay in field work.

The chances of injury and equipment damage can be reduced by following simple safety precautions, and keeping equipment in good condition.

1. Always lock safety lock during transport, if the disk is to be left raised for an extended period, or while working on the machine.

2. Never depend on tractor hydraulic pressure to carry harrow weight in transport—use safety lock, and relieve pressure in cylinders.

3. Lower integral harrows to ground each time tractor engine is shut off, and any time harrow is being serviced or repaired. If it must be raised for repairs, securely block the frame to prevent accidental lowering.

4. Always use lights, reflectors, and SMV emblem when transporting, day or night.

5. Lock the tractor drawbar in fixed position when transporting wheeled disks.

6. Never transport a disk harrow on its own wheels at more than normal tractor speed, and considerably less than that on rough or uneven ground.

7. Never clean, adjust, or lubricate the harrow while it is in motion.

8. Wear protective gloves when working with or near disk blades.

9. Hydraulic fluid escaping under pressure can penetrate the skin and cause serious infection or reactions. Never use hands to locate the source of a small leak which may be nearly invisible. Obtain immediate medical attention if injured by escaping hydraulic fluid.

10. Park or block the harrow so it cannot roll when unhitched.

11. Make sure wings are securely locked in transport position before moving the harrow.

12. Large disk harrows exceed normal vehicle width, so be particularly careful to avoid collisions when meeting other vehicles on the road. Avoid dropping wheels of tractor or harrow into holes, drains, or ditches along the road.

13. Provide adequate tractor ballast for front-end stability and to prevent excessive slippage.

14. Never allow anyone to ride on tractor drawbar or harrow in operation or transport.

15. Never allow anyone but the operator to ride on the tractor.

16. Lower the machine or install safety lock when storing a disk harrow.

17. Never permit children to play on or near a disk harrow while it is in operation, transport, or storage.

18. Stand clear of harrow wings during folding or unfolding.

19. Remove spring-loaded scrapers in proper order to avoid personal injury. Use care in relieving any spring under tension or compression.

20. Do not make sharp turns with blades down.

TROUBLESHOOTING

Recognizing improper performance and its cause makes it easier to keep a disk harrow functioning at its best. Where possible remedies to problems are obvious, based on possible causes, a blank space is left in the "Possible Remedy" column.

TROUBLESHOOTING CHART		
PROBLEM	POSSIBLE CAUSE	POSSIBLE REMEDY
SIDE-DRAFT ON TRACTOR	Disk not running level.	Adjust leveling linkage on hitch or wheels.
		Adjust hitch lift-link for integral harrows
	Gangs improperly angled.	Set right-hand and left-hand gangs the same
EXCESSIVE TRACTOR-WHEEL SLIPPAGE	Disking too deep.	Reduce gang angle or lower transport wheels.
	Inadequate ballast.	
LEAVING CENTER RIDGE	Rear gangs cutting too deep.	Reduce gang angle.
	On integral disks, tractor top link improperly adjusted.	Decrease length to level disk.
	Harrow not level fore-and-aft.	Adjust leveling linkage on hitch or wheels.
		Lower tractor-drawbar height.
	Improper lateral adjustment.	Move rear gangs toward center of machine or put smaller blades on inside of rear gangs.
	Excessive speed.	
LEAVING CENTER VALLEY	Rear gangs cutting too shallow.	Increase rear-gang angle.
		Level harrow fore-and-aft.
		On integral disks, lengthen tractor top link to level harrow.
	Disking too slowly.	
GANGS PLUGGING	Field too wet.	Disk very shallow on first pass to speed drying.

PROBLEM	POSSIBLE CAUSE	POSSIBLE REMEDY
	Excessive gang angle in wet soil.	Reduce gang angle enough to ease plugging.
	Scrapers badly worn or improperly adjusted.	
PLUGGING BETWEEN MAIN GANGS AND WINGS	More space at front edges of main gang and first wing blade than at rear.	Adjust angle of main gang and/or wing.
GANGS DIFFICULT TO ANGLE	Gang-attaching bolts not loose enough; or too loose and gangs tip and bind.	Loosen bolts just enough to allow gangs to slide.
DISKS WOBBLE IN FIELD	Insufficient gang-bolt torque.	Stop immediately and torque to level recommended in manual.
DISK NOT LEVEL IN TRANSPORT	Leveling linkage improperly adjusted (note: some harrows cannot be leveled for transport without changing field setting).	
INADEQUATE TRANSPORT CLEARANCE	Lift cylinders not fully extended.	Extend cylinders; may need to replenish hydraulic oil supply.
	Tire pressure too low.	
	Transport cylinder pins, clevis holes, or brackets worn.	
	Disk not level fore-and-aft.	
DISK NOT LEVEL IN OPERATING POSITION	Unequal depth-stop settings on wheel cylinders (two or more cylinders).	Set depth stops alike on wheel cylinders.
	Leveling linkage improperly adjusted.	
DISK WILL NOT LOWER SUFFICIENTLY	Binding in wheel-frame bearings.	Lubricate bearings.
	Too many depth stops on cylinder rods.	Remove enough stops to obtain desired working depth.
	Transport lock engaged.	
	Cylinder(s) not completely retracted.	
OUTSIDE FURROWS TOO DEEP	Not using furrow fillers.	Install furrow-filler blades on rear gangs.
	Rear of harrow running too deep.	Adjust leveling linkage on hitch or wheels.

PROBLEM	POSSIBLE CAUSE	POSSIBLE REMEDY
FRONT GANGS LEAVING OUTSIDE RIDGE	Outer front blades cutting too deep.	Install spacer between inner end of front gang and frame (or see manual).
	Rear gangs set too close together	Shift rear gangs out on frame.
	Excessive front-gang angle.	
	Excessive speed.	
WINGS NOT DISKING AS DEEP ON OUTER ENDS	Wing-folding cylinder not fully extended.	
	Folding linkage improperly assembled or adjusted, or binding.	
POOR PENETRATION	Insufficient gang angle.	
	Insufficient weight.	

Some disk-harrow performance problems and their solutions are unique to offset disks. Here are some of the causes and remedies.

PROBLEM	POSSIBLE CAUSE	POSSIBLE REMEDY
SIDE DRAFT	Too much left-hand offset.	Swing hitch to left.
	Tractor drawbar pinned.	Use swinging drawbar.
	Tractor drawbar too low.	
	Excessive front-gang angle.	
	Not enough rear-gang angle.	
NOT FILLING FURROW	Too much left-hand offset.	Swing hitch to left.
	Tractor wheel in furrow, enlarging it.	Drive with wheels on land.
	Disk too far from furrow.	Keep left front blade in furrow.
	Wrong lateral position of rear gang.	Shift gang right or left— center first rear blade between two left front blades.
	Not using furrow filler.	Attach furrow filler on right rear.
UNEVEN PENETRATION OF GANGS	Hard ground, front uneven.	Install spacer between left end of front gang and frame (see manual).
	Rear uneven.	Install spacer between right end of rear gang and frame (see manual).

PROBLEM	POSSIBLE CAUSE	POSSIBLE REMEDY
POOR PENETRATION	Extremely hard ground.	Swing hitch to the right. Increase angle of front and rear gangs.
DISK UNSTEADY (MOVES SIDE-TO-SIDE)	Too much angle in gangs.	Reduce front and rear gang angle.
PROBLEMS RAISING AND LOWERING	Same as other disk harrows.	Extend cylinders. Check transport locks and depth stops. Check fluid level in tractor reservoir. Bleed air from cylinders.

Fig. 64—Today Disk Harrows Compete with Plows for Primary Tillage

SUMMARY

Disk plows and disk tillers move soil only in one direction. Most disk harrows, while belonging to the same "family" (i.e., using concave, rotating blades), differ in having opposed tandem gangs. Except for single-action models, the front gang throws soil in one direction, and the trailing gang moves it back, leveling the field, providing double tillage, and assuring more complete mix-down of surface residue.

Almost unknown in North America a century ago, disk harrows have evolved from crude, narrow, black-smith-shop models into efficient, convenient, large-capacity types. Today they often compete with disk and moldboard plows for primary tillage (Fig. 64), and with other implements for secondary tillage. They also are excellent tools for minimum-tillage practices.

Disk harrows may be used in almost every soil condition and cropping plan. Many farmers consider them superior to chisel plows and field cultivators for working chemicals into the soil. They can be used to good advantage before or after moldboard plowing.

Widths vary from integral 5-footers (1.5 m) to large drawn types (including teaming two or more smaller disk harrows behind squadron hitches) whose capacity is limited almost solely by tractor power. There is great variety in blade types, blade size 16 to 36 inches (406 to 914 mm) blade spacing, 7 to more than 13 inches (457 to 838 mm) frame and gang design, and special attachments for special requirements. Adjustments are ample on most models to match desired penetration with optimum field performance and tractor-fuel economy.

CHAPTER QUIZ

1. (True or false) Disk harrows were first used in the United States in the 1870's.

2. What are the major benefits of disking heavy crop residue before plowing?

3. List the major benefits of disking after plowing.

4. What are the four primary disk-harrow types?

5. What is the purpose of a middlebreaker attachment on a disk harrow?

6. What is gained by offsetting and overlapping front and rear gangs of a tandem disk harrow?

7. (Fill in the blanks) Two disk-blade shapes are _____ and _____.

8. (Complete sentence) Notched or cutout blades work better in heavy trash and hard soil because ____.

9. (Choose one) Equally weighted <u>small</u> <u>large</u> diameter blades penetrate better in hard soil.

10. (Fill in the blank) Disk spacing usually ____as blade diameter and machine weight increase.

11. Why does the concave end of a disk gang normally tend to penetrate deeper than the convex end?

12. (Fill in the blanks) To increase offset to the right of an offset harrow, angle hitch to the _____ and _____ cutting angle of the front gang.

13. What is the purpose of increasing disk-gang angle from a line perpendicular to the direction of travel?

14. What is the appearance of a disk blade which has been broken by striking a stone or stump?

15
Roller Harrows
and Packers

Fig. 1—Roller Harrow in Operation

INTRODUCTION

Roller harrows and packers are secondary-tillage tools which are related to various lighter types of field-finishing implements in that one of their purposes is to prepare seedbeds of proper tilth.

These implements employ such seedbed-conditioning devices as heavy, cast, solid-rim, or serrated rollers; crowfoot rollers; sprocket-type rollers; tooth packer or treader wheels; spike, spring, or tine teeth; bladed cutterheads; and smoothing boards.

Differences in construction and work between these tools and tooth-type harrows often are not great, but two generalizations apply: they usually are heavier than tooth harrows, and many of them compact the soil surface (even to a depth of several inches) instead of leaving it loose.

Roller harrows are also known as cultipackers, culti-mulchers, soil pulverizers, and corrugated rollers because of the resulting appearance of the soil surface. Other soil-firming implements which accomplish similar results, but usually under different conditions, are known as treaders, clodbusters, plow packers, and subsurface packers.

Primary purpose of these implements is to crush clods and firm the soil surface, which they do better than any other machines. Roller-type implements are useful for leveling and firming freshly plowed soil.

Roller harrows (Fig. 1) are used to break surface crusts, pulverize clods, firm loose soil, eliminate large air pockets, and leave the ground ready to plant. The small ridges left by rollers may help prevent crusting following heavy rains in some soils. The front roller gang on these harrows crushes clods and levels the surface. Spring teeth then break big lumps in the soil, close large air spaces, and bring up buried clods. The rear roller gang crushes the clods and leaves the surface level. Some roller harrows may be equipped with seeding attachments for once-over tillage and seeding of fall- or spring-plowed land.

Roller packers (Fig. 2) also break soil crust, crush clods, firm the surface, and close air pockets near the surface for fast seed germination. With spring teeth raised, roller harrows may be used as roller packers to repack soil around roots of frost-heaved winter wheat, firm soil over newly-planted crops, or break crusted soil for fast germination and growth if heavy rain immediately follows planting.

Many rotary hoes may also be operated in reverse so that teeth crush and pulverize clods, and firm and level the soil. Rotary-hoe usage for packing is normally recommended where only limited or occasional soil firming of this type is required, and a roller harrow or roller packer would not be a profitable investment.

Firmly-packed, well-pulverized soil provides good seed-soil contact, particularly important for such small-seeded crops as grass and forage plants which must be accurately placed in the soil for good germination and early growth.

Roller harrows and packers are adaptable to many conditions and cropping programs, but cannot be ex-

ROLLER PACKER

Fig. 2—Roller Packer Crushes Clods and Firms Soil Surface

pected to cure problems caused by poor soil structure, working soil too wet, or other unfavorable conditions or management.

TREADERS

Treader packers look like heavy-duty rotary hoes with strong curved teeth on an angled gang. Teeth are normally rotated backwards. That is, the curved backs of the teeth strike the soil first. When angled and used in small-grain stubble or similar conditions, the wheels force some residue into the soil. This anchored residue provides protection from wind and water erosion; trash mixed into the soil also begins to decompose more quickly.

Treaders also break up clods, tear out weeds, firm the soil to reduce moisture loss through evaporation, and provide excellent incorporation of pre-emergence chemicals. Yet, the surface is left open and somewhat rough to promote better water infiltration. These implements are very good for seedbed preparation behind chisel plows and similar tools, and for weed control in summer fallow. Variations in gang angle, working depth, and operating speed provide the desired degree of tillage.

Treader configuration and application depends on the desired tillage results. Separate gangs may be attached to wide-sweep stubble-mulch plows, chisel plows, field cultivators, and rod weeders. Treaders have frames similar to wheeled disk harrows. As with disk harrows, they are available in single-action, double-action or tandem, and offset configurations.

Some treader-type units have been used successfully in stony land, and have outperformed disk harrows because of reduced maintenance requirements. If an individual point is broken from a treader spider, the implement can continue to operate until a convenient time for replacement. However, if a disk blade is cracked or a piece broken out, replacement must be made immediately to prevent damage to remaining

blades and other harrow parts due to progressive deterioration and probable loss of the broken blade.

PLOW PACKERS

Various roller-type packers are also used behind moldboard plows to break clods, close air pockets, and firm the soil. Planters and drills for spring wheat may be pulled behind the packer for a once-over plow-plant operation. In such cases, no other operation is needed which reduces fuel consumption and manpower requirements. This also helps reduce soil compaction from additional traffic at a time when it would be severe due to soil-moisture conditions. Some packer units use tandem, angled gangs with angled, knife-like rollers to slice and mash clods, smooth the surface, and pack soil to a depth of 4 to 6 inches (100 to 152 mm).

FINISHING HARROW

Another packer variation is the finishing harrow (Fig. 3), which has a pair of 5-bladed, horizontal cutterheads to chop and crush clods and firm the soil. Cutterheads are followed by several rows of chain-suspended spike teeth to break additional clods and level the surface. Final finishing and smoothing is done by a hardwood leveling plank mounted across the rear of the machine.

The finishing harrow with angled cutterheads (Fig. 3) can be used for seedbed preparation on flat ground or to firm and level beds. If the machine will be used entirely on flat ground, the cutterheads can be replaced by a double row of spring teeth (Fig. 4) which dig, lift and break clods; close air pockets; and level the soil. Spike teeth and smoothing plank are the same as on the other machine, and leave the same type of firm, well-prepared, ready-to-plant seedbed. Both models have rubber-tired transport wheels, and are raised and lowered by remote hydraulic cylinders.

Other packers use roller wheels similar to those on the roller harrow, but larger in diameter and spaced several inches apart. The V-shaped rim of these wheels presses into the loose, freshly plowed ground to firm the soil several inches under the surface and close many large air spaces. Yet the surface is left somewhat rough and open, and evaporation of soil moisture is retarded by the subsurface packing.

Crowfoot rollers (Fig. 5) are used on some subsurface packers. Such wheels pulverize rather than pack the soil, and leave the surface rougher and pock-marked for better moisture absorption.

Reduction of moisture loss through evaporation after plowing is one of the key advantages of using plow packers. Also, by breaking clods immediately and slowing surface drying, some ground may be spring-plowed which would otherwise quickly dry into hard, almost unbreakable clods which would be very difficult to work into a satisfactory seedbed. If such clods are formed, a roller harrow or roller packer is the only implement which will break them down, and several passes may be needed.

FINISHING HARROW

Fig. 3—Finishing Harrow with Bladed Cutterheads

Fig. 4—Two Rows of Spring Teeth are Available for Deep Penetration

TWO ROWS OF SPRING TEETH FOR DEEP PENETRATION

Fig. 5—Crowfoot Rollers Provide Aggressive Pulverization

CRAWFOOT ROLLER

Working soil too wet and packing immediately after plowing could compact some soils enough to restrict root development and crop growth severely.

ROLLER-HARROW AND PACKER TYPES AND SIZES

The roller harrow or cultimulcher is a combination spring-tooth harrow and roller packer which may be either integral or drawn (Fig. 6). Sizes are approximately 10 to 28 feet (3 to 8 m) wide. Each implement has a front gang or section of independent, heavy cast-Iron rollers mounted on a tubular axle. The rollers are not intended to operate as a unit, but must be free to turn individually. This provides better pulverization, a differential action when turning, and better cleaning of moist soil from the roller unit.

Behind the front roller are two staggered rows of spring teeth which dig up and break clods, break large

DRAWN ROLLER HARROW

Fig. 6—Drawn Roller Harrow in Transport

lumps, and close air pockets in the soil. Teeth also tear out small weeds and break up surface crusts. On the integral model, teeth leave the soil corrugated to resist water and erosion.

These spring teeth are equipped with double-pointed replaceable shovels (Fig. 7), and are hydraulically or lever controlled (Fig. 8). Points of the teeth can be raised to 8 inches (203 mm) above the bottom of the rollers, or set to work down to 6 inches (152 mm) below the rollers.

Behind the spring teeth on some roller harrows, is an optional tine-tooth harrow attachment designed to

break small clods and level the corrugation left by the spring teeth. The flexible tines are coil-spring-cushioned at the upper end to provide a vibrating, soil-shattering action, and protection from soil obstructions. Working depth of the smoothing harrow is controlled by raising or lowering the tine bar on the attaching brackets.

Tine-tooth angle on some leveling harrow attachments is lever-controlled to permit variation of tooth aggressiveness. Tipping teeth back to a 30-degree trailing angle enables them to shed trash easily.

Fig. 7—Double-Pointed Shovels are Reversible for Longer Wear

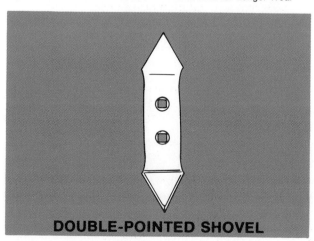

DOUBLE-POINTED SHOVEL

Fig. 8—Spring Teeth are Hydraulically Raised and Lowered

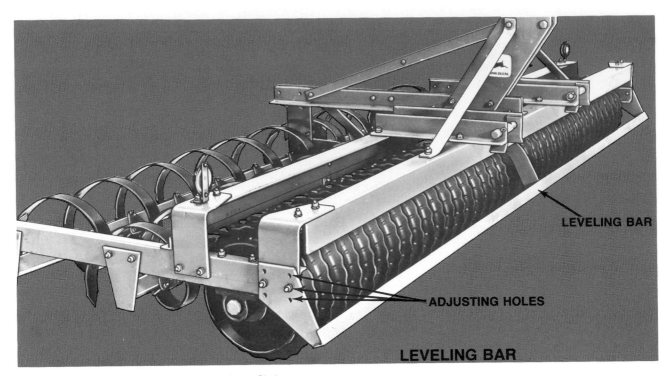

LEVELING BAR

ADJUSTING HOLES

LEVELING BAR

Fig. 9—Leveling Bar Knocks Down Ridges and Large Clods

Fig. 10—Two Rows Each of Rollers and Teeth

TWO ROWS OF ROLLERS AND TEETH

260

DRAWN HARROW TEETH CONTROLLED HYDRAULICALLY

Fig. 11—Drawn Roller-Harrow Teeth may be Controlled Hydraulically or by Hand Lever

An optional leveling bar (Fig. 9) may be attached to the front of the roller-harrow frame to break and level large clods before the rollers pass over them. Bar height may be adjusted to suit soil conditions.

On drawn roller harrows (Fig. 10), a second roller gang works behind the spring teeth to crumble remaining clods, firm the seedbed, and further eliminate air pockets.

Locating transport wheels inside the main frame (Fig. 10) reduces overall width for easier transport. On the other hand, some machines have wheels outside the ends of the main frame to permit reduction in overall machine length, and avoid interference with spring-tooth locations.

Spring teeth on drawn roller harrows may be operated by a remote hydraulic cylinder (Fig. 11) for on-the-go control of working depth, or by hand lever. If hydraulic tooth-control is used, the tractor must have two remote valves to operate tooth and transport-wheel cylinders (Fig. 12) independently. Teeth may be rotated from 8 inches (203 mm) above rollers to clear trash, to 6 inches (152 mm) below rollers to shatter and bring up clods. Or, teeth may be raised completely, and the machine used as a tandem packer only.

Because of the weight of integral roller harrows, they should be used only with tractors having adequate lift capacity and front-end stability for safe operation and transport. See roller-harrow operator's manual for tractor models compatible with each harrow size.

ROLLER PACKER

Roller packers are also called land rollers, cultipackers, corrugated rollers, pulverizers, and packers. They are available in single-gang or triple-gang (Fig. 2) configurations. Typical single-gang sizes range from 9 to 12 feet (2.7 to 3.7 m) while triple-gang sizes may go from 12 to 24 feet (3.7 to 7.3 m) with various combinations of center and wing sections.

Triple-gang roller packers are easily transported by hooking wing units in tandem behind the center sec-

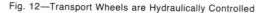

Fig. 12—Transport Wheels are Hydraulically Controlled

WHEELS HYDRAULICALLY CONTROLLED

Fig. 13—Transporting Triple-Gang Packers

Fig. 14—Solid-Rim Packer Wheels

Fig. 15—Serrated Cast Packer Wheel

Fig. 16—Sprocket-Tooth Packer Wheel

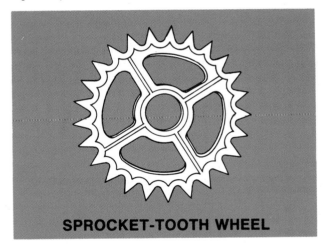

Fig. 17—Scrapers Clean Wheels

Fig. 18—Level Drawn Roller Harrows with Hitch-Leveling Screw

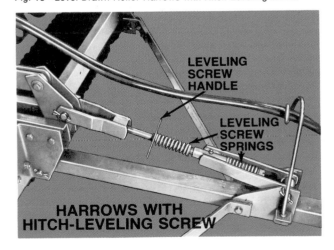

tion and folding extension hitches to the center-gang frame (Fig. 13).

ROLLER WHEELS

Roller-wheel selection depends on soil conditions and tillage objectives. The diameter of packer wheels ranges from 10 to 20 inches (254 to 508 mm) on various machines, depending on wheel type.

Packer wheels are loosely mounted on a tubular axle approximately 2 to 4 inches (50 to 100 mm) in diameter, depending on machine design. This axle is fitted on a gang bolt. Depending on machine width, the gang bolt may be supported by two to five sealed, anti-friction bearings.

If roller or packer wheels become loose on the gang axle due to wheel wear, slack may be taken up without disassembling the gangs by installing split washers on the axle. Packer wheels should never be tight enough to restrict free rotation of wheels.

On roller packers, solid-rim wheels (Fig. 14) or serrated wheels (Fig. 15) are usually provided for normal operation. Serrated cast rollers provide uniform packing and a smoothly corrugated soil surface.

Crowfoot rollers (Fig. 5) may be used in place of solid-rim or serrated rollers on roller harrows and roller-packers. More aggressive clodbreaking is provided by crowfoot rollers, and soil is left with a loosely mulched surface. Crowfoot rollers also firm the soil deeper than solid-rims or serrated rollers. Reversing crowfoot rollers causes them to leave small pockets in the soil which give increased moisture retention, reduced crusting, and better wind and water erosion control. On roller harrows, crowfoot rollers may be used on either front or rear gangs, or both, depending on soil conditions. Note: cast-iron rollers are more susceptible to breakage on rocks or stones than solid packer wheels; many crowfoot rollers are made of nodular iron to reduce breakage.

Some roller packers and roller harrows have alternate solid-rim or serrated packer wheels and sprocket-tooth wheels (Fig. 16). Tooth action of the sprockets helps pulverize soil, and leaves some loose soil on the surface. Sprockets also help keep packer wheels clean of mud and moist soil as they shift and roll between the other wheels.

Optional scrapers may be attached to the frame of roller packers or harrows to clean wet, sticky soil from packer wheels (Fig. 17). Scrapers are set just close enough to clean the wheels without slowing roller rotation (never set scrapers tight against wheels).

Scrapers cannot be used with crowfoot or sprocket-type packing wheels.

PRINCIPLES OF ROLLER-HARROW AND PACKER OPERATION

As explained earlier, roller harrows, roller packers, and similar implements break clods, pulverize and

firm surface soil, and close air pockets. The goal is a firm seedbed of finely divided soil particles to provide good water movement and seed-soil contact for rapid germination and early growth, particularly for small-seeded crops.

Due to their weight, roller harrows will usually operate level. Optimum performance results if these machines have flexible frames, allowing both gangs to follow ground contours. A hitch-leveling screw is provided on drawn roller harrows (Fig. 18) to level the machine fore-and-aft for transport. During operation, this screw should be adjusted to provide equal spring compression as the tractor and roller harrow move over rolling land. Integral harrows are leveled by adjusting tractor top and lift links (Fig. 19).

TRACTOR PREPARATION

Draft per foot of operating width of roller harrows and packers is relatively low, compared to many tillage tools. Although a great deal of power is not required, integral roller harrows are quite heavy, and require a tractor with sufficient front-end weight and stability as well as adequate lift capacity for satisfactory operation. Consult the harrow manual for proper tractor size and the tractor operator's manual for recommended front-end weighting.

To avoid making deep tractor-wheel tracks which cannot be removed by the roller harrow or packer, keep rear-wheel weight to the minimum required for satisfactory traction without excessive slippage. This applies both to integral and drawn machines.

Fig. 19—Level Integral Harrows by Adjusting 3-Point Hitch Linkage

ADJUST TRACTOR LIFT LINKS

Tractor-wheel tread is not particularly important in most cases, and many drawn machines have sufficient tongue length to permit use of either single or dual rear wheels.

3-POINT HITCH

Adjust hitch lift links to equal length in the "float" position so the machine will follow ground contours in operation (Fig. 20). Install sway blocks or chains to provide lateral flexibility in operation, but eliminate side sway when the harrow is raised for transport.

Most current integral roller harrows may be attached directly to the tractor hitch links, or with an implement quick-coupler for fast, no-muscle hookup. Be certain that hitch pins are properly installed on the harrow to match quick-coupler hooks.

Adjust tractor top link to level machine fore-and-aft.

DRAWN TYPE

Set drawbar height as recommended in harrow manual, and pinned in the center for transport. In operation the drawbar may be pinned or allowed to swing, as preferred.

Always use a drawbar pin with adequate strength for the tractor-implement combination, and lock it in palce to avoid possible loss during transport.

Inflate tractor tires to recommended pressure and examine hydraulic couplings, hoses, and cylinders for leaks or damage.

Always raise spring teeth clear of the ground before lowering the roller harrow to the ground for parking or unhitching. Block the machine before unhitching.

Fig. 20—Front Rollers Raise First

MACHINE PREPARATION AND MAINTENANCE

Roller harrows and packers require little preparation or maintenance. However, careful operation and prompt attention to problems will increase productive life of the machine.

Before Each Season:

1. Check for loose bolts and nuts; torque gang-bolts to recommended tension particularly on new machines.

2. Check thoroughly for loose, broken, or worn parts, and replace as necessary.

3. Reverse or replace spring-tooth points when badly worn.

4. Lubricate as recommended (some machines with pre-lubricated bearings require no field lubrication).

5. Clean, repack, and tighten wheel bearings.

6. Adjust scrapers if worn or out-of-line.

7. Inflate tires on drawn machines to recommended pressure.

Daily Before Operation:

1. Visually check tire pressure; look for loose or broken parts, loose nuts and bolts, etc.

2. Lubricate as recommended in manual.

3. After first day's operation of a new machine, check all bolts for tightness—particularly gang-bolts, those holding spring teeth, and the tine-tooth fastening bolts on smoothing-harrow attachments.

Before Storage At End Of Season:

1. Clean dirt and trash from the machine to prevent collection of moisture and rusting.

2. Repaint spots where paint has been scratched or worn away.

3. Coat spring-tooth points with plow-bottom paint or heavy grease to prevent rust.

4. Store the machine inside to prevent damage from weathering.

FIELD OPERATION

Operating the roller harrow in the same direction in which the field was plowed usually provides the best leveling of ridges. In fresh plowing, particularly sod, do not set spring teeth deep enough to drag quantities of trash or chunks of sod back to the surface.

Where trash is not a problem, operating depth is determined by soil conditons and results desired. Normally, set teeth just deep enough to close air pockets and break clods in the planting and early root-growth zones. Working the soil deeper than necessary wastes power and time.

Adjust hitch-leveling screw to provide equal spring compression on drawn machines. Level integral models by adjusting the tractor top-link length.

If soft soil tends to push ahead of the front rollers, hydraulically lowering the transport wheels slightly (Fig. 20) raises front rollers first to allow the harrow to pass over the ridge without seriously disturbing spring-tooth action or raising rear rollers.

Always allow the tractor 3-point hitch to float vertically and laterally when operating integral roller harrows or packers. Do not use load-and-depth or draft control to "borrow" implement weight for traction. Maximum implement weight must be concentrated on the rollers for efficient soil pulverization.

TRANSPORT AND SAFETY

Careful operation and attention to fundamental rules of safety can help reduce accidents, injuries, and costly equipment repairs.

1. Always use reflectors, lights, and SMV emblem as required when transporting equipment—day or night (Fig. 21).

2. Provide adequate front-end weight for tractor stability in operation and transport of integral harrows. Use maximum allowable front ballast if operating integral harrows in lower gears.

3. Never exceed normal tractor speed when transporting drawn roller harrows, and drive considerably slower than that on rough or uneven ground.

4. Do not transport roller packers over hard-surfaced roads—use carriers.

5. Never permit anyone but the driver to ride on the tractor.

6. Never permit anyone to ride on the tractor drawbar or implement during operation or transport.

7. Install transport lock pin before storing, transporting, or parking drawn implement; do not depend on hydraulic pressure to support the weight. Lower machines to the ground whenever the tractor engine is shut off.

8. Pin tractor drawbar in center before transporting.

9. Never lubricate, adjust, or repair the implement while it is in motion or the tractor engine is running.

10. Always raise spring teeth before lowering roller harrow to the ground for parking or storage.

11. Park or block the implement to prevent rolling when it is disconnected from the tractor.

12. Never try to lift or support the roller harrow on the spring teeth for service or repairs.

TRANSPORTING ROLLER HARROW

Fig. 21—Transporting Roller Harrow

TROUBLESHOOTING

Improper adjustment and operation are the most frequent causes of unsatisfactory machine performance. Recognizing the symptoms, causes, and remedies for operating problems permits quick correction and better utilization of equipment, manpower, and fuel. Obvious remedies are not included in the chart.

TROUBLESHOOTING CHART

PROBLEM	POSSIBLE CAUSE	POSSIBLE REMEDY
ROLLERS DON'T TURN FREELY	Rollers plugged with mud.	Allow field to dry; install or adjust scrapers.
	Too many washers installed on gang axles.	
	Scrapers set too close.	
EXCESSIVE SIDE-PLAY IN ROLLERS	Rollers worn on sides.	Install split take-up washers or spacers on gang axle—see manual.
FREQUENT TRASH PLUGGING IN SPRING TEETH	Working wrong direction.	Operate same direction as plowing.
	Teeth set too deep.	
ROLLERS PLUG WITH WET SOIL	Soil too wet to work.	Allow soil to dry; install or adjust scrapers (except on crowfoot packers).
DIFFICULT SPRING-TOOTH PENETRATION	Points worn or broken.	
	Soil too dry and hard.	
INADEQUATE TRANSPORT CLEARANCE	**Integral models:**	
	Inadequate hydraulic pressure.	Check level of hydraulic fluid.
		Check hydraulic-system performance.
		Use larger tractor with greater lift capacity.
	Tractor top link too long.	
	Drawn models:	
	Inadequate hydraulic pressure.	Check level of hydraulic fluid.
		Check hydraulic system performance.
	Machine not leveled.	
SOIL ROLLS UP IN FRONT OF FRONT ROLLERS	Machine low in front.	Lift just enough to pass over ridge of soil.
	Soft, loose soil.	
UNDESIRABLE SOIL RIDGING BY SPRING TEETH	Teeth set too deep.	
	Not using smoothing harrow.	
SMOOTHING HARROW TEETH DRAG TRASH	Teeth set too low.	Raise or angle teeth.

SUMMARY

Roller harrows and roller packers belong to a diversified but functionally related "family" of pre-plant, field-finishing implements whose specific crop-production uses often overlap. Further, many of these various implements are adjustable to a wide latitude of field conditions and desired results. That is why each purchaser should consider all pertinent factors before making a choice.

The primary purpose of roller harrows and roller packers is to crush clods and firm the soil surface and sub-surface soil, which they do better than other implements. They are especially useful for leveling and firming freshly plowed soil. They are adaptable to many conditions and cropping programs, but are not cure-alls for poor soil structure or management mistakes.

Roller harrows and roller packers employ one or more of such seedbed-conditioning devices as heavy, cast, solid-rim, crowfoot, sprocket-type, or serrated rollers; toothed packer or treader wheels; spike, spring, or tine teeth; bladed cutterheads; and smoothing boards. These implements usually are heavier than tooth-type harrows, but draft per foot of width is low compared to many tillage tools.

When deciding what kind of field-finishing implement to choose, each prospective purchaser should consider the nature of soil on his farm (sandy, clay, stony, etc.); his type of primary tillage, probable annual rainfall, soil and wind erosion problems; crops, tractor power and tractor lift capacity, and highway travel between fields. Observing neighbors' success or difficulty will be helpful, and his implement dealer can supply comparative prices of various machines to help assure satisfactory results with minimum investment.

CHAPTER QUIZ

1. What are the primary functions of roller harrows and packers?

2. Why should tractor load-and-depth control not be used when operating integral roller harrows or packers?

3. Why is a treader useful in stubble-mulch farming?

4. (True or false) Treaders are available as attachments for wide-sweep and chisel plows, field cultivators, and rod weeders, as well as with frames similar to disk harrows.

5. Compare the functions of serrated and crowfoot rollers.

6. (True or false) Roller packers and harrows are high-draft implements.

16
Toothed Harrows

Fig. 1—Clod-Buster Attachment for Moldboard Plows

INTRODUCTION

Implements used for secondary tillage differ somewhat by crop and geographic areas, but their wide range almost always includes some form of spike-, spring- or tine-toothed implements.

These machines perform one or more of several functions: break soil crust, shatter clods, smooth and firm soil surface, close air pockets in the soil, kill weeds, loosen and aerate soil, conserve moisture, cultivate small plants, and renovate pastures.

Clods are broken more easily and maximum moisture is saved when secondary tillage immediately follows plowing (Fig. 1). This firms surface soil and closes air spaces from which moisture would evaporate. Almost any tillage of a dry surface exposes moist soil and increases evaporation.

The ultimate goal of most secondary tillage is preparation of a suitable seedbed for a particular crop (Fig. 2) so seeds are closely surrounded by soil for maximum moisture absorption. Small seeds must generally be planted close to the surface for better emer-

gence, but larger seeds usually tolerate coarser soil particles and deeper planting. In any case, seedling growth must be rapid enough for roots to reach deeper, more dependable moisture supplies before soil above the seed dries and kills the new plants.

Timing of tillage and seeding or planting must closely coincide with the availability of proper soil moisture and proper temperature conditions for quick germination and plant growth.

Use of tooth harrows and cultivators depends largely on machine design and desired tillage results. The performance and capabilities of each unit will be discussed in the next section.

TYPES, SIZES, AND PRINCIPLES OF OPERATION

The choice of secondary-tillage equipment depends largely on primary tillage already completed and the degree of added tillage desired. Each tooth harrow and light cultivator serves a slightly different need and the choice should be made accordingly.

Fig. 2—Goal of Most Secondary Tillage is Seedbed Preparation

SPIKE-TOOTH HARROWS

Primitive man learned that seeds he planted were more likely to grow, and less likely to be devoured by foraging birds and animals, if he covered them with soil. So he used the most logical tool available—a tree branch, dragged over the newly scattered seeds. Some branches covered seed better than others. Adding more wooden points made the drag more effective, and so, through many changes, led to the current spike-tooth harrow. It was one of man's first agricultural tools, and is as old historically as the plow.

Spike-tooth harrows (Fig. 3) are also known as peg-tooth, drag, section, or smoothing harrows. They are used to smooth seedbeds, break soft clods, and kill small weeds as they emerge from the soil. Spike-tooth harrows are also used to break rain-crusted soil for fast emergence of seedlings. However, if harrowing is improperly timed—after germination, but before roots are established—the crop may be seriously damaged.

Spike-tooth harrows are low-draft implements and may be attached behind other tools (moldboard plows, disk harrows, and field cultivators (Fig. 4)] to smooth the surface and finish seedbed preparation. They are used to cover broadcast seed, such as forage.

A variation of the spike-tooth harrow, called a chain or flexible harrow (Fig. 5), also is used for seedbed preparation and other spike-tooth harrow functions—scratching up dead grass and loosening surface soil for seeding pasture or meadowland, scattering livestock droppings in pastures, and drying and smoothing dirt feedlots.

The chain-link design provides maximum flexibility and ground-hugging penetration. Sizes range from 4 to 24 feet (1.2 to 7.3 m), made up of 4- to 5½-foot (1.2 to 1.7 m) sections drawn behind a rigid steel evener (Fig. 5). Sections can easily be rolled and loaded or stacked for transportation or storage.

Fig. 3—Final Seedbed Preparation with Spike-Tooth Harrow

Fig. 4—Tine-Tooth Attachment for Mulch Tiller

271

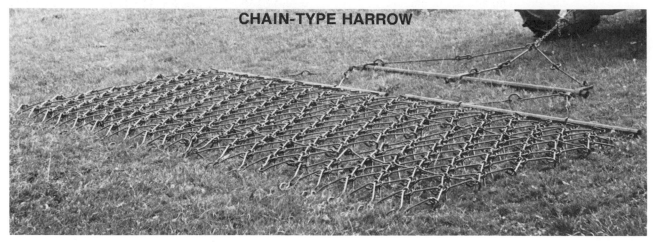

CHAIN-TYPE HARROW

Fig. 5—Flexible or Chain-Type Harrow

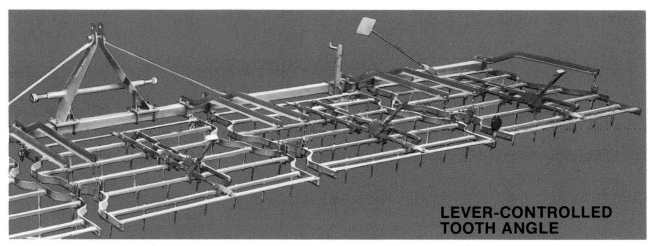

LEVER-CONTROLLED
TOOTH ANGLE

Fig. 6—Lever-Controlled Tooth Angle

Fig. 7—Spring-Cushion Angling Linkage Protects Harrow

SPRING-CUSHION
ANGLING LINKAGE

Aggressiveness

Working depth and aggressiveness of spike-tooth and flexible harrows are controlled by changing tooth angle. Spike-tooth harrows may have from three to nine lever-controlled tooth positions—from vertical to nearly horizontal (Fig. 6). Vertical teeth provide maximum soil agitation and clod breakage, but with maximum draft. Inclining the teeth allows them to slip more easily over trash and around obstacles. Some are equipped with cushion springs in the adjusting linkage (Fig. 7) to absorb shocks and help shed trash. This permits a combination of aggressive tooth action and easy shedding of trash. The tooth-angling lever on many harrows folds flat for easier hauling or storage of harrow sections.

The chain harrow (Fig. 5) can be reversed, or split and reversed, to provide the desired tooth aggressiveness. Maximum penetration is provided by pulling the harrow with teeth pointing forward and down. Reversing the sections and angling teeth to the rear provides better trash shedding and better smoothing. For fine smoothing or careful seedbed preparation, the harrow is turned upside down and dragged with teeth pointing up. For a combination of aggressive digging and smoothing, the sections are split in the middle; the front portion is pulled with teeth pointing forward, and the rear section with teeth pointing back or turned over with teeth up for even better smoothing.

Spike-Tooth Harrow Frames

Most spike-tooth harrows are made in 4- to 6-foot (1.2 to 1.8 m) sections with diamond-shaped teeth either bolted (Fig. 8) or welded (Fig. 9) to a tubular or channel-shaped bar. Diamond-shape teeth are superior to square or round teeth because the sharp leading edge provides better soil penetration and lighter draft. If the leading edges of bolted teeth become worn, the teeth may be reversed for longer wear. Welded teeth do not loosen and get lost, as bolted teeth may.

Each harrow section usually has five lateral toothbars, with teeth staggered on each bar for thorough tillage and optimum trash clearance. With 25 to 45 teeth per section, this provides a working interval of approximately 1½ to 2 inches (38 to 50 mm), depending on tooth spacing and section width.

MAIN-FRAME STYLES

Three main-frame styles are available. Closed-end harrows (Fig. 10) have a steel guard rail on each end to prevent fouling toothbars of adjoining sections or snagging trees and fence posts. Closed ends may be straight or curved; with curved ends, sections fit together to maintain tooth interval from section to section. Open-end sections (Fig. 11) shed trash from each end; they are cross-braced for greater stiffness. Most open- and closed-end sections have sufficient flexibility to permit easy following of ground contours.

Flexible-frame harrows have toothbars individually hinged to permit each bar to rise or drop over obsta-

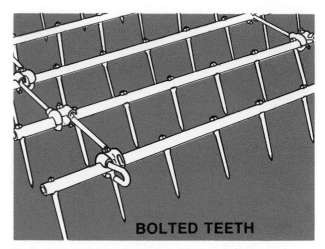

BOLTED TEETH

Fig. 8—Bolted Teeth may be Reversed for Longer Wear

Fig. 9—Welding Teeth to Bar Provides Maximum Durability

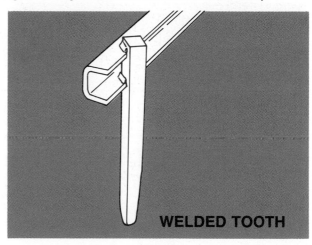

WELDED TOOTH

Fig. 10—Closed-End Spike-Tooth Harrow Section

Fig. 11—Open-End Spike-Tooth Harrow Section

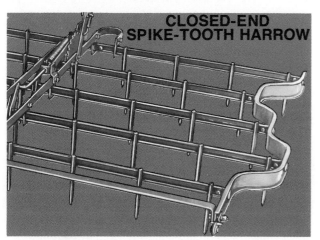

CLOSED-END SPIKE-TOOTH HARROW

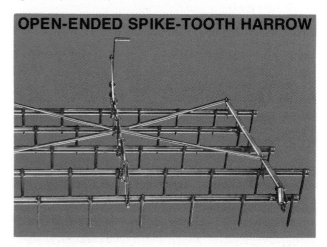

OPEN-ENDED SPIKE-TOOTH HARROW

273

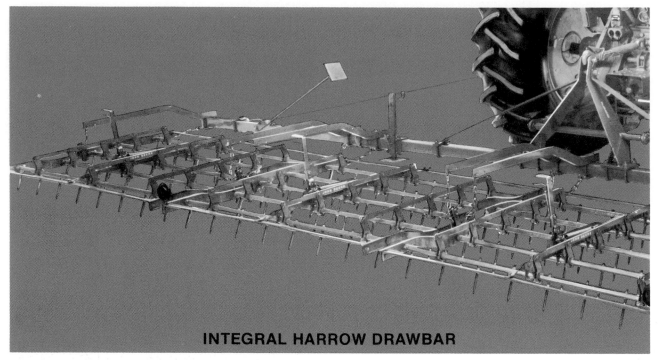

INTEGRAL HARROW DRAWBAR

Fig. 12—Integral Drawbar for Two to Five Sections of Harrow

cles or uneven ground. This provides trash-shedding and protection from soil obstructions.

Two tooth angles are provided by simply reversing the sections of flexible harrows. Pulled in one direction, teeth are held vertically for maximum penetration and soil pulverization. Reversing the section lets teeth slant to approximately a 25-degree trailing angle for better trash clearance and soil leveling.

Flexible-harrow sections may be rolled and stacked for compact storage or hauling.

Fig. 13—Wheeled Carts May Pull Up to Nine Harrow or Field Conditioner Sections

HARROW TRANSPORT

Hitching

Single harrow sections can be pulled behind moldboard plows, and two or more sections with a steel evener are frequently hitched behind disk harrows, chisel plows, or field cultivators to reduce field trips.

When harrowing is a separate operation, sections can be hitched and pulled in three ways, depending on the number of sections used, available tractor power, need for transporting, and operator preference. For simplicity and economy, if the harrow is seldom transported very far, a drag-type steel evener can handle two to six harrow sections, depending on section size. Teeth are adjusted to minimum angle and end sections manually folded over center sections to reduce width for moving short distances.

Integral harrow drawbars handle two to five sections of spike-, spring- or tine-tooth harrow (Fig. 12) with maximum maneuverability and ease in transporting. Wheeled carts (Fig. 13) can be used to pull three to nine tooth-harrow sections (as much as 54 feet; 16.5 m) for big-area harrowing.

Heavy-Duty Harrows

Some heavy-duty spike-tooth harrows, weighing up to 85 pounds per foot of width (125 Kg/m), are available for smoothing and pulverizing soil under very severe conditions. Rugged teeth are mounted on heavy angle crossbars which provide both strength and weight. In operation, previously plowed or disked soil flows over the front two bars as teeth agitate and mulch the soil every two inches. Soil is smoothed by the following bars for final seedbed preparation. Wing sections fold over the center section for transport.

TINE-TOOTH HARROW

Tine-tooth harrows, also called "finger" harrows, are relatives of spike-tooth harrows in appearance and performance under many soil conditions. The additional vibratory action of the tine teeth helps shatter clods, skip around obstructions, rip out small weeds, and smooth soil—and all of these are combined with light weight and low draft. Tine-tooth harrows are sometimes used for fast weeding and thinning of small sugar beet plants after emergence. Depth-control skids are available for many tine-tooth harrows to protect young plants when harrowed shortly after sprouting.

Types And Sizes

Tine-tooth harrows come in many configurations for a wide range of tillage operations.

Five-bar sections (Fig. 14) are available in approximately 5- to 7½-foot (1.5 to 2.3 m) widths for single- or multiple-section use, similar to spike-tooth harrows. These sections may be used individually or attached to drag-type eveners, integral drawbars, or wheeled harrow carts. Tooth interval of these sections is usually 1½ to 2 inches (38 to 50 mm), and teeth may be operated vertically or tilted back as far as 30 degrees for less-aggressive action and better trash flow.

ATTACHMENTS FOR OTHER IMPLEMENTS

Tine-tooth sections are used as smoothing attachments for mulch tillers, roller harrows, spring-tooth harrows, field cultivators, disk harrows, and moldboard plows (Figs. 4, 13, and 15). Very small tine-tooth sections also are available for use on corn planters to provide strip-tillage at planting time in previously worked soil. All these attachments provide ad-

Fig. 15—Tine-Tooth Smoothing Attachments Improve Performance of Other Implements

ditional soil pulverization and leveling for final seedbed preparation and normally eliminate at least one additional pass over the field, reducing soil compaction and saving time and fuel.

Tine-tooth attachments on moldboard plows mulch the surface, crumble clods, and close large air pockets to reduce rapid evaporation of soil moisture (Fig. 1).

Tine teeth are also optional on some field conditioners (Fig. 16) up to 40 feet (12 m) wide for fast crust-breaking, clod-shattering seedbed preparation, and summer fallowing. If teeth are set shallow and the machine is level fore-and-aft, soil is left slightly ridged (Fig. 16). Slightly lowering the rear bar and setting teeth deeper leaves a smoother seedbed.

Depending on tillage objectives, most tine teeth range from 9/32- to ⅜-inch (7 to 9.5 mm) in diameter, and from 10 to 20 inches (254 to 508 mm) in length and are normally coiled at the top (Fig. 17) for added spring-cushion protection and vibratory action. Teeth on some machines are mounted on tubular bars which run through the coil at the top. This may necessitate stripping the entire bar for replacement of individual broken or damaged teeth. In other installations, teeth are attached to an angle or channel-shape bar for easy replacement (Fig. 17).

FLEXIBLE-TINE CULTIVATORS

Breaking hard-crusted soil, preparing fine seedbeds, controlling weeds, renovating pastures, and culti-

Fig. 14—Tine-Tooth Harrow Section

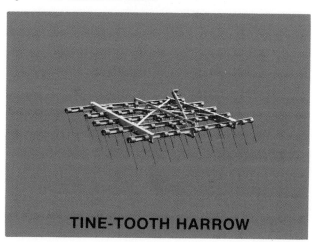

TINE-TOOTH HARROW

CONDITIONERS LEAVE SOIL RIDGED

Fig. 16—Tine-Tooth Field Conditioners May Leave Soil Slightly Ridged

vating row-crops are only a few of the tasks performed by flexible or spring-tine cultivators (Fig. 18). The "S" or double-curved loop in the forged spring-steel tines provides a strong vibrating action (Fig. 19) which shatters soil, breaks clods and pulls weeds out by the roots. These machines may also be used for early cultivation of row crops by relocating tines.

Teeth may do little more than scratch the soil surface at slow operating speeds. As speed is increased to 6 or 7 miles (10 or 11 Km/h) per hour, teeth draw into the soil and perform at their best. Even in hard, dry alfalfa stubble, soil may be broken to full depth and only small

curved ridges left unbroken between tines.

High operating speed and strong tine action combine to make flexible-tine cultivators excellent for tearing out and destroying troublesome weeds and grasses.

In previously worked soil, flexible-tine cultivators break crusts and clods, kill weeds, close air pockets, and leave the ground ready for planting. Working depth is precisely controlled by adjustable gauge wheels or gauge shoes. Tractor load-and-depth or draft control need not be used with flexible-tine cultivators because draft requirement is quite low.

Fig. 17—Coils at Top Permit Flexing and Provide Tooth Protection

Fig. 18—S-Tine Cultivator

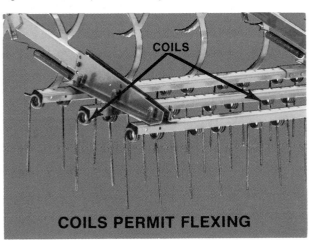

COILS

COILS PERMIT FLEXING

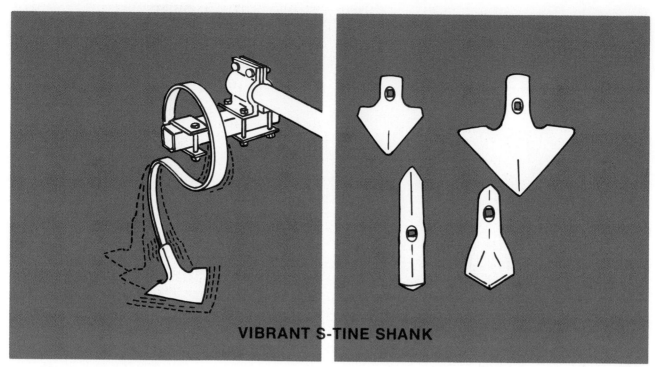

VIBRANT S-TINE SHANK

Fig. 19—Vibrating S-Tine Shank and Some of Available Sweeps

Flexible-tine cultivators pull easily, requiring about 1½ horsepower (1 kW) per tine. With 4-inch (101 mm) spacing, that is 45 horsepower (34 kW) for a 10-foot (3 m) cultivator that can cover as much as 7 acres an hour (2.8 ha/h)—fast, efficient tillage.

Types And Sizes

Flexible-tine cultivators are available in 4½- to 24-foot (1.4 to 7.3 m) integral models. Cultivators 14 feet (4.3 m) wide or wider are normally equipped with extension wings which fold for transport. They have manual, mechanical, or hydraulic wing-folding, depending on size and design.

Frames are usually welded to withstand high operating speeds and tooth vibration. Most cultivators have four tine-bars with tines arranged to work every 4 inches (102 mm). This provides 16 inches (406 mm) between tines on each bar. The wide spacing, plus high operating speed and tine vibration, keeps trash moving through the cultivator with little plugging. In loose soil, or extremely heavy trash, tines may be rearranged for 6-inch (152 mm) working space—24 inches (610 mm) laterally between tines.

Tine locations on most cultivators are marked on the tine-bars to provide 4- to 6-inch (102 to 152 mm) spacing and the properly staggered arrangement for best trash shedding and equal draft load on each side.

Narrow double-point shovels are standard equipment on most flexible-tine cultivators. When these become worn they can be reversed for additional wear. For weeding, or more complete soil working, 2½- to 4-inch (64 to 102

mm) sweeps (Fig. 19) may also be used. Sweeps are not used in hard-soil conditions because of increased drag and resistance to penetration.

Tine-tooth and drag-type smoothing attachments (Fig. 20) are available for some cultivators to provide additional soil smoothing.

SPRING-TOOTH HARROW

The spring-tooth harrow (Fig. 21) works 3 to 6 inches (76 to 152 mm) deep to loosen soil crust, dig, lift, and break clods which are not too hard, and pull many roots of quack grass and similar plants. Used immediately after plowing, it closes air pockets in the soil, breaks up clods, and levels the surface to make it ready to plant. In freshly plowed sod, operating in the direction of plowing with a shallow setting, the first time over, will help prevent dragging chunks of sod to the surface. The spring-tooth harrow is an excellent light-draft tool for summer-fallow and orchard and vineyard work.

Deeper penetration in crusted soil and more-aggressive action on sprouting weeds makes the spring-tooth harrow much better suited for seedbed preparation than the spike-tooth harrow. However, large disk harrows and field cultivators have replaced spring-tooth harrows on many farms because of their greater ability to penetrate hard ground and kill tough weeds, particularly if the field cultivator has wide sweeps.

The spring-tooth harrow is better than the disk harrow for stony ground, but plugs badly in heavy trash due to the limited clearance of most spring-tooth harrow sections.

SMOOTHING BOARD ATTACHMENT

Fig. 20—Smoothing-Board Attachment

Depth of penetration is controlled by adjusting the angle of teeth as they enter the ground and by raising or lowering skid shoes or runners under each section. Tooth angle is controlled by hand lever or hydraulically, depending on machine design and operator preference.

Spring-tooth harrows are sometimes used for renovating alfalfa sod, but are not as well-suited for this work as flexible-tine harrows and field cultivators, particularly in heavy soil.

Types And Sizes

Spring-tooth harrows are available in 3- to 6-foot (0.9 to 1.8 m) sections (Fig. 22) for use with drag-type eveners, integral drawbars, or wheeled harrow carts. When attached to drag-type eveners, teeth may be rotated clear of the ground and the harrow transported short distances on the skid shoes. Hard-surfaced roads should be avoided to prevent excessive skid wear.

Most sections have three toothbars with teeth spaced approximately 12 inches (305 mm) apart on each bar— a tooth working every 4 inches (102 mm). Alternate 6-inch (152 mm) working interval provides additional trash clearance.

Typical spring-teeth are 5/16-inch (8 mm) thick, 1¾ inches (44 mm) wide, and available in three general types (Fig. 23). Utility or seedbed teeth are recom-

Fig. 21—Spring-Tooth Harrow in Operation

SPRING-TOOTH HARROW

278

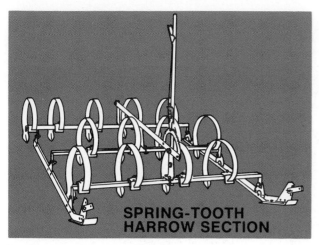

Fig. 22—Spring-Tooth Harrow Section

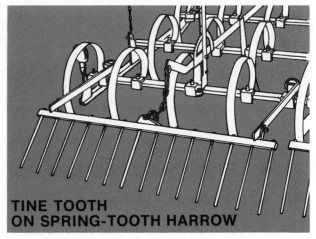

Fig. 24—Tine-Tooth Smoothing Attachment on Spring-Tooth Harrow

mended for most soils and general seedbed preparation. Sharp-pointed, weeding teeth are excellent for destroying quack grass, Johnson grass, and other persistent weeds. For seedbed preparation or weed control in highly abrasive soils, use reversible-point teeth for double wear and easy, economical replacement.

A tine-tooth smoothing attachment (Fig. 24) is available for some spring-tooth harrows to level and firm the soil prior to planting.

Spring-Tooth Planter Attachment

A spring-tooth harrow attachment on the corn planter will save time, fuel, and an extra trip over the field at planting time. It provides final seedbed preparation, last-minute weed control, and freshly worked soil for easier planting. These units usually have two toothbars and are attached directly to the front of the planter frame. A longer planter hitch may be necessary to per-

mit short turns, particularly if the tractor has dual wheels. Some are raised and lowered with the planter; others have separate hydraulic depth control.

If strip-tillage in the row area is desired, remove some of the spring-teeth and arrange two or three teeth directly in front of each row unit. Make sure the same number of teeth are used on each side of the planter and that spacing is balanced to prevent side draft.

Wheeled Spring-Tooth Harrow

Compared to regular spring-tooth harrow sections, more trash clearance and increased frame strength are key features of wheeled spring-tooth harrows (Fig. 25). Vertical clearance on typical units is 16 inches (406 mm) from tooth-point to underside of toothbar. A long hitch helps eliminate seesaw action in the field and permits short turns, even with dual tractor wheels. Either 4- or 6-inch (102 or 152 mm) tooth spacing may be used, depending on trash conditions and desired tillage results.

Fig 23—Utility or Seedbed Tooth, Weeding Tooth, and Reversible Point Tooth

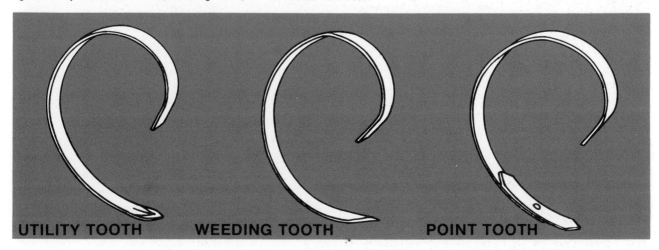

Fig. 25—Wheeled Spring-Tooth Harrow

Longer-than-normal spring-teeth are made of 1-3/4 × 3/8 inch (44 × 9.5 mm) high-carbon steel with forged points. Two holes in tooth points permit attachment of doublepoint shovels when original tooth points become worn (Fig. 26).

Widths range from 8 to 20 feet (2.4 to 6 m); larger models have manually or hydraulically folded wings to reduce transport width. Hydraulic folding permits the operator to change from field to transport or back and never leave the tractor seat (Fig. 27).

Depth control for wings is provided by a gauge shoe at the outer end of each unit. A flat wear-pad, which acts as a gauge shoe, is used to keep wings from digging too deep (Fig. 28) when operating against terraces or rolling ridges.

Wheeled spring-tooth harrows are not intended to replace field cultivators, but work very well for seedbed preparation in previously worked soil, light summer

fallow, or weed control. Tooth breakage and bending often occur when users operate spring-tooth harrows too fast or too deep, instead of employing field conditioners or field cultivators.

FIELD CONDITIONERS

Field conditioners are essentially heavy-duty wheeled spring-tooth harrows—built wider and stronger. These lightweight machines are designed to make seedbeds, break soil crust, kill grass and weeds, control volunteer grain, incorporate chemicals and fertilizer,

Fig. 27—Wings Fold Hydraulically

Fig. 26—Spring-Teeth Have Forged Points and Holes for Replacement Shovels

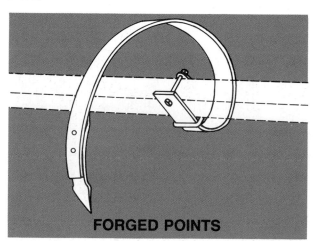

FORGED POINTS

280

and work air pockets out of the soil. Working depth of most units is down to 6 inches (152 mm), depending on soil conditions. Vertical clearance is about 20 inches (508 mm).

Normal tooth-working interval on field conditioners is 6 inches (152 mm), which may be increased to 9 or even 12 inches (229 or 305 mm) in extremely heavy trash, or decreased to 4 inches (102 mm) with some units for seedbed work. Most machines have a standard 2-bar frame, with an optional third bar available to provide wider tooth spacing on each bar and still work soil completely.

Several means are offered to provide increased trash clearance with a 2-bar frame and retain close tooth spacing. For instance, one machine has four rows of teeth on two bars with alternate long and short teeth on each bar. Another alternates long and short teeth on the rear bar only to provide three rows of teeth on two bars (Fig. 29).

Types And Sizes

Working width of field conditioners ranges from 12 to 40 feet (3.6 to 12 m). Machines more than 15 feet (4.5 m) wide usually have folding wings for flexibility in the field and reduced transport width. At 5 miles an hour (8 Km/h), a 40-foot (12 m) conditioner can cover as much as 20 acres an hour (8 ha/h)—fast tillage to permit timely completion of field operations. Hydraulic wing-folding attachments are available for some large sizes. Others are designed for easy manual folding when the implement is lifted to full height hydraulically.

Typical field-conditioner teeth (Fig. 30) include the sharp-pointed weeding tooth (left), recommended for tight, moist soils, which can usually be operated faster than general-purpose teeth; general-purpose, broad-point (or spade point) tooth (center) for average conditions; and heavy-duty tine-tooth adapted to rocky conditions and very light soil. The two teeth on the left have holes for attaching reversible double-point shovels when original teeth become worn. Teeth on some machines may be equipped with sweeps for very light soils, but their use is definitely not recommended in hard or heavy soil.

Other tooth shapes include "S" curves and full loops for greater vibrating action, more trash clearance, or to meet specific operating conditions.

A remote hydraulic cylinder controls raising and lowering of the conditioner. Precise working depth is varied on most machines by individual wheel adjustments (Fig. 31). Vertical hitch and tractor drawbar height adjustments permit leveling the machine fore-and-aft; the hydraulic cylinder is also used for this purpose.

Tine-tooth attachments are available for most field conditioners to help break clods, shake soil from weed roots, and level the surface for final seedbed preparation. Smoothing attachments are designed to flex over uneven ground, and have adjustments to angle tine teeth to control aggressiveness.

GAUGE SHOE

Fig. 28—Gauge Shoe Controls Depth

Fig. 29—Alternate Long and Short Teeth on Each Bar Provide Better Trash Clearance

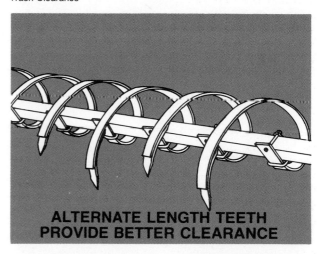

ALTERNATE LENGTH TEETH PROVIDE BETTER CLEARANCE

Fig. 30—Choose Teeth to Match Soil Conditions

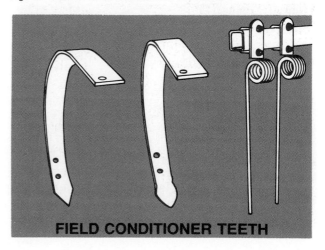

FIELD CONDITIONER TEETH

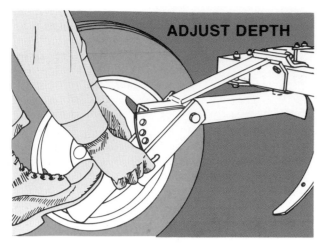

Fig. 31—Wheel Adjustments Set Working Depth

ROLLER HARROWS

Roller harrows (Fig. 32) are a specialized type of field-finishing implement which employ both heavy rollers and heavy-duty spring teeth. They are considered ideal for spring working of fall-plowed ground which has been compacted by snow and rain, or for working spring-plowed ground which has dried to form clods.

See Chapter 15 on Roller Harrows and Packers

Fig. 32—Roller Harrow with Two Rows of Serrated Rollers and Two Rows of Deep-Probing Spring Teeth

POWER HARROW

Utilizing tractor engine power through the PTO, rather than drive wheels, reduces or eliminates wheel slippage, reduces required tractor weight, and permits close control of the degree of tillage. The reciprocating action of the power harrow (Fig. 33) breaks clods and crusts without completely pulverizing soil when operated properly. Pointed tines work through the soil with no inversion and very little trash incorporation. Power harrows may overwork the soil unless properly used.

Moist lower layers of soil are not exposed, which means more moisture is retained for crop production. This also prevents exposure of large chunks of moist soil to rapid surface drying and formation of hard clods. By carefully controlling working depth, only drier surface soil is disturbed. Nor is there any slicing or sealing between layers of worked and unworked soil to restrict later water and root penetration.

Maximum working depth is approximately 8 inches and widths of 10 to 14 feet (3 to 4.3 m) are available. Wider units may be transported endways on a wheeled carrier.

Power requirements vary with soil type and condition, working depth, desired tillage results, and tractor speed. A 10-foot (3 m) harrow normally requires about 40 PTO horsepower (30 PTO kW)—a 14-foot (4.3 m) harrow, 55 horsepower (41 kW). However, adequate power must be available to provide ample forward speed at all times to prevent over-pulverization of the soil.

A rolling crumbler attachment can be fitted to the rear of the power harrow to smooth and firm the top 2 to 3 inches (51 to 76 mm) of soil when working in very loose seedbeds. The crumbler is made of lateral bars welded to circular spacers; it can be adjusted vertically to control the amount of pressure applied to the soil.

Scuffing boards (Fig. 34) can be used to help prepare extra-fine seedbeds for sugar beets and similar crops. These boards are clamped to the reciprocating harrow

Fig. 33—Reciprocating Tines on Power Harrow Shatter Soil

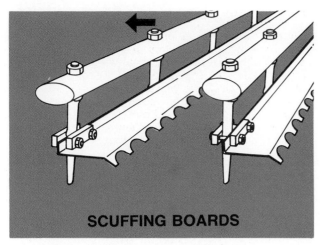

SCUFFING BOARDS

Fig. 34—Vertically Adjusted Scuffing Boards Provide Extra Pulverization and Depth Control

tines and help control depth as well as providing additional pulverization. One board may be set vertically (Fig. 35) as a leveling board, and the second horizontally to permit shallow operation in very rough soil.

Both 2- and 4-bar power harrows are available to provide different degrees of tillage. The 2-bar machine is driven from the PTO through a flywheel and eccentric yoke, with harrow-bar pivot arms free-floating to prolong bearing life. Crank arms on 4-bar harrows are designed to move the front bar about 4 inches (102 mm) side-to-side at a relatively low speed to start the soil-breaking process. The second bar, moving in the opposite direction, shifts 9 inches (229 mm) at a faster speed. The third bar is attached to the ends of crank arms carrying the front bar and has a 14-inch (356 mm) movement. The fourth bar is attached to the same arms as the second bar and has a high-speed throw of 20 inches (508 mm). Alternate bars always work in opposite directions to balance load, provide better pulverization, and reduce trash plugging.

Fig. 35—Setting One Board Vertically Levels Rough Soil

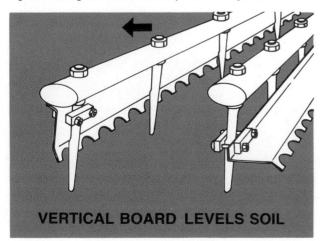

VERTICAL BOARD LEVELS SOIL

The implement produces progressively finer soil particles as it moves forward. Faster forward speeds reduce the extent of pulverization.

A PTO slip clutch is provided to prevent damage from soil obstructions.

SPIRAL CUTTER BLADES

Some field-finishing tooth-type harrows intended for knocking down old beds are made more effective by use of spiral cutter blades which work ahead of the teeth (Fig. 36) and provide an aggressive mixing, stirring, clod-breaking action.

Sharpened cutter blades are attached to supporting reels on each one-row unit (Fig. 37). Axles run in triple-sealed, flange-type bearings to resist abrasion. These implements are made in 4-, 6-, and 8-row sizes for 38- to 40-inch (965 to 1016 mm) rows.

SAVING FUEL AND TIME

Tooth harrows and light cultivators are all relatively light-draft implements, even in the larger sizes, and will seldom require full power of large tractors needed for other farm operations. This provides opportunities to conserve fuel and manpower and reduce repair bills—for instance, by pulling two implements in tandem to save a field trip.

TRACTOR AND IMPLEMENT PREPARATION

Here are suggestions for tractor and implement preparation:

1. Tune engine for optimum fuel consumption.

2. For most integral implements, set 3-point hitch sway blocks or chains for lateral flexibility in operation and to lock out side-sway in transport. For power-harrow operation, eliminate linkage side-sway in operation and transport.

3. Adjust hitch lift links to provide level operation of integral equipment. Adjust top-link length to level machine fore-and-aft.

4. Provide adequate front-end weight, particularly with integral machines, for front end stability in transport and operation (Fig. 38). Use maximum allowable front-end weight if working in lower gears.

5. With drawn machines, adjust drawbar height to level recommended in implement manual. Always pin drawbar for transport, but allow it to swing in operation with some implements (see manual). If implement tends to seesaw, pin the drawbar.

6. Set tractor wheel tread equidistant from tractor centerline, although tread measurement is generally not critical unless wheels interfere with implement in turning or extend beyond implement width. Use widest possible tread when working on steep slopes.

Fig. 36—Heavy-Duty Finishing Harrow with Cutter Blades, Teeth, and Smoothing Board

7. Inflate tires to recommended pressure. To reduce soil compaction, use only enough ballast to prevent excessive slippage. See tractor and implement manuals.

IMPLEMENT MAINTENANCE

Timely servicing and prompt repair or replacement of worn or damaged parts prolongs implement life and often prevents small problems from causing major repair bills.

Fig. 37—Cutter Blades

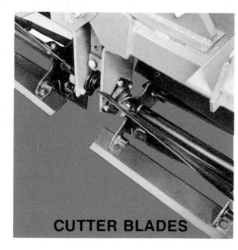

CUTTER BLADES

Before Each Season:

1. Inspect entire machine for worn, broken, or missing parts, and replace as necessary.

2. Check for loose bolts, nuts, and cotter pins.

3. Clean, repack, and tighten wheel bearings.

4. Lubricate as recommended; don't overlook wing hinges and wheel frames (see the manual).

5. Inflate tires to recommended pressure.

6. Reverse badly worn shovels or replace dull teeth or points.

Daily Before Operation:

1. Visually check for loose bolts and nuts, missing or damaged parts, and under-inflated tires.

2. Lubricate as recommended.

3. Replace or reverse worn teeth.

Before Storage At End Of Season:

1. Clean all trash and dirt from machine to prevent collection of rust-causing moisture.

2. Repaint spots where paint is scratched or worn away.

3. Cover sweeps, reversible shovels, spring-teeth, etc., with grease or plow-bottom paint to prevent rusting.

4. Lubricate machine as recommended.

Fig. 38—Provide Adequate Front-End Weight

5. Store inside if possible and remove weight from tires.

6. Protect soil-engaging parts from sinking into the ground and rusting during storage.

7. Lower implements or install transport locks and relieve hydraulic pressure in cylinders; fully retract cylinders to protect cylinder rods from rusting.

Fig. 39—Harrows May be Operated in Almost Any Desired Pattern

SPRING-TOOTH HARROW

FIELD OPERATION

Tooth harrows and light cultivators are quite simple to operate. In freshly plowed soil, spring-tooth and flexible-tine units should follow the direction of plowing to prevent dragging chunks of sod to the surface. Otherwise, these machines can be operated in nearly any desired pattern (Fig. 39). Following contours and working across slopes helps reduce water runoff and soil erosion, and working perpendicular to prevailing winds reduces soil blowing.

Operating the long way of a field, instead of making short passes across the field, is more efficient because fewer turns are required and more time is spent in productive work.

Working diagonally across passes made by other tillage tools—plows and disks, for instance—provides better leveling and more complete working of the soil.

Faster operating speed with these tools means more tooth vibration, better tilling, more weed pulling, and better trash movement. This is particularly true of flexible-tine cultivators which function best at speeds of 6 to 7 miles an hour (10 to 11 Km/h). If implements tend to bounce in some soil conditions, speed should be reduced. Excessive speed also may cause needless tooth failure.

Remember: Increasing speed usually gets more work done in a given time. But it also increases draft, wear-and-tear on equipment, and the potential danger of accidents. Increasing speed beyond recommended limits for a given machine usually results in poor performance, rapid equipment deterioration, and unsafe operating conditions.

OPERATING TIPS—SECTION-TYPE HARROWS

Chain-type flexible harrows are generally limited to operation with drag-type eveners. Most spike-tooth,

CARTS EASY TO TRANSPORT AND MANEUVER

Fig. 40—Wheeled Harrow Carts are Easy to Transport and Maneuver

tine-tooth, and spring-tooth harrow sections can be operated with drag-type eveners, integral drawbars, or wheeled harrow carts (Fig. 40). For optimum harrow performance, follow these operating tips as well as instructions in the operator's manual.

1. Securely attach all sections to evener or drawbar.

2. Obtain desired tillage results by controlling working depth of teeth.

3. Set teeth perpendicular for maximum aggressiveness—slanted back for better trash shedding and improved leveling.

4. Adjust tooth angle of all sections exactly the same for uniform penetration and stable implement operation.

5. Setting teeth to run deeper than they are designed to work causes unstable harrow operation, uneven performance, and excessive wear and tooth failure.

6. Avoid extremely sharp turns in which evener draft rods may catch on tires. If rods catch on lugs, harrow could be pulled onto the tractor very quickly causing serious equipment damage and possible operator injury.

7. Match tractor wheel tread, center-to-center, with harrow cart wheels—or adjust tread so cart wheels cannot run partly in and partly out of tractor tracks.

OPERATING TIPS—INTEGRAL HARROWS AND FLEXIBLE-TINE CULTIVATORS

1. Level machines laterally by adjusting hitch lift links.

2. Level fore-and-aft by adjusting top-link length so all teeth penetrate equally.

3. Provide adequate front-end weight for stability.

4. Install hitch-pin adapters before using quick coupler.

5. Lower integral drawbar enough for harrow sections to float free of drawbar. Never allow pull chains to pull up or down on front of sections—chains should pull level.

6. Adjust all gauge wheels to exactly the same depth.

7. Allow gauge wheels, skid shoes, or tooth angle to control working depth. Do not use tractor load-and-depth or draft control.

8. Raise machine from ground before backing to prevent tooth damage.

9. Reduce speed when turning or crossing rough areas.

OPERATING TIPS—WHEELED SPRING-TOOTH HARROWS AND FIELD CONDITIONERS

1. Use same number of equally spaced teeth on each side of implement to balance draft load and assure straight, stable operation. See manual for recommended tooth patterns.

2. Level machine fore-and-aft for uniform soil penetration.

3. To reduce ridging, or if machine seesaws, run rear bar slightly lower than front bar.

4. Set all gauge or transport wheels to operate at the same level for uniform penetration across the implement.

5. Save time and reduce fuel consumption by not overworking soil. Limiting soil manipulation helps preserve structure and reduces tendency for soil to crust.

6. Promptly reverse or replace dull points for better penetration, reduced draft, and improved weed control.

7. Do not operate at excessive speed—bouncing and/ or damage may result.

OPERATING TIPS—POWER HARROWS

1. Level harrow laterally to provide uniform penetration.

2. Adjust top link to level fore-and-aft.

3. If clods build up in front of first bar, lengthen top link slightly.

4. Because PTO shaft is very short, do not run PTO with shaft angled more than 30 degrees.

5. Always start PTO before lowering harrow to working position. Never engage PTO when harrow tines are in the soil.

6. Do not overwork the soil.

TRANSPORT AND SAFETY

It is the operator's responsibility to know and abide by local and state regulations when transporting equipment on a road or highway. Every operator must also follow the fundamental rules of safety for operation and transport of equipment.

1. Provide sufficient front-end weight for safe, stable operation and transport.

2. Use widest practical wheel tread to improve tractor stability, especially when working on steep slopes.

3. Reduce implement to narrowest possible width for transport on roads or highways.

4. Lock wings or folded sections securely in place before transporting equipment (Fig. 41).

5. Use lights, reflectors, and SMV emblem as required by law when transporting equipment—day or night.

6. Schedule moves for least hazardous periods; avoid transporting equipment on busy roads, during peak traffic periods, or after dark.

7. Never transport wheeled harrows, harrow carts, or field conditioners at more than tractor transport speeds; transport considerably slower on rough or uneven terrain.

Fig. 41—Lock Folded Sections Securely for Transport

8. Never make sharp turns at high speeds.

9. Never allow anyone to ride on the tractor but the operator.

10. Never allow anyone to ride on the tractor drawbar or implement in operation or transport.

11. Always stop implement and tractor engine to adjust, repair, or lubricate.

12. Lower implements to the ground before stopping tractor engine, before servicing or repairing equipment, or at any time the machines are left unattended.

13. Never make extremely short turns with drag-type eveners which could foul tractor tires.

14. Never park implements where they could be hidden by tall or growing crops, grass, or weeds.

15. Keep PTO shaft properly shielded. Never get off tractor without disengaging PTO and stopping engine.

TROUBLESHOOTING

Immediate recognition and correction of equipment malfunctions will increase equipment productivity, save time, and often reduce repair costs. Where possible remedies to problems are obvious, a blank space is left in the right-hand column.

TROUBLESHOOTING CHART		
PROBLEM	**POSSIBLE CAUSE**	**POSSIBLE REMEDY**
ONE END OF HARROW DRAGS BACK	Trash plugging on one end.	Check for improper tooth arrangement.
	Gauge wheels not set alike.	
	Teeth improperly set.	

PROBLEM	POSSIBLE CAUSE	POSSIBLE REMEDY
MACHINE SEESAWS IN OPERATION	Front teeth working deeper than rear. Tractor drawbar swinging.	Level fore-and-aft, or let rear bar run slightly deeper.
SEVERE TRASH PLUGGING	Trash and soil too wet. Too much trash present. Working too deep, dragging trash out of plowed ground. Spike or tine teeth set too steep.	Work very shallow on first pass to speed drying. Shred or disk trash rather than harrowing.
MACHINE BOUNCES IN OPERATION	Front teeth working deeper than rear. General-purpose teeth not suitable. Working too fast. Soil too hard.	Level fore-and-aft, or with rear slightly lower. Use pointed teeth.
INSUFFICIENT SOIL PULVERIZATION	Soil too wet. Insufficient working speed. Soil too hard.	
INADEQUATE PENETRATION	Teeth or points worn or broken. Speed too slow. Soil too hard.	

SUMMARY

Despite their wide variety, the tooth-type harrows discussed in this chapter have several things in common.

• Their main use is preparation of germination-hastening, seedling-nurturing seedbeds atop the rootbeds created by primary tillage.

• They do this by breaking soil crust and clods, smoothing and firming the soil, and closing air pockets, all with a minimum of moisture loss from evaporation.

• Their other uses, depending on type and circumstances, include killing small weeds, covering broadcast seeds, summer fallowing, renovating pastures, thinning sugar beet seedlings, fighting wind erosion, and pulling up quack grass roots.

• Most modify the top layer of soil without mix-down of trash.

• All employ rigid, flexible, or vibrating teeth, which usually are adjustable for field conditions and desired results.

• All have much lighter draft than primary-tillage tools, and so can be used in conjunction with other implements, such as plows and planters.

• All are sophisticated and specialized descendants of tree branches dragged by primitive farmers in the dawn of agriculture to smooth seedbeds and cover seeds.

CHAPTER QUIZ

1. What are the primary functions of tooth harrows and light cultivators?

2. How is spike- and spring-tooth aggressiveness controlled?

3.   (Fill in blanks) Spike teeth may be _____ or _____ to the frame bars.

4.   (True or false) Vibratory tooth action of tine-tooth harrows helps break clods and pull small weeds.

5.   Flexible-tine harrows perform best at (choose one) high or low speed.

6.   (Fill in blanks) Spring-tooth harrows are normally operated from _____ to _____ inches deep.

7.   (True or false) Spring-tooth harrows can penetrate hard ground and kill weeds better than a disk harrow or field cultivator.

8.   What are the major advantages of wheeled spring-tooth harrows over regular spring-tooth harrow sections?

9.   (Fill in blanks) Compared to wheeled spring-tooth harrows, field conditioners are built _____ and _____.

10. How can trash clearance of field conditioners be increased?

11. Utilizing tractor power through the PTO with a power harrow has what effect on wheel slippage and tractor ballast?

12. (True or false) A power harrow exposes moist lower soil for rapid drying on the surface.

17
Field Cultivators

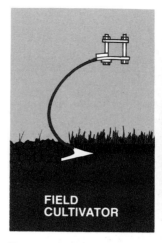

Fig. 1—Field Cultivators are Intended for Less-Severe Conditions

INTRODUCTION

Field cultivators are widely used across North America for seedbed preparation, weed control, stubble-mulch tillage, summer fallow, and roughing fields to increase moisture absorption and control wind and water erosion.

The field cultivator does not compact soil as a disk harrow does, and when equipped with wide sweeps is much better for controlling persistent weeds such as quack grass, Johnson grass, Canadian thistle, and wild morning-glory.

Fig. 2—Integral Field Cultivators are Very Maneuverable

INTEGRAL FIELD CULTIVATOR

Field cultivators and chisel plows have similar appearance and operating characteristics, and in some areas chisel plows are known as field cultivators. But the field cultivators described in this chapter are lighter than most chisel plows and are intended for less-severe operating conditions (Fig. 1).

When soil permits, field cultivators occasionally replace chisel and moldboard plows for primary tillage as well as secondary tillage. Usually, however, they are intended for secondary tillage of previously worked soil, such as seedbed preparation in fall- or spring-plowed fields, use after stalks or stubble have been chisel-plowed or disked, summer fallowing after use of a disk tiller, chisel plow, or wide-sweep plow, and similar operations.

Field cultivators leave most of the residue on top or mixed into the upper few inches of soil. When using spike points the surface is usually left ridged, rough, and open so moisture infiltration is increased and soil blowing and runoff are reduced.

Field cultivators may be used to renovate pastures and meadows, and some can be equipped with unit planters for simultaneous tilling and planting.

TRASH CLEARANCE

Trash clearance of field cultivators is generally much better than that of spring-tooth harrows, but somewhat less than chisel plows because of the normally narrower tooth spacing—6, 9, or 10 inches (152, 229, 254 mm) in most cases, compared to 12 inches (305 mm) or more on chisel plows. Vertical clearance is usually 19 to 22 inches (483 to 559 mm) compared to 26 to 32 inches (660 to 813 mm) for chisel plows.

Working depth depends on soil conditions, desired tillage results, and soil-engaging tools used. Double-pointed shovels and similar narrow points may be used down to 5 inches (127 mm) or more in lighter soil with relatively little residue. When sweeps are used for weed control or seedbed preparation, depth is usually about 4 inches (102 mm). If working freshly plowed soil, particularly sod, depth must be limited to prevent dragging chunks of sod or trash to the surface.

TYPES AND SIZES

Integral field cultivators (Fig. 2) range from 7½ to 24½ feet (2.3 to 7.5 m) wide. Those wider than 15 feet (4.5 m) are equipped with folding wings to reduce transport width. Drawn models (Fig. 3) are available from about 8½ to 60 feet (2.6 to 18 m) or more to match tractor power and acreage. Various sizes may be assembled by adding stub bars, extensions, or folding wings to basic integral or drawn center sections. Extremely wide cultivators (Fig. 3) may have two wing sections on each side to provide maximum flexibility and reduce overall height of folded wings.

Integral cultivators may be attached directly to the tractor 3-point hitch, or to an implement quick-coupler for pick-up-and-go operation.

DRAWN FIELD CULTIVATOR

Fig. 3—Drawn Field Cultivators Match Tractor Power

Most current field cultivators have three shank bars and 18-inch (457 mm) lateral spacing between shanks for 6-inch (15 cm) working interval. The third bar is optional on some units and shanks are spaced 12 inches (305 mm) for the 6-inch (152 mm) interval. This reduces cost and implement weight, and is satisfactory so long as trash is not a problem. Shanks on a 2-bar machine may be respaced at 9-, 10-, or 12-inch (229, 254, or 305 mm) working intervals to increase trash clearance without adding the third bar.

Increasing working interval to 9 or 12 inches (229 or 305 mm) on 3-bar cultivators provides 27- or 36-inch (686 or 914 mm) lateral spacing between shanks for maximum trash flow and minimum plugging. Fore-and-aft bar spacing ranges from 20 to 32 inches (508 to 813 mm) on different machines, which means some are better adapted to handling heavy trash.

TRANSPORT WHEELS

Transport wheels on field cultivators may be located inside the main frame (Fig. 4) or in front of the frame. Each method offers advantages. All shanks operate at

Fig. 4—Transport Wheels may be Located Inside the Frame above or in front of the Front Bar

TRANSPORT WHEEL INSIDE FRAME

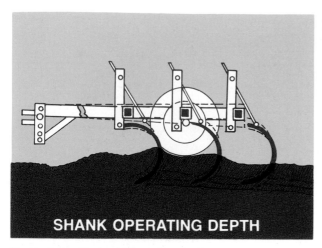

SHANK OPERATING DEPTH

Fig. 5—Center-Mounted Wheels Provide More Uniform Operation—
Forward-Mounted Wheels Avoid Shank Interference

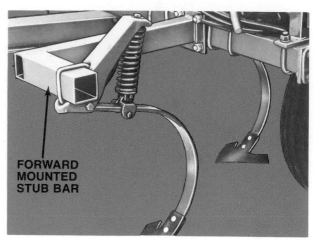

FORWARD
MOUNTED
STUB BAR

Fig. 6—Forward-Mounted Stub Bar Moves Shanks Away from Wheels

more uniform depth over uneven ground if wheels are inside the frame (Fig. 5). Locating wheels ahead of the frame prevent wheel interference with any desired shank pattern or sweep size. The choice depends on operating conditions and operator preference. If wheels interfere with trash flow, sweep usage, or shank pattern, a forward-mounted stub bar (Fig. 6) may be used on some machines to position the shank away from tires or areas where plugging might occur.

Offset tandem wheels mounted on opposite sides of a

walking beam are optional for the center section of some field cultivators (Fig. 7). These wheels help maintain uniform cultivator operation, even if one wheel of each pair drops into a ditch or passes over a ridge. Either wheel can rise or drop with little effect on machine operation.

FOLDING WINGS

Field-cultivator wings may be folded hydraulically for transport, mechanically with a winch and cable, or

Fig. 7—Offset Tandem Wheels under Center Section Help Maintain Uniform Penetration

OFFSET TANDEM WHEELS

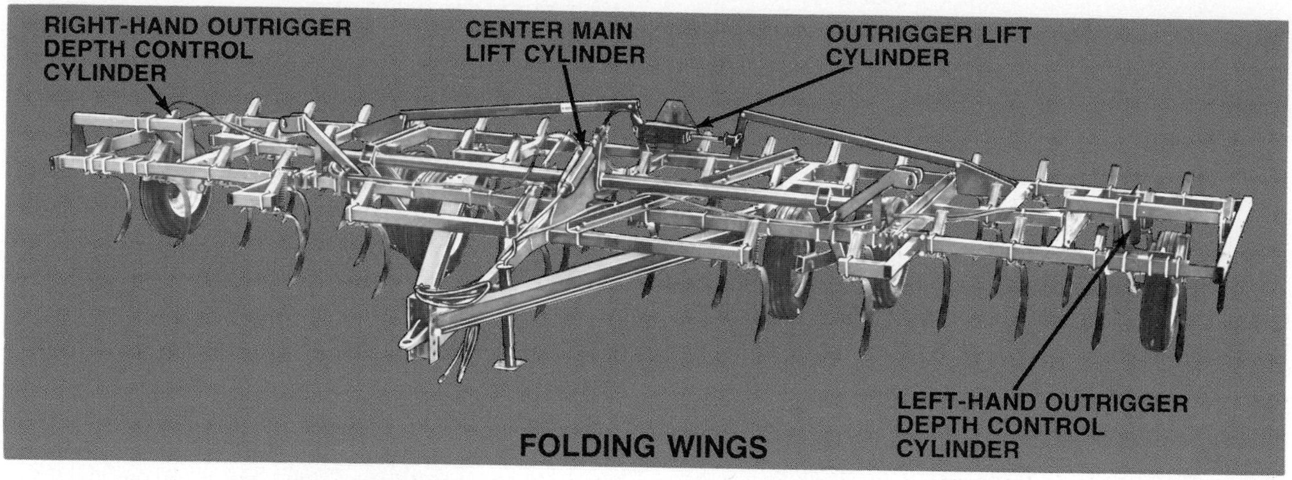

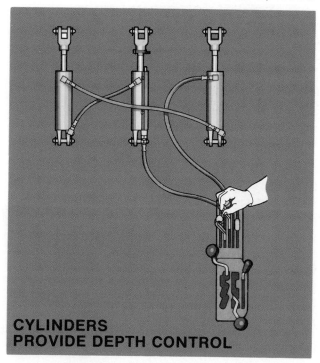

RIGHT-HAND OUTRIGGER DEPTH CONTROL CYLINDER

CENTER MAIN LIFT CYLINDER

OUTRIGGER LIFT CYLINDER

LEFT-HAND OUTRIGGER DEPTH CONTROL CYLINDER

FOLDING WINGS

Fig. 8—Hydraulic Power Folds Wings Easily

manually, depending on machine design and size. Some small cultivators and those with specially angled hinges may be folded manually with little effort. Larger wings are usually folded hydraulically (Fig. 8) by one or two cylinders, again depending on design and size. Wings must be securely locked in folded position before transporting or storage to protect equipment and people from possible damage or injury. Restrictors must usually be placed in hydraulic lines to control the rate of wing raising and lowering within safe limits.

Fig. 9—Three Remote Cylinders in Series Can Provide Depth Control

CYLINDERS PROVIDE DEPTH CONTROL

Wide flexible wings with transport wheels follow ground contours and provide more uniform tillage of uneven ground, compared to rigid-frame models. Hydraulic depth control may be provided by a single remote cylinder and special linkage connecting center and wing wheel rockshafts, or by three hydraulic cylinders in series (Fig. 9). A single cylinder is used on rigid-frame cultivators and those wide machines on which wings are folded manually.

SHANKS

Spring-cushion shank mounting (Fig. 10) permits point to lift from 8 to 11 inches (203 to 279 mm), if they strike obstructions. Spring-cushioning also provides additional shank vibrating action for better soil pulverization, particularly in hard soil. Some spring-cushion mountings permit increasing shank rigidity for working extremely hard soil. Others permit adjusting shank pitch for normal, hard, and toughest soil conditions.

Spring-cushion shanks are standard equipment on some cultivators and optional on others, which may use fixed-clamp shanks (Fig. 11) for economy in soils with few obstructions. These shanks depend on flexing of the shank for protection from obstacles.

Another shank-cushion design uses elastomer pads between shank bar and mounting clamp (Fig. 12) to protect equipment and provide increased shank vibration. There is no metal-to-metal contact between clamp and bar.

SOIL-ENGAGING TOOLS

A wide variety of soil-engaging tools is available for field cultivators. Double-point shovels, which can be reversed for longer wear, are most common in many areas. Sweeps from 4½ to 12 inches (114 to 305 mm) wide are favored for other conditions. For economy, on some machines optional shank ends are forged into a chisel-type

295

SPRING-CUSHION SHANKS PROTECT EQUIPMENT

Fig. 10—Spring-Cushion Shanks Protect Equipment from Soil Obstructions

Fig. 11—Fixed-Clamp Shanks for Economy and Obstruction-Free Soil

Fig. 12—Elastomer Cushions Are Used for Protection on Some Clamps

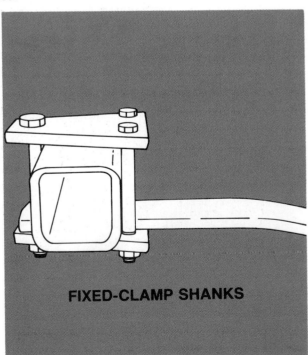

FIXED-CLAMP SHANKS

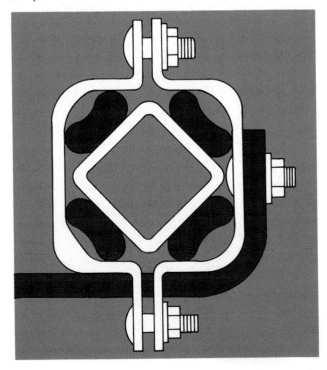

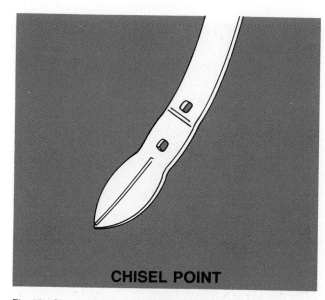

CHISEL POINT

Fig. 13—Shank End Forged Into Chisel Point

Fig. 14—Typical Soil-Engaging Tools

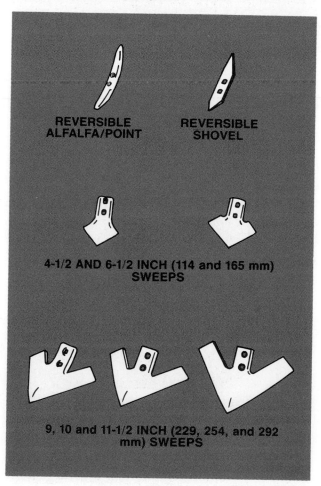

REVERSIBLE
ALFALFA/POINT

REVERSIBLE
SHOVEL

4-1/2 AND 6-1/2 INCH (114 and 165 mm)
SWEEPS

9, 10 and 11-1/2 INCH (229, 254, and 292
mm) SWEEPS

point (Fig. 13). When the original point wears down, reversible points may be attached for continued use.

Typical soil-engaging tools (Fig. 14) include:

● *Narrow, reversible alfalfa point for renovation of alfalfa stands.*

● *Double-point, reversible shovel (1¾ × 10¼-inch; 44 × 260 mm, shown), recommended for general tillage, hard ground, fall tillage or deep penetration.*

● *4½- and 6½-inch (114 and 165 mm) sweeps for seedbed preparation and general tillage.*

● *9-, 10-, and 11½-inch (229, 254, and 292 mm) sweeps for killing persistent weeds and for fine seedbed preparation; sweeps should overlap for complete cutting of weed roots.*

ATTACHMENTS

Field-cultivator versatility is increased by adding various attachments, such as pull-behind implement hitches for section-type harrows or corn planters; packer wheels for firming soil in the row area ahead of a planter hitched to the cultivator, and a hitch jack. Tine-tooth smoothing attachments (Fig. 15) are available for many cultivators, as are support stands to steady the parked cultivator if equipped with a smoothing harrow. There are also alternate hitches to match tractor-drawbar type and height, and gauge wheels for integral and drawn cultivators where wings are not carried on transport wheels.

Fig. 15—Tine-Tooth Harrow Attachment Levels and Firms Soil

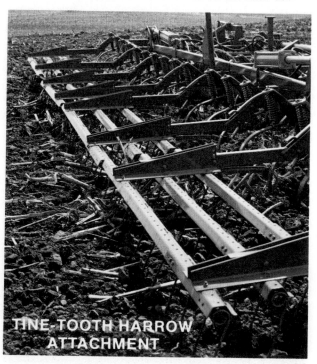

TINE-TOOTH HARROW
ATTACHMENT

FIELD CULTIVATOR AND PLANTER AS A UNIT

Fig. 16—Unit Planters on Field Cultivator for Once-Over Operation

Fig. 17—Lift-Assist Wheel Helps Support Planter Weight

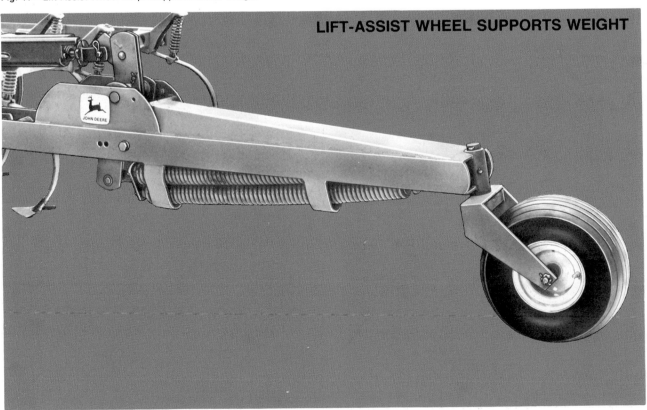

LIFT-ASSIST WHEEL SUPPORTS WEIGHT

298

On some integral field cultivators, four, six or eight unit planters may be attached for once-over seedbed preparation and planting (Fig. 16). If tractor hydraulic-lift capacity is adequate, or a lift-assist wheel is added, granular herbicide and insecticide attachments may also be included to eliminate additional trips over the field. Spray tanks may be mounted on the tractor and a boom used in front of or behind the field cultivator, depending on the chemical used. Markers permit accurate control of row spacing.

Lift-assist wheels (Fig. 17) are hydraulically and automatically raised and lowered by directly connecting the single-acting remote lift cylinder to the 3-point hitch hydraulic circuit. One or two lift-assist wheels improve machine stability for turning and transport and provide easier lifting.

COMBINATION MACHINES

With the growing acceptance of conservation tillage where much of the residue from previous crops is left on the surface, many combination machines have been developed to work in heavy residue. Among these are field cultivators with disk gangs or coulters across the front. These units appear very similar to stubble mulch tillers (See Chapter 10). However, they are built for secondary tillage and seedbed preparation. Frames, shanks, and other parts are much lighter than stubble mulch tillers. These field cultivators, with disks or coulters, are not intended for primary tillage. Their operation and adjustment is very similar to regular field cultivators, except disks or coulters must be set as specified in the operator's manual for each machine.

PRINCIPLES OF FIELD CULTIVATOR OPERATION

Field-cultivator design, function, and performance are similar to those of chisel plows. (See Principles of Chisel Plow Operation, Chapter 7). Both can penetrate hard soil, break up large clods, and leave the surface open and broken to absorb moisture and resist erosion. The field cultivator is an excellent tool for seedbed preparation (Fig. 18) when equipped with sweeps or smoothing-harrow attachment and working light or previously-tilled soil.

Varying shank spacing from the usual 6 inches (152 cm) to 9, 10, or even 12 inches (229, 254 or even 305 mm) permits using field cultivators in heavy-trash or hard-soil conditions normally limited to chisel-plow operation. In extremely hard conditions, some shanks can be removed from each end of the cultivator to reduce draft load and still maintain desired operating depth.

If shank spacing is altered, or shanks are removed from the frame, the same number of equally-spaced shanks must be retained on each side of the centerline to provide uniform load and prevent side-draft.

HITCHING

Nearly all integral field cultivators can be attached to the tractor with or without an implement quick-coupler. Quick-coupler use (Fig. 19) makes good sense for tractor power does the work. The coupler can compensate for minor deviations in alignment.

Fig. 18—Field Cultivators Prepare Seedbeds

Fig. 19—Quick-Coupler for Integral Cultivators

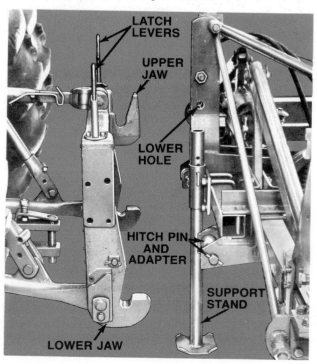

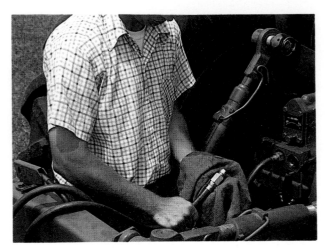
Fig. 20—Examine Couplings for Leaks or Damage

If 3-point-hitch lift capacity is marginal for the cultivator, quick-coupler use is not recommended. The quick-coupler places the center of implement weight a few inches (centimeters) farther behind the tractor, making the cultivator a little harder to lift and control. If quick-coupler convenience is important, cultivator size should be reduced to provide satisfactory operation and stability.

Tractor load-and-depth or draft control normally regulates operating depth of field cultivators and provides some weight transfer to the tractor for improved traction. In some soil conditions it may be desirable to use gauge wheels to limit cultivator penetration without frequent changes in 3-point-hitch setting.

Never depend on gauge or stabilizer wheels fully to regulate working depth of integral cultivators. Most of the control must be supplied by the tractor 3-point hitch. Properly adjusted wheels should carry little weight when cultivating at the desired depth in heavy soil when load is highest. Gauge wheels pick up more load in lighter soil when draft is reduced and normal control-system response calls for deeper penetration to maintain a uniform load. Gauge wheels must be set equally to assure uniform operation.

Drawn field cultivators are attached to the tractor drawbar, and are normally designed to permit vertically adjusting the hitch point to level the cultivator fore-and-aft. Means also are generally provided for lateral leveling to assure uniform penetration.

If a cultivator does not operate level from side-to-side, first be sure all tires are of equal size and equally inflated as recommended. Consult the manual for additional leveling instructions if required.

TRACTOR PREPARATION

Draft per foot (meter) of width of field cultivators is generally less than disk harrows or chisel plows. However, the total draft and weight of large integral and drawn cultivators can require full power output from even large tractors. Therefore, the tractor must be tuned for maximum power output and optimum fuel consumption for economical operation.

Tires must be inflated to recommended levels and adequate ballast added to prevent excessive wheel slippage. Front-end weight must be adequate to provide tractor stability in operation and transport, particularly with integral cultivators or when the tractor is operated in the lower gears.

The tractor hydraulic system must function properly to provide satisfactory depth control and wing folding. Check reservoir for correct oil level after filling new cylinders. Examine cylinders, hoses, couplings, and lines for leaks or damage and repair as needed (Fig. 20). Adjust cylinder depth stops equally for uniform penetration and install restrictors if required to control speed of wing raising and lowering.

PINNING DRAWBAR

Adjust tractor drawbar to height recommended in the cultivator manual. Allowing the drawbar to swing will aid in turning with wide cultivators, but pinning the drawbar in the center provides more stable operation, particularly in hard soil, or if the machine tends to see-saw. Always pin the drawbar when transporting equipment.

Adjust tractor tread to keep tires off worked soil when using small cultivators (not likely a problem unless using dual wheels). If using planter attachments on integral cultivators, center the wheels between the rows. Adjust 3-point-hitch sway blocks or stabilizers to permit lateral flexibility in operation, but lock out sway for transport of integral cultivators.

FIELD CULTIVATOR PREPARATION AND MAINTENANCE

"Off-season" equipment preparation means implements are ready to go when fields are ready to work. Needed repair parts may be ordered and installed when there's more time for careful attention to details and necessary servicing.

Careful preparation and servicing can reduce later downtime by avoiding some equipment failures in the field as well as by reducing draft, improving performance, and cutting time and fuel consumption.

Before Each Season:

1. Check for loose bolts and nuts and broken, worn, or missing parts and tighten or replace as needed.

2. Check soil-engaging tools for excessive wear or damage; reverse or replace as needed.

3. Lubricate as recommended in manual.

4. Clean, repack, and tighten wheel bearings.

5. Check tires for proper inflation and uniform size.

6. Check lift and wing-folding linkage and cylinders for proper alignment and operation.

Daily Before Operation:

1. Lubricate as recommended in manual.

2. Visually check for tire inflation and loose, broken, or missing bolts and parts.

Before Storage At End Of Season:

1. Clean trash, soil, and dirty grease from cultivator to prevent rusting.

2. Repaint spots where paint is scratched or worn.

3. Coat soil-engaging tools with heavy grease or plow-bottom paint to prevent rust.

4. Lubricate entire machine as recommended in manual.

5. Store inside if possible; remove weight from tires.

6. Install safety lock and relieve pressure in hydraulic cylinders, or lower machine for storage; fully retract cylinders to protect cylinder rods from rust.

7. Place boards under soil-engaging points, if lowered, to prevent settling into the ground and rusting.

FIELD OPERATION

If all soil is evenly tilled, no particular field patterns need to be followed for field cultivators. Working on contours or across slopes helps reduce water runoff and erosion. Working at right angles to prevailing winds helps reduce soil blowing. Operating diagonally or at right angles to previous tillage operations provides better leveling and more uniform tillage.

If tillage and planting are done at the same time with either unit-type or trailing planters, follow a normal planting pattern. Adjustable markers attached to the cultivator frame will help maintain proper row-spacing.

Observe the following tips, plus detailed instructions in the cultivator operating manual, for easier operation and better performance.

● *Level the cultivator fore-and-aft (a) when changing working depth, (b) when first starting to work a field, and (c) when changing tractors or drawbar height.*

● *Work at the speed (usually 5 to 7 mph; 8 to 11 Km/h) which produces the best tillage results (Fig. 21).*

● *Be sure wing-folding cylinders are properly operated to permit wing flexing over uneven ground.*

● *If master and wing-lift cylinders operate in series, keep cylinders synchronized by cycling the unit several times when first starting and holding the valve open for 10 to 15 seconds each time as the cylinders are extended.*

● *Follow manual instructions for leveling wing sections with the center frame.*

● *Before raising wings, prevent possible equipment damage by making certain that lock-up pins are in the storage position, not in the "wing locked" position (Fig. 22).*

● *Use recommended restrictors in hydraulic lines to wing-folding cylinders.*

● *Never try to move a wide cultivator, with wings folded, under low telephone or electric power lines. If cultivator catches power lines NEVER attempt to disengage it or touch cultivator or tractor from the ground. If dismounting from the tractor, jump clear to avoid electrical shock. Then call power company for assistance.*

● *Match soil-engaging tools to working depth and desired results—sweeps for shallow tillage and weed control, points for deep penetration and a rough surface.*

● *Always raise the cultivator when making sharp turns to avoid twisting shanks.*

Fig. 21—Work at Speed that Produces Best Results

SPEED A FACTOR IN FIELD CULTIVATOR EFFICIENCY

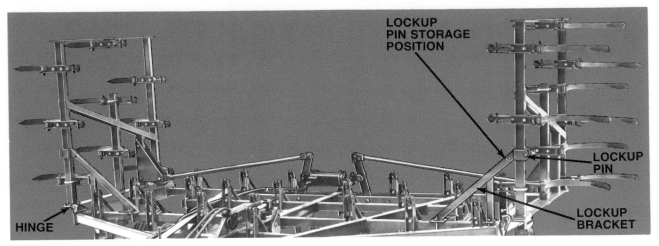

Fig. 22—Never Raise Wings if Pins Are in "Lock" Position

• *Use smoothing harrow to provide fine-finished seedbed.*

TRANSPORT AND SAFETY

Safety features and simple adjustments are built into field cultivators and tractors wherever possible to encourage safe, proper use. Most accidents are caused by failure to follow fundamental rules of safety.

1. Never exceed recommended transport speed for the cultivator used; if speed is not stated, do not exceed maximum tractor speed.

2. Reduce speed for turning and travel over rough or uneven ground.

3. Use transport locks and relieve pressure in cylinders when transporting field cultivators. Do not depend on hydraulic pressure to carry the weight. Always lock wings in transport position and relieve pressure in cylinders.

4. Never walk or work under wings when they are in the folded position.

5. Follow state and local regulations regarding lights, reflectors, SMV emblem, and maximum width when transporting on roads or highways.

6. Transport width of most field cultivators exceeds normal vehicle width. Therefore, use extreme caution when meeting other vehicles and avoid the possibility of dropping tractor or implement wheels into holes, drains, or ditches along the road edge.

7. Never permit anyone to ride on the tractor drawbar or cultivator in transport or operation, or to stand near the machine while it is operating—particularly when raising or lowering wings.

TROUBLESHOOTING

Improper adjustment and servicing are the primary causes of field-cultivator performance problems. Recognition of possible causes is the first step toward solution of the problem. This chart gives suggestions (remedies are not included if they are obvious).

TROUBLESHOOTING CHART		
PROBLEM	**POSSIBLE CAUSE**	**POSSIBLE REMEDY**
POOR PENETRATION	Depth-control cylinders not functioning properly.	Synchronize cylinders. Level center frame and outriggers. Check for proper assembly of cylinders and hoses. Check depth-stop adjustment. Check for cylinder or hose leakage.
	Field cultivator not leveled.	

PROBLEM	POSSIBLE CAUSE	POSSIBLE REMEDY
ERRATIC DEPTH CONTROL	Air in hydraulic system.	Bleed cylinders and hoses (see manual).
	Insufficient tractor hydraulic pressure.	Check system for proper pressure. Repair if needed. Use correct-size lift cylinders in relation to tractor system pressure.
	Hoses and cylinders improperly connected.	Reconnect hoses. Be certain couplings on hoses match tractor couplings.
	Series cylinders not synchronized.	
WING-LIFT MALFUNCTIONS	Wings rise too slowly.	Remove hose restrictors or increase hose diameter.
	Wings rise too fast.	Install hose restrictors or decrease hose diameter.
	Leakage in cylinder.	
	Lift linkage out of adjustment.	
	Lift cylinder too small.	
CULTIVATOR DOESN'T TOW STRAIGHT WHEN OPERATING	Outrigger lift linkage restricting movement.	Fully retract lift cylinder or adjust linkage (see manual).
	Depth-control cylinders leaking.	
	Improper tire inflation.	
	Frame not leveled side-to-side	
	Standards not properly arranged.	
PLUGGING	Plugging around wheels.	Remove standard from wheel area and place on forward-mounted stub bar.
	Wrong tool equipment.	Match shovels, sweeps, or spikes to operating conditions.
	Standards too close.	Respace to 9, 10, 12 inches (23, 25 or 30 cm).
	Third bar not used.	Add third bar and spread standards.
	Trash too heavy.	Disk or shred trash first.
	Trash too wet.	

8. Never permit anyone but the operator to ride on the tractor.

9. Keep children away from the cultivator during operation and storage.

10. Provide adequate front tractor weight for safe, stable operation and transport, especially with integral models.

11. Lower the cultivator to the ground any time the tractor engine is shut off, when the machine is being serviced or repaired, or when it will be left unattended. If the cultivator must remain raised for service or repairs, install transport locks or securely block the frame to prevent accidental lowering.

SUMMARY

Field cultivators have more general utility than many others in the wide variety of secondary-tillage implements. Their numerous uses range from seedbed preparation to weed and erosion control and pasture renovation. Sometimes they even are satisfactory for primary tillage in light soils.

Wide adaptability to available tractor power, soil conditions, and work to be done is a major reason for the way field cultivators fit so neatly into many cropping operations. Soil-engaging tools for the vibrant shanks include interchangeable types from spike points and shovels to comparatively wide sweeps.

Field cultivators usually are lighter in construction and lower in cost than their close "relatives," chisel plows, and are intended for less-deep work and less-severe operating conditions. They normally have less vertical frame clearance and narrower shank spacing than chisel plows, and so are less capable of steady operation in heavy trash.

Adaptability and the range of practical working widths is increased by the following attachments: gauge and transport wheels, lift-assist wheels, add-on wing sections with hydraulic or manual folding for transport, stub bars to permit greater latitude of shank spacing, pull-behind hitches for tooth-type smoothing attachments and planters, and insecticide and herbicide applicators.

CHAPTER QUIZ

1. (Fill in blanks) Field cultivators are widely used for _____, _____, and _____.

2. (Fill in blank) Compared to other implements, field cultivators are most like _____ in design and appearance.

3. What are the advantages of locating field-cultivator wheels inside of the main frame? In front of the main frame?

4. Name two advantages of spring-cushion shanks.

5. What type of soil-engaging tool is normally used on field cultivators for deep cultivation? For shallow work?

6. (Fill blanks) Working depth of integral field cultivators is normally controlled by _____ and _____.

18
Rod Weeders

Fig. 1—Weeder Rods Overlap for Complete Coverage

INTRODUCTION

Rod weeders are widely used in western wheat-growing areas of the United States and Canada, primarily for weed control in summer fallow and prior to seeding. Weeding can be done shortly after seeding, before germination and root development, by setting the rod far enough above planting depth to root out weeds without disturbing seed.

The rotating weeder rod is operated from just under the soil to several inches deep to pull weeds out by the roots and work them to the top, along with coarser soil particles, to provide a surface mulch.

Rod weeders normally only cover about 10 percent of the crop residue each pass, which makes them excellent for stubble-mulching where maximum residue should be left exposed to catch and hold moisture and resist runoff and soil blowing.

Draft is relatively light, about one-fourth that of a disk tiller, partly because of the rod rotation in the soil.

Fig. 2—Drive Wheels are Interconnected

Draft varies from about 40 to 120 pounds per foot (580 to 1750 N/m) of width, depending on soil type and working depth. This permits efficient use of wide multiple units for fast ground coverage.

TYPES AND SIZES

Width of individual rod-weeder units ranges from 8 to 14 feet (2.4 to 4.3 m). Two to eight weeders may be pulled with special hitches to cover up to 350 to 400 acres (140 to 160 ha) a day. Multiple units are hinged to follow ground contours, and weeder rods overlap to assure complete coverage. For instance, on the 10-foot (3 m) sections (Fig. 1) each rod is 10½ feet (3.2 m) long. This keeps tough weeds from slipping around the ends of the rods and prevents missed streaks if sections flex in irregular ground.

RODS DRIVEN FROM WHEELS

Rods are driven by chains, or a combination of chains and gears, from rubber-tired drive wheels at the front of the weeder. Drive wheels are interconnected (Fig. 2) by telescoping drive axles with ratchets and universal joints. This provides uniform rod rotation regardless of changes in terrain, poor traction on one wheel, and on turns.

Each weeder rod is mounted on three beams or pendants (Fig. 3) which locate the rod between the drive and gauge wheels to maintain uniform penetration. Two types of shoes are available (Fig. 4). The pointed type penetrates best in hard soil and normal operating conditions with few if any soil obstructions. Stub-pointed shoes are designed to absorb shocks and ride over rocks and stones.

Additional protection from soil obstructions is provided by shear-pin pendants (Fig. 5). If the rod strikes something solid, the pin breaks before the rod does; the rod lifts over the obstacle and must be manually reset with a new shear pin.

ROD LOCATED MIDWAY BETWEEN WHEELS

Fig. 3—Rod is Located Midway Between Wheels to Average Ground Variations

HYDRAULIC RESET

If rocks are a serious problem and replacing shear pins is a nuisance, the pendants may be equipped with hydraulic-reset protection. Individual hydraulic cylinders on each drive boot (Fig. 6) permit the rod to lift, up to 13 inches (330 mm) if necessary, on-the-go, to clear obstacles. When the obstruction is cleared, the rod is instantly and automatically returned to operating position and held there by hydraulic pressure.

The hydraulic-reset system is similar to that on some moldboard plows, and is available for direct operation from closed-center tractor hydraulic systems with adequate pressure, or by pre-charged hydraulic accumulators (Fig. 7). A bladder in the accumulator is filled with compressed nitrogen gas. When the rod rises, the pendant cylinder is closed, forcing oil into the accumulator. As the obstruction is passed, the pressurized nitrogen gas forces oil back into the cylinder to return the rod to working position.

Fig. 4—Pointed Shoes Help to Penetrate Hard Ground; Stub Shoes are Recommended for Rocky Soil

Fig. 5—Shear-Pin Protects Rod from Solid Objects

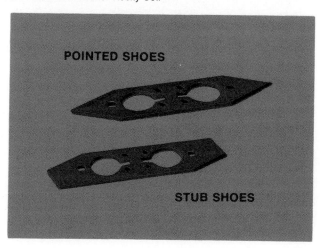

POINTED SHOES

STUB SHOES

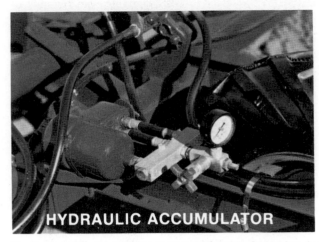

Fig. 6—Power-Reset Cylinders Cushion Rod from Damage

Fig. 7—Hydraulic Accumulator or Tractor Hydraulic System can be Used to Reset Pendants

Varying the nitrogen pressure in the accumulator, or changing relief valves in the tractor-controlled system, permits matching trip resistance to soil conditions. For instance, in very light soil full pressure is not required to maintain penetration, so reset pressure can be reduced to provide better rod protection.

The accumulator system is recommended where tractors have an open-center hydraulic system, if there is inadequate pressure in a closed-center system, or in absence of necessary hydraulic outlets to operate a tractor-controlled system.

OPERATING DEPTH

Rod operating depth is hydraulically controlled by individual remote cylinders on gauge wheels (Fig. 8). All cylinders are connected in parallel to one remote valve on the tractor, and gauge wheels are free to flex independently over uneven ground. Parallel connection also prevents any one gauge wheel from controlling depth of penetration for the entire implement.

An adjustable depth stop (Fig. 9) is optional on some rod weeders to permit returning the rod to the same

Fig. 8—Individual Hydraulic Cylinders Control Gauge Wheels

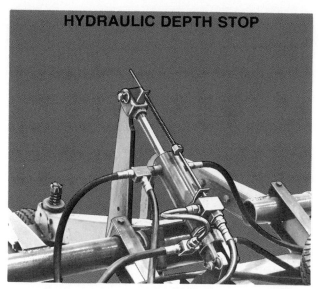

HYDRAULIC DEPTH STOP

Fig. 9—Hydraulic Depth Stop Returns Rod to Same Position Each Time

working depth every time the machine is lowered. This is particularly helpful when weeding is done after seeding, to keep from disturbing seed placement.

In some cases chisels have been installed ahead of the weeder rod to loosen compacted soil for better rod-weeder operation. To accomplish similar goals, rod-weeder attachments are available for some chisel plows (Fig. 10). These tear out weed roots possibly missed by the chisels or sweeps, and help pulverize and level the soil.

To improve rod penetration in hard soil, additional weights may be added to gauge wheels (Fig. 11), gauge-wheel tires may be filled with calcium chloride solution, or both.

TRANSPORT ATTACHMENT

An optional transport attachment permits pulling rod weeders endways for fast movement between fields. By folding or removing the regular hitch, transport width is reduced to approximately 8 feet (2.4 m) (Fig. 12) for safe movement on roads, across bridges, and through gates.

Fig. 10—Rod-Weeder Attachment on Chisel Plow Provides Better Weed Control

ROD WEEDER ATTACHMENT

Fig. 11—Extra Weight Improves Penetration

PRINCIPLES OF ROD-WEEDER OPERATION

Weeder rods may be round or square, and more than one rod size is available for each machine — a smaller one for economy and a heavier one for increased strength for severe conditions. Typical options are 7/8- and 1-inch-square (22.2 and 25.4 mm) high-carbon steel rods.

The front side of the rod moves upward as it revolves, to pull up and get rid of roots while leaving most of the surface residue intact (Fig. 13). This requires a reversing drive in the boot (Fig. 14). As the machine moves forward, the drive axle rotates in the direction of the arrow (left), turning the double sprocket at the top of the boot. This powers the rod drive chain which runs down through the boot to rotate the rod in the proper direction. Square rods simply slide through square holes in the rod sprocket (Fig. 15) and are clamped in white-iron bearings on each pendant. Round rods must be drilled or keyed to hold the sprockets in place.

For better drive-wheel traction, install tires with lugs pointing toward the rear of the rod weeder when looking down on the tire (Fig. 16).

Multiple cable hitches are used on some rod weeders, particularly older, wider machines. However, the combination of rigid hitch and cables (Fig. 17) provides better hillside control and permits backing the weeder. Cables supply draft support for outer ends.

The two rear gauge wheels are pinned to prevent swiveling and provide stability in transport without fishtailing. All other wheels are free to caster for easy turning and maneuvering.

Use of transport attachments and rubber-tired gauge wheels permits easy movement of equipment between fields, and eliminates loading and hauling weeders, or buying separate sets of equipment as some farmers have done in the past.

Fig. 12—Transport Width is about 8 Feet (2.4 m), Regardless of Length

TRANSPORT WIDTH 8 FEET (2.4 m)

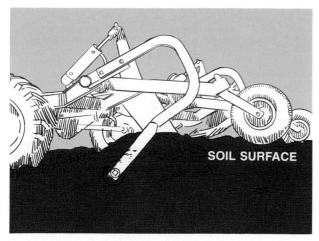

Fig. 13—Rod Revolves to Pull Up Roots

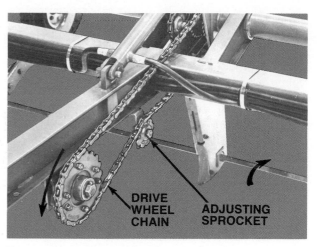

DRIVE WHEEL CHAIN

ADJUSTING SPROCKET

Fig. 14—Reverse Rod Rotation Helps Pull Weeds

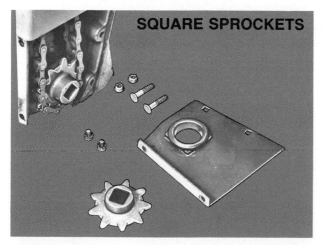

SQUARE SPROCKETS

Fig. 15—Sprockets Have Square Holes To Match Rod Size And Provide Positive Drive

REAR

LUGS ON TOP POINT TO REAR

Fig. 16—Lugs On Top Of Tire Should Point To The Rear For Best Traction

Fig. 17—Rigid Hitch Permits Backing Rod Weeder

RIGID HITCH PERMITS BACKING WEEDER

TRACTOR PREPARATION

Rod weeders may not require full power of tractors used for other tillage and farm operation, but the tractor should be in good mechanical condition and tuned for economical operation. If full power is not required, shift to the highest gear which will pull the rod weeder at a satisfactory speed (usually 4 to 6 mph; 6.5 to 10 Km/h) and throttle back the engine. This reduces fuel consumption and provides quieter operation for greater operator comfort.

1. Provide sufficient ballast on rear wheels to prevent excessive slippage (Fig. 18). Or, remove unnecessary ballast to reduce soil compaction.

2. See tractor manual for recommended ballasting of front end. Added weight may not be necessary unless full tractor power will be required in lower gears.

3. Inflate tires to recommended pressure.

4. Set drawbar height as recommended in rod-weeder manual. Drawbar may be pinned in the center or allowed to swing. A swinging drawbar aids turning with wide implements, but fixing the drawbar provides better stability. See the rod-weeder manual.

5. Use widest practical wheel-tread for work on steep slopes. Precise setting is usually immaterial, so long as both wheels are equidistant from the tractor center.

6. Check hydraulic system for proper operation, adequate pressure, and reservoir oil level. Recheck fluid level each time hydraulic-reset system is charged.

7. Check hydraulic couplings, hoses, and cylinders for leaks or damage and repair as needed.

8. Be sure hoses are properly connected if using hydraulic-reset rod protection.

ROD-WEEDER PREPARATION AND MAINTENANCE

Before using a rod weeder, carefully check the implement to be sure it is ready to go. Time spent making repairs in the field is wasted and could delay other work, such as seeding. Such delays are even more wasteful if they could have been avoided by making the same repairs during the off-season.

Before Each Season:

1. Check for loose bolts and nuts and broken, worn, or missing parts. Replace as needed.

2. Check tightness of setscrews on weeder rod to avoid lateral movement and rod and bearing damage.

3. Inflate all tires to recommended pressure.

4. Lubricate entire implement as recommended.

5. Check nitrogen pressure in hydraulic-reset accumulator, if used. Recharge if needed.

6. Obtain extra shear pins for manual-reset pendants, if used.

Fig. 18—Provide Sufficient Ballast on Rear Wheels

Daily Before Operation:

1. Lubricate as recommended in manual.

2. Inspect hydraulic lift and reset systems for leaks or damage.

3. Visually check for loose, broken, or missing bolts, nuts, and parts, and underinflated tires.

4. Check all drive chains for proper tension to prevent undue wear on chains and sprockets.

Before Storage At End Of Season:

1. Clean all trash, soil, and dirty grease from the rod weeder to prevent collection of moisture and rusting.

2. Repaint spots where paint has been scratched or worn off.

3. Coat soil-engaging parts with heavy grease or paint to prevent rusting.

4. Relieve hydraulic pressure in power-reset and depth-control systems when storing the machine for a week or more. See manual for instructions.

5. Fully retract cylinders to protect cylinder rods from rust.

6. Store inside, if possible, to prevent weathering, and remove weight from tires.

7. Lubricate entire machine as recommended.

FIELD OPERATION

Careful consideration of field size, shape, and working pattern is required to make efficient use of such wide implements as multiple rod weeders.

Wherever possible lay out fields in multiples of machine width, or consider field dimensions when selecting equipment. Try to avoid an extra pass to work a 10-foot (3 m) strip with a 50-foot (15 m) rod weeder, ending the last pass at the far side of the section, and similar situations. Try to make each pass as long as possible to avoid unnecessary turns. In most cases, work is started at one edge of the field and adjacent passes made until the job is completed. The rod may be raised for turning or allowed to remain in the ground and continue operating. If weeding progresses around the field, an extra pass must be made from the center to each corner after the field is finished to till the unworked arc left on each turn.

To reduce water runoff and soil blowing, work across slopes or at right angles to prevailing winds (Fig. 19).

OPERATING TIPS

1. Always relieve pressure in depth-control cylinders and reset system before detaching hoses from tractor. This prevents difficulty in reattaching hoses and possible leaks caused by high pressure in the system.

2. For maximum protection from the hydraulic-reset system, maintain the minimum pressure required to hold the weeder rod in the ground.

3. Follow instructions in operator's manual and periodically check pressure in the reset system to assure correct functioning and optimum pressure.

4. If rod weeder has shear-pin pendant protection, use only replacement pins provided by the manufacturer. Common bolts or other pins could be too hard and not fail when they should, or fail prematurely and cause unnecessary delays.

5. Keep rod-drive chains correctly tensioned. Excessive tension damages chains and sprockets. Chains running too loose may jump off or wrap on sprocket and break. See manual for proper tensioning procedure.

6. Check oil level in tractor hydraulic system after recharging reset system. Replenish if necessary.

7. Be sure hoses are properly connected when using tractor-powered reset system. Never operate weeder without attaching hoses to the tractor if it is equipped with a tractor-powered reset system.

TRANSPORT AND SAFETY

Individual attention to proper operation and transport can reduce accidents. Observation of these basic rules and specific instructions in tractor and rod-weeder manuals will provide a safer environment in which to work.

• Allow only the operator to ride on the tractor during operation and transport.

• Never permit anyone to ride on the rod weeder.

• Never exceed recommended transport speed or, if not stated, maximum tractor speed. Reduce speed on rough or uneven terrain or when turning.

• Use lights, reflectors, and SMV emblem as required by state and local regulations when transporting equipment, day or night.

• Avoid busy highways and peak traffic periods if possible. Move equipment only in daytime.

• Never attempt to repair, adjust, or lubricate the rod weeder while it is in motion.

• Shut off the engine and relieve hydraulic pressure in hoses before disconnecting them.

• Hydraulic fluid escaping under pressure can penetrate the skin and cause serious infections or reactions. Seek immediate medical attention if injured by hydraulic fluid. Small hydraulic leaks under pressure may be nearly invisible; never attempt to locate them by hand—use paper or metal.

• Never stand between the rod weeder and tractor when someone is backing the tractor for hitching.

• Never allow children to play on or near the rod weeder during operation, transport, or storage.

Fig. 19—Work at Right Angles to Prevailing Winds

WORK AT RIGHT ANGLES TO PREVAILING WINDS

TROUBLESHOOTING

Immediate correction of operating problems can pay off in better performance, time saving, and reduced operating costs. Failure to correct some malfunctions or making incorrect adjustments can lead to implement breakdowns and lost time. Where Possible Remedies are obvious, a blank space is left in that column.

TROUBLESHOOTING CHART		
PROBLEM	**POSSIBLE CAUSE**	**POSSIBLE REMEDY**
INADEQUATE ROD PENETRATION	Insufficient weight on rod.	Add weights on gauge wheels or put fluid in gauge-wheel tires.
	Rod weeder not operating level.	Inflate all tires to proper pressure.
	Using stub-point shoes in hard soil.	Use pointed shoes unless soil is too rocky.
	Hitch cables improperly adjusted.	
ROD FLOATING	Insufficient hydraulic pressure in reset system.	Tractor-powered system: be sure hoses are correctly attached; check tractor pump for correct pressure.
		Accumulator system: check to see if system is properly charged.
	Air in lines.	See manual for bleeding instructions; recharge hydraulic system.
RODS FAIL TO TRIP	Improve accumulator charge.	Check nitrogen pressure in accumulator. Reduce pressure or recharge system if needed. See manual.
	Hoses improperly attached to tractor.	
	Faulty relief valve.	
	Improper shear pins used.	
LOSS OF NITROGEN PRESSURE FROM ACCUMULATOR	Loose core valve.	Tighten core valve and recharge accumulator, or see dealer.
	Faulty core valve.	Replace core valve and recharge system, or see dealer.
	Faulty accumulator.	Replace accumulator and recharge system.
RODS BENDING AND SPOOLS BREAKING IN LIGHT SOIL	Relief valve pressure rating too high (tractor system)	Replace relief valve with lower psi (K Pa) relief valve.

PROBLEM	POSSIBLE CAUSE	POSSIBLE REMEDY
	Nitrogen pressure or hydraulic pressure too high (accumulator system).	Reduce nitrogen pressure in accumulator. Reduce hydraulic pressure in system to 200 psi (1380 kPa) above nitrogen pressure in accumulator.
	Improper shear pins used.	
UNEVEN ROD PENETRATION	Inner drive wheels running in tractor tracks.	Add liquid weight (calcium chloride and water) to the outer drive wheels not in the track.
	Outrigger hitch cables not properly adjusted.	

SUMMARY

Rod weeders perform an important double function—weed control with moisture conservation—in major wheat-growing areas where rainfall is light. Not only do they pull weeds out by the roots, they work the weeds—and coarser soil particles—to the surface to provide an erosion-resistant mulch.

The rod-weeder principle is simple—⅞- or 1-inch (22.2 or 25.4 mm) round or square rods turning backward under the surface. The rods are powered from drive wheels connected by chains to the rod sprockets. Drive wheels are interconnected on multiple units so all rods keep turning even if one drive wheel temporarily loses traction.

Rod-weeder draft is comparatively light, so multiple hitches of 8- to 14-foot (2.4 to 4.3 m) units capable of working 350 or more acres (140 ha) a day behind proper power are not uncommon. When several units are used together, hinged end-to-end attachment permits the wide span to follow ground contours.

CHAPTER QUIZ

1. When and where are rod weeders normally used?

2. (Fill blank) Rod weeders are excellent for stubble-mulch farming because they cover only about _____ percent of the residue on each pass.

3. Why do weeder rods on multiple-unit rod weeders overlap?

4. (Fill blanks) Two types of shoes are used on rod weeders. _____ shoes penetrate better in hard soil and normal operating conditions, but _____ shoes are better for rocky conditions.

5. Which way does the weeder rod rotate?

6. Why must pendant shear pins be replaced only with those recommended by the manufacturer?

19
Row-Crop
Weed-Control
Equipment

Fig. 1—Weed Control Is a Primary Goal of Cultivation

INTRODUCTION

Complete weed-control programs must include primary tillage, secondary tillage, planting practices, cultivation of row crops, and use of herbicides. They also involve general farm housekeeping—weed control along fencerows and ditch banks, around farmsteads, and in pastures and fallow ground as well as any other unused land, to reduce the spread of weed seed.

Chemical herbicides are a vital part of most weed-control programs. Careful choice of selective herbicides, applied at the right time and correct rate, will kill certain kinds of weeds and grasses without harming crop plants. In many cases these herbicides may make the difference between satisfactory yield and crop failure.

The margin of safety between adequate weed control and crop damage from some herbicides is very slim. Carryover of some herbicides in the soil from one season to the next may limit crop choices on some fields. At higher application rates some herbicides are quite expensive, and none are completely reliable under every combination of soil, moisture, and weather conditions. Therefore, a combination of mechanical and chemical weed-control methods is generally recommended, even by many herbicide manufacturers.

This chapter will concentrate on mechanical weed-control equipment, with some references to relationships to chemical measures.

ROW-CROP CULTIVATION

Weed control and preparation of the land for irrigation are considered the primary purposes for row-crop cul-

ROTARY HOE

Fig. 2—Rotary Hoe Provides Fast Early Cultivation

tivation (Fig. 1), although there are other, harder-to-define benefits. Even with chemical weed control, cultivating corn and soybeans at least once, whether weeds are present or not, is generally worthwhile.

Tests in many areas have shown corn-yield increases of 5 to 20 bushels an acre (315 to 1250 Kg/ha) where corn was cultivated at least once and herbicides also used, versus herbicides or cultivation alone. Such increases assure return from the investment in equipment and time required to cultivate.

Soils which tend to crust seem to benefit most from cultivation. Stirring the soil surface breaks up the crust, improves aeration and moisture absorption, and kills weeds growing between rows. Cultivation also helps control weeds which may be resistant to most herbicides, and may discourage germination of some late-starting weeds, such as fall panicum.

ROTARY HOEING

About the fastest, economical means of cultivation is rotary hoeing (Fig. 2), which can be done prior to crop emergence, or after plants are at least 2 inches (50 mm) tall, to break up soil crust. One or two passes with the rotary hoe when the crop is small often eliminates later cultivations with other equipment.

The rotary hoe picks out small weeds and grasses and covers some which are not yet firmly rooted. It may be

used only once, or whenever a new crop of weeds emerges, until crop plants are 6 to 8 inches (152 to 200 mm) tall. The hoe will not destroy weeds or grasses after their roots become established. There may be slight crop mortality from hoeing, but the benefits more than offset any losses and no yield reduction is normally noticed.

Tine-tooth and spike-tooth harrows may also be used to weed small row crops and provide an action similar to the rotary hoe. But the spike-tooth harrow cannot be used after the crop is more than 3 or 4 inches (76 to 100 mm) tall because of minimal clearance.

OTHER CULTIVATORS

Sweep, shovel, or rotary cultivators may be used from the time crops emerge until they are 30 to 36 inches (762 to 914 mm) tall, depending on cultivator and tractor clearance (Fig. 3). Suitable shields must be used when plants are small to prevent covering crop plants with soil. Speed must also be reduced, often to 1½ or 2 miles an hour (2.4 to 3.2 Km/h), to reduce covering crop plants, so progress is slow. Rolling shields permit higher speeds. Early cultivation with the rotary hoe usually per-

Fig. 3—Cultivators may be used until Crop Height Exceeds Equipment Clearance

CROP HEIGHT DETERMINES CULTIVATOR USE

mits delaying other cultivation until most danger of crop coverage is past and cultivating speed can be increased.

In some areas cotton is cultivated four or five times, even with herbicide usage, to control problem weeds and vines. Corn and soybeans are generally cultivated only once, when herbicides are used.

FLAME CULTIVATION

Selective flame cultivation provides effective weed and grass control in cotton and many other crops. Flaming does not kill weeds by actually burning or consuming them, but causes moisture in the plant cells to boil, expand, and rupture the cell walls. This blocks the flow of moisture and nutrients, killing the plants. As plants mature they develop a resistance to such heat exposure, so flaming must be done after crop plants become resistant, and before weeds become immune.

Crops and weeds usually germinate and grow in about the same time, but a good pre-emergence herbicide delays weed growth in the row area to give plants a chance to get a headstart in building resistance. Normally, flaming of cotton starts when plants are 6 to 8 inches (152 to 203 mm) tall, corn 8 to 10 inches (203 to 254 mm), and grain sorghum about 9 to 10 inches (229 to 254 mm).

Soybeans are more sensitive and must be 10 to 12 inches (254 to 305 mm) tall before flaming. Extra care must be taken when flaming soybeans to avoid damaging lower leaves and buds. If these are overheated during flaming, yields may be reduced.

Flame cultivation, properly used, can control many broadleaf weeds and vines for which chemical herbicides are generally ineffective. However, it is not a cure-all weed-control method. Interest in flaming has declined in recent years as more effective herbicides have been developed. But there remains a place for flame cultivation as a portion of a complete weed-control system, particularly in cotton. Two more benefits of flaming—there is no residue or carryover, and no pruning of crop roots.

WEED-CONTROL EQUIPMENT TYPES AND SIZES

Weed-control equipment includes various types and sizes of machines.

ROTARY HOE

Rotary hoe sizes are generally measured by row capacity, and most of these implements are made with one section per row (Fig. 4), particularly older models. Row sections are coupled together to work the desired width.

For many years each section was about 40 to 42 inches wide (1016 to 1067 mm), which was fine for that width of row, but then reduced row width in many crops began causing some problems. This was solved in some cases by making each hoe section slightly wider, up to 48 inches (1219 mm), and using four sections for six 30-inch (762 mm) rows.

Section frames are usually constructed of steel angles, channels, or tubing (Fig. 4), with hinged joints front and rear between sections for flexibility.

Fig. 4—One Section of Rotary Hoe is used Per Row

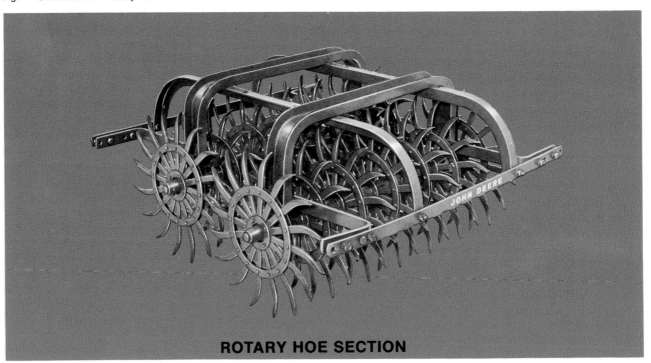

ROTARY HOE SECTION

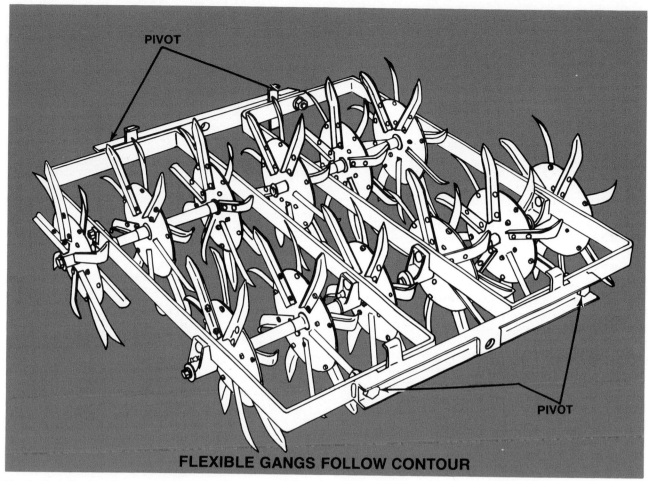

PIVOT

PIVOT

FLEXIBLE GANGS FOLLOW CONTOUR

Fig. 5—Two Flexible Gangs on Each Row Unit Follow Beds or Ground Contours

Some row sections are divided into two flexible gangs which are pivoted on the front and rear frame members to permit the hoe to follow beds and uneven ground (Fig. 5).

Axles are normally mounted in white-iron bearings which provide excellent wear resistance in the extremely dirty working conditions. Hoe-axle bearings usually are not lubricated, preventing accumulation of abrasive material in the bearing area.

FORE-AND-AFT ADJUSTMENT

Fore-and-aft axle spacing is adjustable on many hoes. Wider spacing is recommended for normal operation, while closer spacing permits more cross-action between front and rear gangs to keep hoes free of trash.

Individual hoe wheels, often called spiders, have 10 to 16 teeth. Most wheels are about 18 to 21 inches (457 to 533 mm) in diameter. Lateral wheel spacing is usually 6 to 7 inches (152 to 178 mm) on each axle, and front and rear wheels are staggered to provide 3- to 3½-inch (76 to 89 mm) working interval.

Cable-drawn hitches are available for two to eight hoe sections (Fig. 6) for fast coverage of large acreages. As many as 24 sections with custom-made hitches are sometimes used behind large wheel tractors. Transport width of many drawn 6- and 8-row hoes may be reduced to four rows by swinging outer sections to the rear and pinning them behind the center sections.

Integral hitches for two to six hoe sections provide greater maneuverability than drawn hoes, but cannot match the capacity of drawn units of great width.

SPRING-LOADED SHANKS

A different rotary-hoe design has hoe wheels attached in pairs to spring-loaded shanks. One wheel works slightly ahead of and 3½ inches (89 mm) to one side of its mate. The arm connecting the two wheels permits them to rise and fall individually over uneven ground or field obstructions, or to move together on the shank (Fig. 7).

Spring-loaded shanks are mounted on a common 15- to 30-foot (4.5 to 9 m) integral toolbar (Fig. 8) to provide

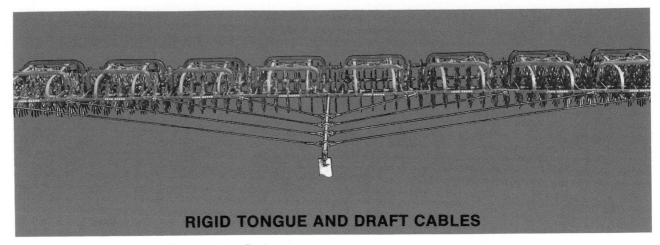

RIGID TONGUE AND DRAFT CABLES

Fig. 6—Rigid Tongue and Draft Cables Permit Easy Towing

Fig. 7—Two Spider Wheels are Mounted on Each Shank

Fig. 8—Toolbar Rotary Hoe has Spring-Loaded Shanks

full coverage for up to a dozen 30-inch (762 mm) rows. When the hoe is raised to transport position, spring pressure on the shanks is relieved and shanks drop almost straight below the bar. A steel-mesh screen is attached to the toolbar directly behind the tractor (Fig. 9) to catch stray clods and reduce the amount of dirt thrown on the tractor and driver.

An endways transport attachment permits quick change-over from working position to transport and reduces width to less than 8 feet (2.4 m). Optional frame-stabilizer wheels are available for outer ends of the toolbar to maintain proper height in operation. These same wheels can be used for the transport attachment.

ROW-CROP CULTIVATORS

Size of row-crop cultivators is counted by the number of rows covered, and ranges from single-row units, for very small tractors, up to eight 40-inch (1016 mm) rows, twelve 30-inch (762 mm) rows, and even sixteen 20-inch (508 mm) rows. Size depends on acreage, tractor power, and tractor lift capacity. Also considered is the period during which timely cultivation may be completed without

interruption by weather or interference with farm operations.

Tricycle-type tractors with adjustable-tread rear wheels were developed primarily for use with front-mounted cultivators and other row-crop equipment. In recent years the safer, more stable wide-front-end tractors with adjustable front and rear tread have taken over most cultivating chores, and cultivators have largely switched from front-mounted to 3-point-hitch-mounted units.

Currently, the most-common type of front-and rear-mounted cultivators has either one or two independent gangs or rigs between each pair of rows (Fig. 10). These rigs are attached to the toolbar with parallel linkage and each has an adjustable gauge wheel which timely cultivation may be completed without interruption by weather or interference with farm operations

For economy and relatively uniform terrain, a single gang with up to five shovels or sweeps (Fig. 11) may be used for each row-middle from 28 to 40 inches. (711 to 1016 mm). For more varied soil contours, two gangs with two and three shanks each (Fig. 12) may also be used for 28- to 40-inch (711 to 1016 mm) rows. When working 20- to 30-inch (508 to 762 mm) rows, a single rig with one to three sweeps (Fig. 13) is more suitable. Many farmers prefer a single 10- to 12-inch (254 to 305 mm) sweep for 20-inch (508 mm) rows for fast, clean cultivation, especially if heavy surface residue is present.

Fig. 9—Steel-Mesh Screen Protects Operator

Fig. 10—One or Two Independent Gangs Cultivate Each Row Middle

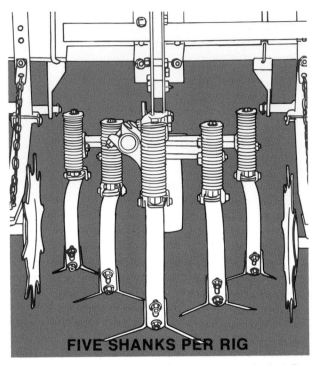

FIVE SHANKS PER RIG

Fig. 11—Single Rig with up to Five Shanks may be used for Each Row

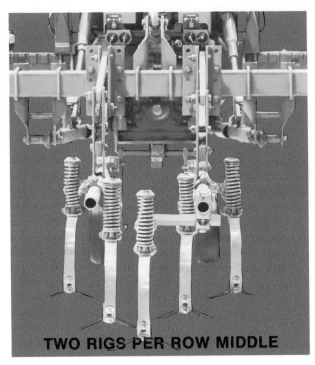

TWO RIGS PER ROW MIDDLE

Fig. 12—Two Rigs Per Row-Middle Follow Ground Contours Better

Fig. 13—Single Rigs are Recommended for 20- to 30-Inch (508 to 762 mm) Rows

Fig. 14—Guide Coulters may be Rigid or Spring-Loaded and Flexible, as Shown

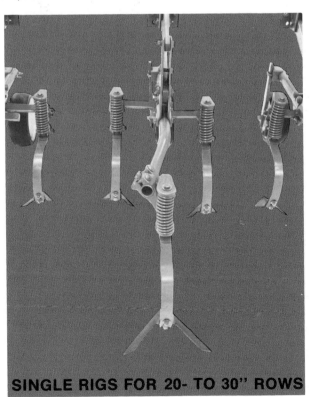

SINGLE RIGS FOR 20- TO 30" ROWS

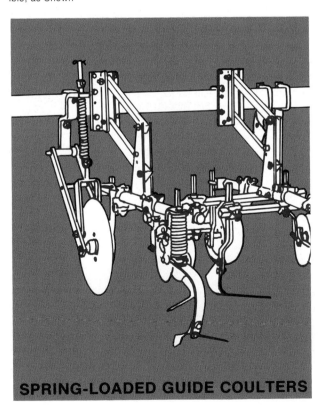

SPRING-LOADED GUIDE COULTERS

REAR-MOUNTED CULTIVATORS

One or two guide fins or coulters (Fig. 14) on the cultivator toolbar provide lateral stability and help average out minor steering corrections as the cultivator follows the tractor.

A steering guide is usually attached to the tractor front axle or frame directly over one row to permit easier steering.

Rear-mounted cultivators for up to four 40-inch (1016 mm) rows are generally transported on the tractor 3-point hitch without change. Many larger units and wide rotary hoes may be equipped with an end-transport attachment and reduced to less than 8-foot (2.4 m) road width (Fig. 15). Others are mounted on folding toolbars (Fig. 16), so end sections may be folded for transport. Hydraulic cylinders are located on the inside of this bar to provide maximum versatility in attaching equipment and varying row spacings (Fig. 17).

Split-rockshaft lift attachments are available for some rear-mounted cultivators (Fig. 18); they permit lifting rigs on either half of the cultivator without disturbing

remaining units, for better cultivation of point rows and waterways, etc.

Economical 2- and 4-row cultivators with double-bar angle frames (Fig. 19) adapt to rows from 28 to 42 inches (711 to 1067 mm). The continuous row of holes in each bar permits quick width adjustment of the rigid beams. Guide coulters in the center or on each end stabilize operation, particularly on slopes or contours.

FRONT-MOUNTED CULTIVATORS

Cultivators of up to 12 rows may be front-mounted on many tractors, giving the operator greater opportunity to observe each row unit, compared to rear-mounted cultivators.

Front-mounted cultivators were, for many years, difficult to attach and detach. However, design improvements now make their installation almost as easy and fast as hookup of rear-mounted cultivators. After the mounting frame is attached to the tractor, simply drive up to the cultivator, attach cylinder hoses, and pin the unit in position after hydraulic pressure has adjusted cultivator height (Fig. 20).

Fig. 15—Endways Transport Takes Less than 8 Feet (2.4 m) of Road

FOLDING TOOLBAR REDUCES WIDTH

Fig. 16—Folding Toolbar Reduces Transport Width

Fig. 17—Folding Cylinders are Inside Bar for Easier Row-Space Adjustment

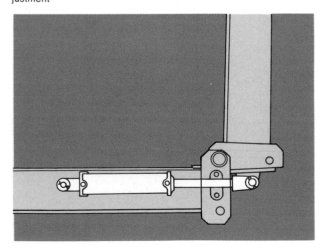

Six-row and wider cultivators are often hinged at the tractor and use frame-mounted castering gauge wheels to support the outer ends when folded forward for transport (Fig. 21) and to permit vertical flexing in operation.

Most front-mounted cultivators may be equipped with individual hydraulic lift cylinders on each side to provide selective lift. This permits easier finishing of point rows and working across or parallel to waterways.

A hitch-mounted rear rig assembly is used with most front-mounted cultivators (Fig. 22) to remove wheel tracks, reduce erosion from water following tracks downhill, and to remove any unworked strip under the center of the tractor. One or more sweeps or shovels may be used behind each wheel.

BEET-BEAN-VEGETABLE CULTIVATORS

Continuous-bar cultivators (Fig. 23) provide the necessary versatility for cultivation of beets, beans, vegetables, and small specialty crops. These crops are

SPLIT LIFT
ROCKSHAFT

SPLIT LIFT
ROCKSHAFT

FRAME GAUGE
WHEEL

SPLIT LIFT PERMITS EASIER CULTIVATION

Fig. 18—Split Lift Permits Easier Cultivation of Point Rows with Wide Rear-Mounted Cultivators

Fig. 19—Angle-Steel Frame Bars are Perforated for Easy Row-Width Adjustment

Fig. 20—Drive-Up-and-Attach Cultivator Designs Save Time

PERFORATED ANGLE-STEEL FRAME

Fig. 21—Frame-Mounted Gauge Wheels Carry Outer Ends for Transport and Operation

Fig. 22—Rear Rig for Front-Mounted Cultivators Removes Wheel Tracks

Fig. 23—Continuous-Bar Beet-Bean-Vegetable Cultivator

Fig. 24—Optional Fourth Bar Increases Tool Capacity

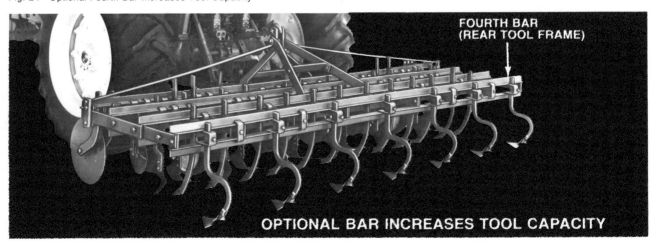

grown in varying row widths and bed configurations, and with different cultural practices. This calls for easy cultivator adjustment and an almost endless variety of possible combinations.

Three double bars on most cultivators permit quick lateral adjustment of shanks, disk weeders, and other tools to match specific operating conditions. An optional fourth bar provides additional tool capacity (Fig. 24). Either stabilizer fins (Fig. 25) or guide coulters (Fig. 24) may be used to keep the cultivator in line behind the tractor.

Cultivator width varies, but the typical model shown (Figs. 23 and 24) can be used for six 20- to 24-inch (508 to 610 mm) rows or four 26- to 36-inch (660 to 914 mm) rows. Beet-bean-vegetable cultivators may be either front- or rear-mounted.

LISTER CULTIVATORS

Lister-planted crops require special cultivation when small to prevent covering the new plants. Most lister cultivators have hooded shields (Figs. 26 and 28) for maximum plant protection, and use disk hillers to pull soil away from the row on the first cultivation. Pointed shovels or other tools remove weeds close to the row. On later cultivations, disk hillers are reversed and soil is turned from the ridge into the row area to provide greater plant support and, if desired, furrow the row middle for irrigation.

Row-spacing is easily adjustable and down-pressure is varied by changing the crank setting on each row unit (Fig. 27).

Lister cultivators (Fig. 28) are available in 4- or 6-row sizes to match 4-, 6-, 8-, or 12-row planting capacity. Most current lister cultivators are rear-mounted, although some are front-mounted. In the past many were drawn behind the tractor on their own wheels, but these are no longer common.

ROTARY CULTIVATOR

Rotary cultivators (Fig. 29) are versatile implements made of a series of spider gangs which look much like rotary-hoe wheels. Contrary to rotary-hoe operation, rotary cultivator wheels turn backwards (Fig. 30) and have curved slicing teeth. Teeth are designed for either right- or left-hand cutting. Gangs are individually angled for moving soil toward or away from rows, working on the sides and tops of beds, or in combination with sweeps and shovels. Rotary gangs are also used for band-incorporation of chemicals at planting time (Fig. 31). By arranging gangs for complete soil coverage, they can be used to incorporate broadcast chemicals.

The curved rotary teeth slice and twist as they pass through the soil, uprooting small weeds, cutting large ones, and breaking soil crust. Gangs can be set close to rows for clean cultivation of small crops, and later turned to throw soil into the row to cover late weeds there. Gangs usually have from two to five spiders, depending on design and application.

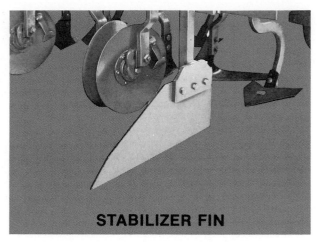

STABILIZER FIN

Fig. 25—Stabilizer Fins (Above) or Guide Coulters Help Steer the Cultivator

Fig. 26—Lister Cultivators are Designed to Protect Small Plants

Fig. 27—Crank Adjusts Shovel Height and Down-Pressure

CRANK ADJUSTMENT

ADJUSTING CRANK

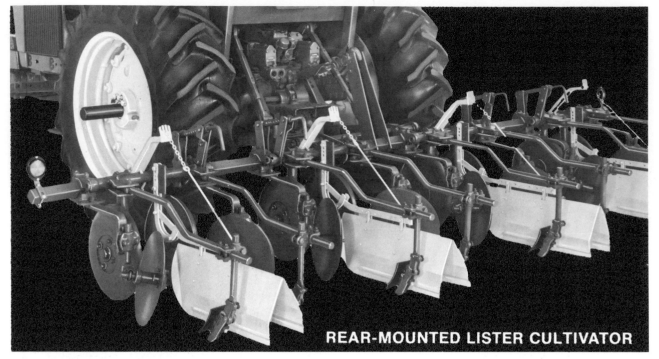

Fig. 28—Rear-Mounted Lister Cultivator

The cultivators can:

1. Operate in a wide variety of crops, from very small to toolbar-high.

2. Work flat, bedded, or lister-planted crops.

3. Provide fast destruction of weeds, from small seedlings to well-rooted weeds and grasses.

4. Provide fast cultivation—up to 7 or 8 miles an hour (11 to 13 Km/h) when properly adjusted and equipped with shields for small plants.

CULTIVATOR ATTACHMENTS

Most front- and rear-mounted cultivators use the same soil-engaging attachments, with only slight variations for some applications.

Several shank types are available, including spring-trip (Fig. 32), which trip and automatically reset when an obstacle is encountered, and quick-adjustable and friction-trip shanks (Fig. 33) for soils with fewer obstacles. There are many shapes and designs of spring-trip and rigid flat standards (Fig. 34) for beet-bean-

Fig. 29—Rotary Cultivators can Work up to 7 or 8 Miles Per Hour

Fig. 30—Rotary Cultivator Wheels have Sharp Teeth for Slicing Soil and Weeds

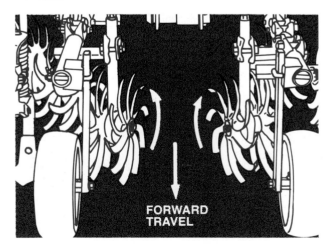

FORWARD TRAVEL

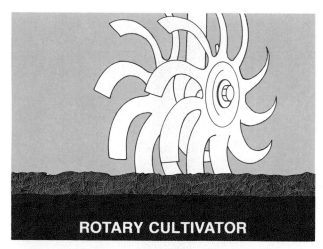

ROTARY CULTIVATOR

Fig. 31—Rotary Cultivator Gangs Provide Excellent Chemical Incorporation

Fig. 32—Spring-Trip Shanks Automatically Reset after Passing over Obstructions

Fig. 33—Quick-Adjustable and Friction-Trip Shanks are in Common Use

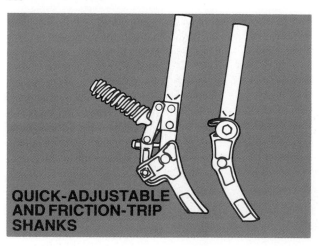

QUICK-ADJUSTABLE AND FRICTION-TRIP SHANKS

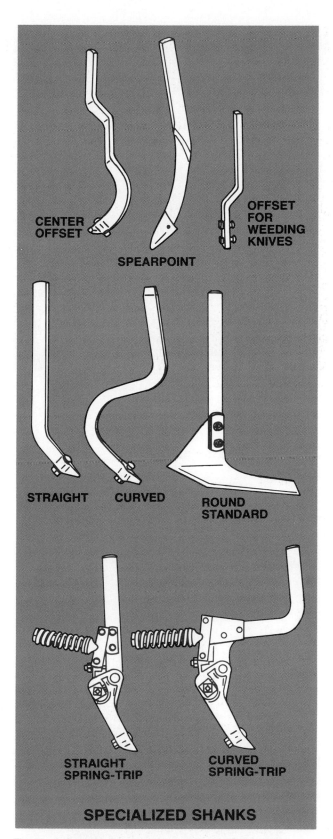

CENTER OFFSET

SPEARPOINT

OFFSET FOR WEEDING KNIVES

STRAIGHT

CURVED

ROUND STANDARD

STRAIGHT SPRING-TRIP

CURVED SPRING-TRIP

SPECIALIZED SHANKS

Fig. 34—Specialized Shanks for Beet-Bean-Vegetable Cultivators

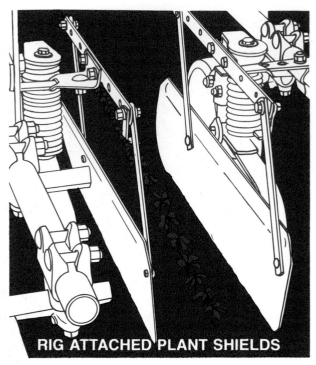

RIG ATTACHED PLANT SHIELDS

Fig. 35—Rig-Attached Plant Shields are Adjustable for Row Spacing and Crop Conditions

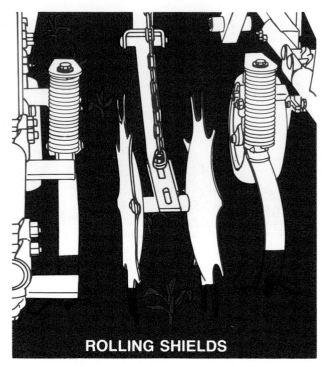

ROLLING SHIELDS

Fig. 36—Rolling Shields Work Well in Trash and for High-Speed Cultivation

Fig. 37—Hooded Shields Provide Maximum Crop Protection

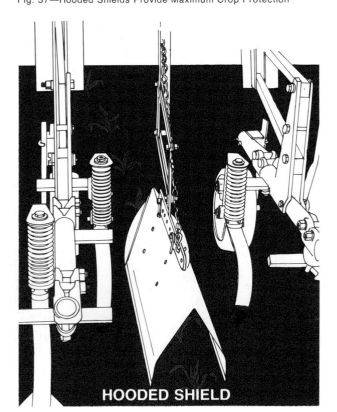

HOODED SHIELD

vegetable cultivators to match soil conditions, bed shapes, and row spacings.

Row shields are required when cultivating small crops to prevent covering plants with soil. Some shields are attached directly to the rig (Fig. 35) on parallel linkage to permit vertical flexing over obstructions and maintain level operation.

Rolling shields (Fig. 36) are particularly helpful when cultivating small plants in heavy trash, because they roll over instead of dragging trash. They're ideal for high-speed cultivation.

Hooded shields (Fig. 34) are fully adjustable and provide maximum possible plant protection. Rotary cultivator shields (Fig. 38) protect bushy crops, such as soybeans, and may be moved up or down to control flow of soil to the plants.

Spray shields (FIg. 39) include two nozzles for directed herbicide application in the row area while middles are being cultivated. Spray is held within a controlled area.

Cultivator soil-engaging tools range from 24-inch (610 mm) wide sweeps to 16-inch (406 mm) diameter disk hillers, their use depending on crop, soil and moisture conditions. Some tools are available with hard-facing for longer wear in abrasive soils. Typical cultivator sweeps (Fig. 40) include:

• *Universal sweeps, 5 to 24 inches (127 to 610 mm)— low crown, narrow shank, low wing angle, and quick scouring for high-speed cultivation*

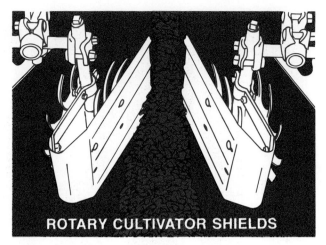

ROTARY CULTIVATOR SHIELDS

Fig. 38—Rotary Shields Protect Bushy Plants

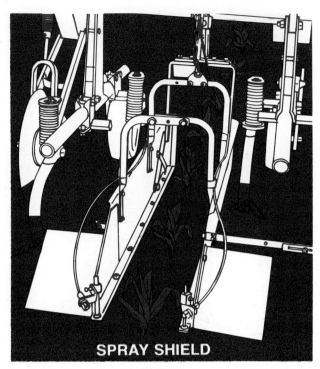

SPRAY SHIELD

Fig. 39—Spray Shields Permit Herbicide Application while Cultivating

• *Peanut sweeps, 6 to 20 inches (152 to 508 mm), for high-speed cultivation with minimum soil movement*

• *Mixed-land sweeps, 4 to 16 inches (102 to 406 mm)— high, broad crown, high wing angle, and broad shank to throw soil and leave a mulched surface*

• *Blackland sweeps, 4 to 20 inches (102 to 508 mm)— high crown and wing angle for difficult scouring conditions*

• *General-purpose sweeps, 4½ and 6 inches (114 to 152 mm)—low-pitched wings for good soil flow*

• *Solid triangular sweeps for hilling cotton beds and cleaning irrigation furrows*

Many sweeps are also available in half-sweep and three-quarter-sweep styles for close-to-row cultivation without throwing soil on small plants; they are usually used only on front shanks next to the rows.

Cultivator shovels (Fig. 41) include:

• *Non-reversible, spear-point shovels, 2 × 7 to 5½ × 7¾ inches (51 × 178 to 140 × 197 mm)—excellent for killing weeds in small crop plants*

• *Reversible, double-point shovels, 2 × 11 to 3 × 11 inches (51 × 280 to 76 × 280 mm)—ideal for deep cultivation*

Fig. 40—Typical Range of Cultivator Sweeps

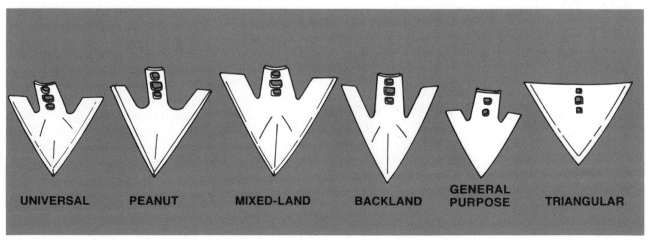

UNIVERSAL   PEANUT   MIXED-LAND   BACKLAND   GENERAL PURPOSE   TRIANGULAR

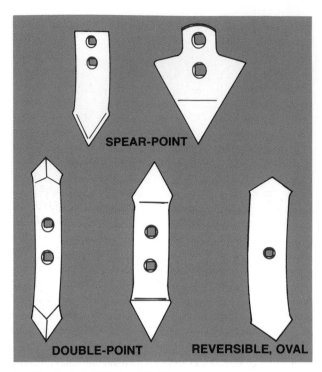

Fig. 41—Typical Cultivator Shovels

SPEAR-POINT

DOUBLE-POINT   REVERSIBLE, OVAL

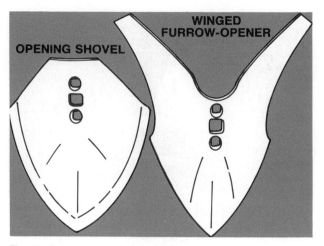

OPENING SHOVEL   WINGED FURROW-OPENER

Fig. 42—Typical Furrow Shovel and Opener

● *Reversible, oval, double-point shovel for grooving soil*

Furrow openers and shovels (Fig. 42) leave ground rough and grooved to absorb and retain moisture with reduced danger of erosion, or to make or clean irrigation furrows. They come in a variety of sizes and shapes such as:

● *Plain 8×10-inch (203 × 254 mm) opening shovel*

● *Winged 8×11-inch and 8½×13-inch (203 × 279 mm and 216 × 330 mm) furrow openers*

Fig. 43—Typical Beet-Bean-Vegetable Tools

Special tools (Fig. 43) are available for close weeding of beets, beans, and vegetables, such as:

● *Extra-flat duckfoot sweeps cut weeds without ridging; 6- to 12-inch (152 to 305 mm) sizes*

● *Irrigation shovels make small furrows between narrow-planted crops without burying seedlings*

● *Reversible, oval-shaped, diamond-point shovels*

● *Spear-point shanks are beveled for better cutting and deep penetration in hard soil; point is replaceable*

Weeding knives (Fig. 44) are used in beet-bean-vegetable cultivating. Typical knives include:

● *Round-turned knives shaped to protect shallow roots; 6 to 10 inches (152 to 254 mm)*

● *Square-turn knives cut through packed soil; 6 to 8 inches (152 to 203 mm)*

● *Low-point, square-turn knives permit very close cultivation of young plants; 6- to 8-inch (152 to 203 mm) sizes*

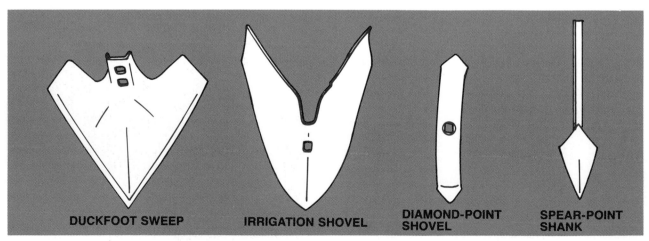

DUCKFOOT SWEEP   IRRIGATION SHOVEL   DIAMOND-POINT SHOVEL   SPEAR-POINT SHANK

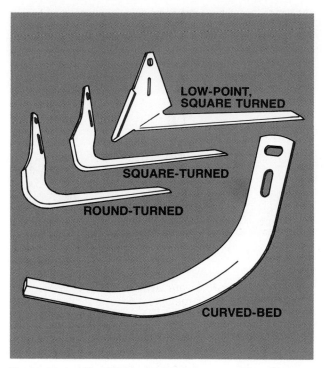

Fig. 44—Typical Weed Knives for Beets, Beans, and Vegetables

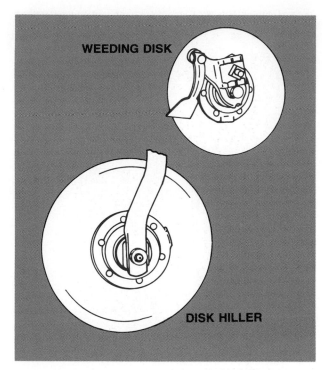

Fig. 45—Typical Cultivator Disks

● *Curved bed knives to cut through tough, heavy weed growth without damaging small plants*

Disks (Fig. 45) may be set to move soil toward or away from rows, depending on crop size and desired results.

● *Weeding disks for beet-bean-vegetable work have adjustable scrapers; 9 to 11 inches (229 to 279 mm), right- and left-hand units*

● *Disk hillers are sometimes used instead of sweeps or shovels, especially for lister-planted or ridged crops. They also work well in heavy trash where sweeps would plug. Diameters range from 12 to 16 inches (305 to 406 mm)*

Either rotary-hoe (Fig. 46) or rotary-cultivator gangs can also be used on most cultivators.

Rotary-hoe gangs mount directly over the row for early cultivation of small crops and to break soil crusts.

Rotary-cultivator gangs with two to five wheels provide excellent high-speed cultivation of many crops and are frequently used in combination with sweeps. Adjustments permit moving soil to or away from rows, on beds or flat land. Wheels can also be used for incorporation of chemicals.

Fast, simple cultivation is provided by attaching flexible-tine shanks to parallel-mounted rigs (Fig. 47). High-speed vibrating tine action shatters crust, kills weeds, and levels soil. These are usually available with reversible points or small shovels.

PRINCIPLES OF OPERATION

Speed, timing, and soil conditions can affect the optimum performance of a machine in weed-control operations.

ROTARY HOE

The points of rotary hoe wheels enter the soil almost vertically and emerge pointing rearward with a lifting action which shatters crust and pulls small weeds.

Fig. 46—Typical Rotary-Hoe Gang

Fig. 47—Flexible-Tine Cultivators Shatter Crust and Kill Weeds

Fig. 48—Hoe Sections can be Pulled Backwards to Firm Soil

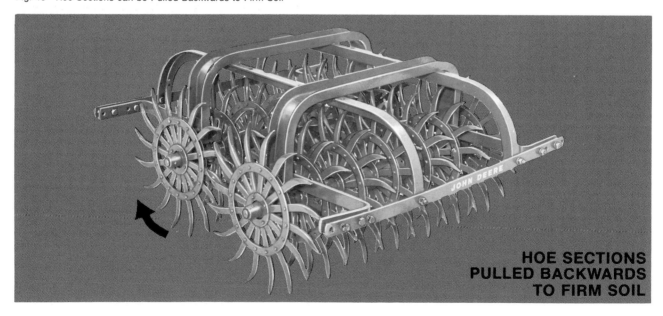

**HOE SECTIONS
PULLED BACKWARDS
TO FIRM SOIL**

The high-speed action throws many weeds free of the soil and leaves them exposed to die.

Successful rotary-hoe performance depends on timing the operation so weeds and grasses are just coming up and crop plants are developing well-established root systems. As soon as weeds or crops are firmly rooted the rotary hoe will do them little harm.

If soil is extremely hard, weight may be added to most hoes for better penetration. Optimum performance occurs at operating speeds of 7 to 12 miles an hour (11 to 19 Km/h) with a minimum of 4½ (7 Km/h) recommended for many hoes. Recommended top speed varies by manufacturer, but extremely high speeds may tear leaves from crop plants and cause some uprooting.

Pulling section-type rotary hoes backwards (Fig. 48) by reversing the hitch allows them to be used for pulverizing and packing the soil. As the curved backside of each point strikes the soil it breaks clods and closes air pockets.

ROW-CROP CULTIVATION

Sweeps, shovels, and rotary-cultivator wheels slice off weeds, pull small weeds, and shatter the soil surface to expose additional weed roots to die in the sun. When using small sweeps and shovels, some well-rooted weeds are apt to slip around the tool without being cut. This can be minimized by overlapping large sweeps or weeding knives, and must be considered in arranging tools on the cultivator.

Pre-emergence chemical herbicides are usually incorporated into the upper 2 or 3 inches (50 to 76 mm) of soil to kill susceptible weeds as they germinate. If later cultivation is required to control other weeds and grasses, stirring soil deeper than the chemical was incorporated simply exposes new weed seed to favorable germinating conditions and counteracts the effectiveness of the chemical.

Similarly, if only a band of herbicide is applied in the row, covering that treated area with soil from the row middles again brings up more fresh weed seed.

Weeds usually germinate in the upper inch or two (25 to 50 mm) of soil, so shallow cultivation will normally control those weeds and reduce germination of additional seeds which are buried deeper.

Evidence is mounting that deep cultivation of most crops, particularly soybeans and corn, may cause severe root pruning (Fig. 49). Although soybean roots have long been considered to consist primarily of taproots, high concentrations of feeder roots have been found in the top 2 to 4 inches (50 to 100 mm) of soil. Corn roots will quickly spread across the entire inter-row area in uncompacted soil. Cutting the plant's roots with deep cultivation reduces the crop's ability to gather moisture and nutrients from the soil and can seriously reduce yields.

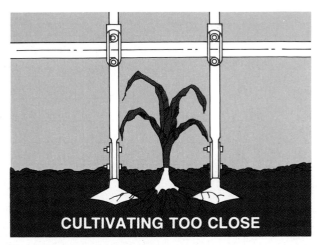

CULTIVATING TOO CLOSE

Fig. 49—Cultivating too Close or too Deep Cuts Roots

CULTIVATOR MOUNTING

When lower draft links of most tractor 3-point hitches are allowed to flex laterally during cultivation, the guide coulters and converging hitch linkage cause the cultivator to follow the tractor on turns and through steering corrections.

If sway blocks or stabilizers prevent sidesway of hitch links, steering corrections will be amplified and the cultivator will move in the opposite direction to front wheel movement resulting in crop damage.

In contrast to rear-mounted cultivators, front-mounted units are rigidly attached to the tractor laterally and shift directly with changes in steering direction (Fig. 50). More-continuous steering attention is required with these cultivators because they do not have the lateral flexibility of rear-mounted equipment to average out steering corrections or deviations. Visibility of the row units is better with front-mounted cultivators, but is often overstressed in relating one type to the other, particularly on wider cultivators. The operator simply cannot continuously monitor all row units on 6-, 8- or 12-row cultivators, front or rear. It becomes even more difficult when speeds approach 5 to 6 miles an hour (8 to 10 Km/h) and crops are 2 to 3 feet high (60 to 90 cm). An occasional glimpse of outer front units and any of the rear rigs may be all the operator gets as he concentrates on steering. This makes a strong argument in favor of automatic guidance systems for tractors to permit more operator attention to be focused on cultivator operation than on steering.

FLAME CULTIVATORS

One burner is usually mounted on each side of each row with the flame directed at about a 45-degree angle across the row area (Fig. 51). The flame usually strikes the soil about 2 inches (50 mm) from the near side of the row and passes through the plant area. Burners are staggered

Fig. 50—Front-Mounted Cultivators "Steer" with the Tractor

to prevent collision of the flames, and the row area must be smooth and free of clods or stones which could deflect flames upward into crop leaves.

Fig. 51—Flame Cultivator

Parallel flaming of very-small crop plants may control weeds nearly the size of the crop which is not yet able to withstand normal flaming. Burners are directed to the rear with only the edge of the flame coming close to the crop row. As crops develop flame tolerance the burners may be turned more toward the row.

Shielding crop plants from flame damage has been done by placing a fan-type nozzle about 2 inches (50 mm) above the burner to direct a curtain of water parallel to the flame. This cools the air above the flame and prevents flame and heat from rising into the crop foliage.

Flame cultivators usually burn LP-gas, normally propane or a mixture of propane and butane.

ROTARY-CULTIVATOR GANGS

Rotary-cultivator wheels cut and pull weeds as they slice through the soil. Wheel aggressiveness is controlled by angling gangs to or away from the direction of travel. Setting the axle perpendicular to direction of travel provides minimum soil movement; weed cutting and soil movement increase as the gang axle is angled from the perpendicular.

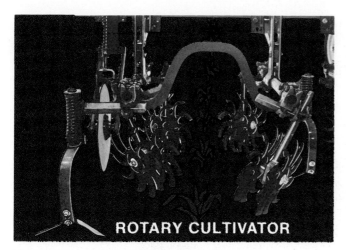

Fig. 52—Rotary Cultivators can be used on Beds or Flat Land

Different gangs are used to move soil to the right and left, according to the shape of wheel teeth. Tilting the gangs on the cultivator frame permits cultivation of beds or ridging crops for later irrigation (Fig. 52).

Recommended operating speeds, usually 5 to 8 miles an hour (8 to 13 Km/h), will vary with soil conditions, crop size, and operator experience.

TRACTOR PREPARATION

Cultivation has relatively low draft per foot of machine width, but a large tractor with adequate lift capacity, strength, and stability is required for operation of 8- to 18-row cultivators. Proper hydraulic-system performance is a vital part of cultivator operation and must be given proper priority in preparing the tractor.

1. Check hydraulic reservoir for adequate oil supply; examine hoses, couplings, and cylinders for leaks or damage.

2. Set hitch lift links to equal length as recommended in operator's manual. Adjust tractor top link to level cultivator fore-and-aft.

3. If gauge wheels are used on outer ends of toolbar, set links to allow bar to float from end to end.

4. Adjust sway blocks to permit lateral flexibility in operation and lock out sway in transport.

5. Position drawbar to clear cultivator when raising or lowering.

6. Set rockshaft control in depth or position range to hold toolbar at fixed height, or allow gauge wheels to control height. Do not use load-and-depth control.

7. Attach hitch-pin adapters on cultivator if using quick coupler.

8. Adjust wheel tread so tires are centered between rows. Wheels on both sides must be equidistant from tractor centerline.

9. Inflate tires to recommended pressure.

10. Supply adequate front ballast to keep front wheels on the ground at all times—in operation, transport, or when raising the cultivator. Do not exceed manufacturer's maximum weight recommendations for tires or tractor frame. Rear wheels normally require no added ballast, and weights used for other work may need to be removed to reduce compaction.

11. Cultivation requires accurate steering. Be sure tractor steering system is responsive and functioning properly.

12. Install steering guide on frame or front axle for rear-mounted cultivators.

PREPARATION AND MAINTENANCE

Breakdowns during cultivation may delay work until weeds are out of control or crops become too large, so equipment must be prepared ahead of time.

Before Each Season:

1. Check entire machine for loose, worn, broken, or missing bolts, nuts, and parts.

2. Lubricate as recommended. Do not lubricate white-iron bearings on rotary hoes for lubrication will encourage collection of abrasive material in bearings.

3. Check parallel-rig linkage for lateral movement; tighten if too loose.

4. Check soil-engaging tools for dull or broken points.

5. Check rotary-hoe and rotary-cultivator wheels for bent or broken points.

6. Check rotary-hoe bearings for excessive wear and replace as needed.

7. Inflate gauge-wheel tires to recommended pressure.

8. Check position of all cultivator rigs for proper location to match desired row spacing.

9. Attach plant shields for protection of small plants.

10. Check the operator's manual for specific instructions for preparation and maintenance of flame cultivators.

Daily Before Operation:

1. Check for loose or missing bolts, nuts, and parts.

2. Check tires for proper inflation.

3. Lubricate as recommended.

4. After each day's work coat cultivator soil-engaging tools with heavy oil to prevent rust.

Before Storage At End Of Season:

1. Thoroughly clean implement and repaint spots where paint is scratched or worn.

2. Lubricate entire implement.

3. Coat soil-engaging parts with heavy grease or plow-bottom paint to prevent rust.

4. Use proper storage stands, securely positioned, to prevent equipment from overturning during storage.

5. Store inside and remove weight from tires.

6. Remove hydraulic cylinders or fully retract them to prevent rusting of cylinder rods.

FIELD OPERATION

Cultivator size must be compatible with planter size. That is, the number of cultivator rows must equal planter rows, or be equally divided into planter size; use a 2-, 4-, 6-, or 12-row cultivator if a 12-row planter was used, but not an 8-row cultivator. If the acreage justifies a 12-row planter, 2- or 4-row cultivators are probably economically unfeasible because of excessive labor involved and low capacity. Even the 6-row unit may be marginal. (See FMO Manual, Machinery Management, for further discussion.)

To avoid possible crop damage from row-width variations between planter passes, always cultivate only rows planted at the same time. The indiscriminate coverage of rotary hoes eliminates the necessity of matching hoe size to specific planter size (Fig. 53).

The key to successful weed control with rotary hoes and row-crop cultivators is uniform and adequate penetration. This includes adding weight to rotary hoes where needed, and keeping soil-engaging tools sharp, rust-free, and properly adjusted. Sufficient soil must be moved into the row area to cover weeds which cannot be mechanically removed by the cultivator.

Uniform and adequate penetration means not cultivating deep to cause crop-root damage, not covering small plants with soil, and not unnecessarily exposing weed seeds to favorable germinating conditions.

Built-in adjustments permit angling, raising, and lowering tools to fit almost any crop and soil condition. When starting a new cultivator there may seem to be more wrong adjustments than right ones unless the entire procedure is done systematically as outlined in the operator's manual.

For instance, shanks must be set at the proper angle for sweeps to penetrate correctly (Fig. 54). Running points too high can ride the sweep out of the ground and cause unnecessary wear on sweep wings. Tipping points down causes too much digging and extra wear on points.

Adjusting too much suction into soil-engaging tools increases the load on gauge wheels, causing extra wear on tires and bearings. It also increases the difficulty of maintaining accurate depth control, because tools keep trying to go deeper and gauge wheels try to resist. This results in bouncing.

Setting guide coulters (Fig. 55) straight behind helps the cultivator follow the tractor on turns and stabilizes the unit laterally. Two coulters set differently will fight each other as soil conditions change, resulting in un-

Fig. 53—Rotary Hoes do not Need to be Matched to Planter Size

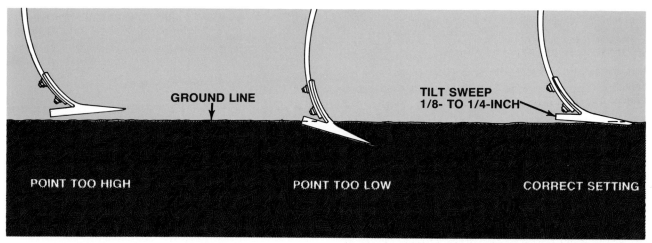

GROUND LINE

TILT SWEEP
1/8- TO 1/4-INCH

POINT TOO HIGH   POINT TOO LOW   CORRECT SETTING

Fig. 54—Adjust Shanks for Correct Sweep Penetration

stable cultivator operation. Don't set coulters too deep. They don't cut well if too deep, and may drag trash on hubs. Set just deep enough to maintain steering control.

Adjust tractor rockshaft or frame gauge wheels to maintain toolbar at recommended height. Running the bar too high or too low may restrict range of vertical flexibility of rigs as they pass over uneven ground. Running toolbar too low may also damage tall crops.

PRELIMINARY ADJUSTMENTS

Uniform adjustment of all gauge wheels, shanks, sweeps, shovels, and coulters before taking a cultivator to the field can save time and improve cultivator performance. To simplify making these adjustments, attach the cultivator to the tractor on which it will be used and park the outfit on a smooth, level surface, preferably a concrete floor or driveway. Time spent rechecking used cultivators can be just as worthwhile as with a new machine.

1. From the centerpoint of the cultivator toolbar, measure and mark the location on the bar of each row.

2. For quick visual checking of results, measure and mark locations on the ground or floor under the toolbar. Always measure from the centerpoint to avoid cumulative error.

3. Cut boards to length recommended for toolbar height and place under ends of toolbar. Adjust gauge wheels or set rockshaft control and hitch lift links to that height.

4. See manual for recommended height of main rig beam and cut a board to that length. Place board under front end of beam (Fig. 56) and adjust parallelogram linkage so that beam is level. Repeat on all rigs.

5. Set shanks to recommended height (Fig. 56) and rotate clamps on beam so shanks are vertical. Shift

Fig. 55—Set Guide Coulters for Proper Cultivator Operation

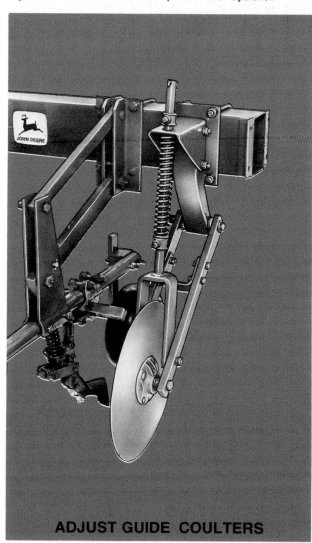

ADJUST GUIDE COULTERS

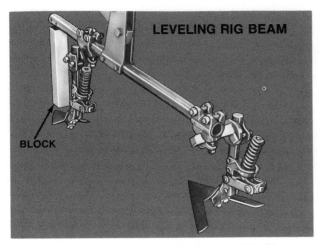

Fig. 56—Use Block of Wood to Level Rig Beams and Set Shanks

shanks laterally to provide complete soil coverage between rows—see manual for recommended settings.

6. Shift rigs laterally to provide desired clearance between soil-engaging tools and rows.

7. Adjust shanks for proper sweep or shovel angle (Fig. 54).

8. Adjust all rotary-cultivator gangs to the same angle (Fig. 57)—see manual.

9. Place a block of wood under each rig gauge wheel to provide uniform wheel setting in relation to soil-engaging tools. Use a block slightly thinner than the desired cultivating depth to compensate for compression of the soil under the gauge wheel tire.

10. For beet-bean-vegetable cultivators, level frame laterally and fore-and-aft on tractor hitch. Block frame at recommended height and position shanks, standards, and disks at the correct height.

Fig. 57—All Rotary Cultivator Gangs must be Set to the Correct Angle

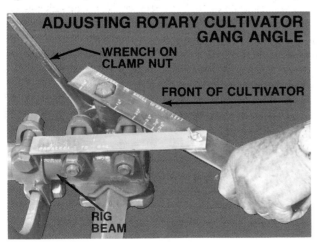

11. Retighten all bolts after making adjustments.

12. Consult operator's manual for additional specific adjustments and then make final settings in the field according to crop size and soil and moisture conditions.

TRANSPORT AND SAFETY

Attempting to save a few seconds or minutes by "short-cutting" the rules of safety could result in a much longer delay for repair of equipment or recovery from injury.

Some "shortcuts" are errors in judgment, others are the result of haste or fatigue, and some indicate lack of knowledge. Operator training cannot stop with explanation of the guidance and mechanical manipulation of equipment functions. It must include all aspects of safe operation and transport—safe for the operator, the equipment, and other people.

1. Always stop the engine and lower all equipment before adjusting, repairing, lubricating, or leaving the tractor and implement unattended.

2. Securely block the frame and stay clear from under the equipment if it must be raised for adjustment or service.

3. Use proper tools when making adjustments to prevent possible equipment damage or personal injury.

4. Allow only the driver to ride on the tractor at any time.

5. Never allow anyone to ride on equipment.

6. Never exceed recommended transport speed, and reduce speed when turning or crossing rough areas or slopes. Reduce width by transporting cultivator endways (Fig. 58).

7. Use reflectors, lights, and SMV emblem as required by law whenever transporting equipment.

8. Whenever possible, avoid road transport during hazardous periods—after dark or during peak traffic flow.

9. Be extremely careful when working close to the sharp edges of sweeps, shovels, and disks.

10. Lower parking stands or securely block equipment to prevent overturning during storage.

11. Provide adequate front-end weight to assure safety and stability when using integral equipment.

12. Raise cultivator before reversing the tractor to avoid possible equipment damage.

13. Before operating equipment after installation or adjustment, be certain there is ample clearance between cultivator and tractor tires.

Fig. 58—Transporting Cultivator Endways

TROUBLESHOOTING

Recognition of problem symptoms, possible causes, and potential solutions are essential for any field operation. Where possible remedies to problems are obvious, based on possible causes, a blank space is left in the "Possible Remedy" column:

TROUBLESHOOTING CHART		
PROBLEM	**POSSIBLE CAUSE**	**POSSIBLE REMEDY**
SOIL COVERING SMALL PLANTS	Rotary gangs set to move soil toward plants.	Switch gangs to other side to move soil away from row.
	Plant shields not used, or set too high.	
	Driving too fast.	
	Rotary-cultivator gangs angled too much.	
	Rigs set too close to row.	

PROBLEM	POSSIBLE CAUSE	POSSIBLE REMEDY
INADEQUATE PENETRATION	Dull soil-engaging tools. Gauge wheels set too low. Rotary gangs not angled enough. Improper sweep or shovel angle. Inadequate weight on rotary hoe.	
EXCESSIVE PENETRATION	Gauge wheels too high. Improper sweep or shovel angle. Soil too wet. Rotary gangs angled too much.	
UNEVEN PENETRATION FRONT-TO-REAR	Tractor top link set wrong. Rig frame not leveled. Shanks not adjusted to equal height.	Adjust top link to level cultivator. Level beam with ground— see manual.
UNEVEN PENETRATION SIDE-TO-SIDE	Tractor lift links not set equally. Toolbar gauge wheels set wrong. Rigs not all set the same.	Level toolbar by adjusting lift links. Set wheels the same on each end of bar.
DIFFICULTY IN SCOURING	No land-polish on tools. Improper tool selection for soil conditions. Paint or rust on tools Soil too wet.	Scrape tools frequently till scoured. Select tools for easy scouring.
TRASH PLUGGING	Poor choice of tools. Too many shanks per row. Tools not scouring Rotary-hoe axles too far apart. Conditions too wet.	Use disk hillers or single wide sweeps in heavy trash. Reduce number of shanks and use wider tools. Scrape tools for easy scouring. Move axles closer together for mutual trash-clearing action.

SUMMARY

Successors to wooden-handled hoes, modern multi-row row-crop cultivators can accomplish more, and probably do it better, in a few minutes than a dozen men could in a day with those old-time tools.

There are a variety of row-crop cultivators specialized for certain crops and specific field conditions. They include rotary hoes, sweep, shovel, knife, and rotary cultivators, tooth-type harrows, and flame throwers.

Despite value of chemical herbicides, and few weeds in the crop, cultivation is important for its soil-conditioning action—crust-breaking and improved aeration and moisture absorption.

But all row-crop cultivators have a distinct potential for reducing yields. Too-deep cultivation stunts some crop roots. And remember each cultivator rig constantly runs within inches (centimeters) of wiping out portions of the rows being cultivated.

Improper adjustment, careless driving, or failure to stop and clear any buildup of trash could destroy parts of one or several crop rows in seconds.

For instance, if 90 feet each from a couple of 30-inch rows is lost, that is slightly more than one percent of acre—and, at 5 miles an hour, 90 feet is covered in about 12 seconds.

If 20 meters of crop is lost from each of six rows spaced 75 cm apart, that equals almost 1 percent of a hectare. At 8 Km/h, 20 meters is traveled in about 9 seconds.

As with all mechanical equipment, the short time a farmer spends reading and understanding cultivator manufactorers' instructions is one of the most profitable investments he can make.

CHAPTER QUIZ

1. Why is mechanical cultivation recommended even when few weeds are present in the crop?

2. (True or false?) Rotary hoes kill weeds up to 6 inches tall.

3. (True or false?) Rotary-hoe axles must be kept well greased to permit high-speed cultivation.

4. What is the purpose of pulling a rotary hoe backwards?

5. Why are most beet-bean-vegetable cultivators made with continuous-bar frames?

6. Choose one: rotary-cultivator wheels appear similar to those on rotary hoes and turn in (1) the same (2) opposite direction.

7. (True or false?) Rotary-cultivator wheels are shaped differently for right- and left-handed soil movement.

8. What controls the aggressiveness of rotary-cultivator gangs?

9. (True or false?) Sway blocks or stabilizers must not be used to prevent lateral movement of rear-mounted cultivators when in operation.

10. What are the dangers of cultivating too deep?

11. (True or false?) Lister cultivators always turn soil into the row area.

12. Why should cultivator and planter sizes be matched?

13. (Complete sentence) The key to successful weed control with rotary hoes and row-crop cultivators is _____.

14. (True or false?) Flame cultivators kill weeds by burning off the leaves.

Appendix

SUGGESTED READINGS 349
USEFUL INFORMATION 349
GLOSSARY 355
INDEX 361

SUGGESTED READINGS

TEXTS AND BULLETINS

Farm Machinery and Equipment; Smith, H.P.; McGraw-Hill Book Company, New York, 1964.

Farm Power and Machinery Management; Hunt, D.; Iowa State University Press, Ames, Iowa, 1977.

Fundamentals of Machine Operation: Agricultural Machinery Safety; John Deere Service Publications, Dept. F., John Deere Road, Moline, Illinois 61265, 1974.

Fundamentals of Machine Operation: Machinery Management; John Deere Service Publications, Dept. F., John Deere Road, Moline, Illinois 61265, 1981.

Fundamentals of Machine Operation: Planters; John Deere Service Publications, Dept. F., John Deere Road, Moline, Illinois 61265, 1981.

Fundamentals of Machine Operation: Tractors; John Deere Service Publications, Dept. F., John Deere Road, Moline, Illinois 61265, 1981.

Machinery Performance Data; Agricultural Engineers Yearbook, American Society of Agricultural Engineers, St. Joseph, Michigan, 1974.

Men, Machines and Land; Farm and Industrial Equipment Institute; Chicago, Illinois, 1982.

Modern Corn Production; Aldrich, S.R. and Leng, E.R.; The Farm Quarterly, Cincinnati, Ohio, 1965.

Modern Soybean Production; Scott, W.O. and Aldrich, S.R.; The Farm Quarterly, Cincinnati, Ohio, 1970.

Plowman's Progress; Nichols, M.L. and Cooper, A.W.; Yearbook of Agriculture, USDA, Washington, D.C., 1960.

Plows and Plowing; Ridenour, H.E.; The Ohio Agricultural Education Curriculum Materials Service, Ohio State University, Columbus, Ohio, 1969.

Power Requirements of Tillage Implements; Promersberger, W.J. and Pratt, G.L.; North Dakota Agricultural College, Fargo, North Dakota, 1958.

Preparing the Seedbed; Lovely, W.G., Free, G.R. and Larson, W.E.; Yearbook of Agriculture, USDA, Washington, D.C., 1960.

Principles of Farm Machinery; Kepner, R.A., Bainer, R., and Barger, E.L.; The AVI Publishing Company, Inc., Westport, Connecticut, 1972.

Fundamentals of Machine Operation: Crop Chemicals; John Deere Service Publications, Dept. F., John Deere Road, Moline Illinois 61265, 1981.

Rotary Hoe and Its Uses; Vogel, S.L.; North Dakota Agricultural College, Fargo, North Dakota, 1957.

Rotary Hoes . . . For Fast Low-Cost Cultivation; Hull, D.O.; Iowa State College, Ames, Iowa, 1956.

The Operation, Care and Repair of Farm Machinery; Deere & Company, Moline, Illinois, 1957.

MANUFACTURER'S LITERATURE

Performance-improving suggestions and safety tips are contained in the operator's manual supplied with each tractor and implement. If the manual for a particular piece of equipment is missing, contact the manufacturer for a replacement giving the make, model, year of manufacture (if known) and serial number of the unit.

VISUALS

Tillage Slide Set (FMO-112S) 35 mm. Color. Matching set of 200 slides for illustration in FMO *Tillage* text. John Deere Service Publications, Dept. F., John Deere Road, Moline, Illinois 61265.

INSTRUCTOR'S GUIDE

FMO 11502T, teaching tips, class activities, quiz answers, masters based on FMO *Tillage* text. John Deere Service Publications, Dept. F., John Deere Road, Moline, Illinois 61265.

WORKBOOK

FMO 11602TW, chapter quizes, activities, exercises. John Deere Service Publications, Dept. F., John Deere Road, Moline, Illinois 61265.

USEFUL INFORMATION

The following tables and charts are designed to serve as a quick reference to useful information related to Tillage.

WEIGHTS AND MEASURES

The following two charts show conversions of weights and measures from **U.S. systems** to **metric** and vice versa.

WEIGHTS AND MEASURES—U.S. TO METRIC

U.S. System			Metric Equivalent
LENGTH			
Unit	*Abbreviation*	*Equivalents in Other Units*	
Mile	mi	5280 feet, 320 rods, 1760 yards	1.609 kilometers
Rod	rd	5.50 yards, 16.5 feet	5.029 meters
Yard	yd	3 feet, 36 inches	0.914 meters
Foot	ft. or ′	12 inches, 0.333 yards	30.480 centimeters
Inch	in or ″	0.083 feet, 0.027 yards	2.540 centimeters
AREA			
Unit	*Abbreviation*	*Equivalents in Other Units*	
Square Mile	sq mi or m²	640 acres, 102,400 square rods	2.590 square kilometers
Acre	A	4840 square yards, 43,560 square feet	0.405 hectares, 4047 square meters
Square Rod	sq rd or rd²	30.25 square yards, 0.006 acres	25.293 square meters
Square Yard	sq yd or yd²	1296 square inches, 9 square feet	0.836 square meters
Square Foot	sq ft or ft²	144 square inches, 0.111 square yards	0.093 square meters
Square Inch	sq in or in²	0.007 square feet, 0.00077 square yards	6.451 square centimeters
VOLUME			
Unit	*Abbreviation*	*Equivalents in Other Units*	
Cubic Yard	cu yd or yd³	27 cubic feet, 46,656 cubic inches	0.765 cubic meters
Cubic Foot	cu ft or ft³	1728 cubic inches, 0.0370 cubic yards	0.028 cubic meters
Cubic Inch	cu in or in³	0.00058 cubic feet, 0.000021 cubic yards	16.387 cubic centimeters
CAPACITY			
Unit	*Abbreviation*	*U.S. Liquid Measure*	
Gallon	gal	4 quarts (231 cubic inches)	3.785 liters
Quart	qt	2 pints (57.75 cubic inches)	0.946 liters
Pint	pt	4 gills (28.875 cubic inches)	0.473 liters
Gill	gi	4 fluidounces (7.218 cubic inches)	118.291 milliliters
Fluidounce	fl oz	8 fluidrams (1.804 cubic inches)	29.573 milliliters
Fluidram	fl dr	60 minims (0.225 cubic inches)	3.696 milliliters
Minim	min	1/60 fluidram (0.003759 cubic inches)	0.061610 milliliters
		U.S. Dry Measure	
Bushel	bu	4 pecks (2150.42 cubic inches)	35.238 liters
Peck	pk	8 quarts (537.605 cubic inches)	8.809 liters
Quart	qt	2 pints (67.200 cubic inches)	1.101 liters
Pint	pt	½ quart (33.600 cubic inches)	0.550 liters
MASS AND WEIGHT			
Unit	*Abbreviation*	*Equivalents in Other Units*	
Ton	tn (seldom used)		
short ton		20 short hundredweight, 2000 pounds	0.907 metric tons
long ton		20 long hundredweight, 2240 pounds	1.016 metric tons
Hundredweight	cwt		
short hundredweight		100 pounds, 0.05 short tons	45.359 kilograms
long hundredweight		112 pounds, 0.05 long tons	50.802 kilograms
Pound	lb or lb av also #	16 ounces, 7000 grains	0.453 kilograms
Ounce	oz or oz av	16 drams, 437.5 grains	28.349 grams
Dram	dr or dr av	27.343 grains, 0.0625 ounces	1.771 grams
Grain	gr	0.036 drams, 0.002285 ounces	0.0648 grams

*In Canada meter and liter are spelled metre and litre.

WEIGHTS AND MEASURES—METRIC TO U.S.

Metric System			U.S. Equivalent		
LENGTH					
Unit	Abbreviation	Number of Meters			
Kilometer	km	1,000	0.62 mile		
Hectometer	hm	100	109.36 yards		
Decameter	dkm	10	32.81 feet		
Meter	m	1	39.37 inches		
Decimeter	dm	0.1	3.94 inches		
Centimeter	cm	0.01	0.39 inch		
Millimeter	mm	0.001	0.04 inch		
AREA					
Unit	Abbreviation	Number of Square Meters			
Square Kilometer	sq km or km^2	1,000,000	0.3861 square mile		
Hectare	ha	10,000	2.47 acres		
Are	a	100	119.60 square yards		
Centare	ca	1	10.76 square feet		
Square Centimeter	sq cm or cm^2	0.0001	0.155 square inch		
VOLUME					
Unit	Abbreviation	Number of Cubic Meters			
Stere	s	1	1.31 cubic yards		
Decistere	ds	0.10	3.53 cubic feet		
Cubic Centimeter	cu cm or cm^3 also cc	0.000001	0.061 cubic inch		
CAPACITY					
Unit	Abbreviation	Number of Liters	Cubic	Dry	Liquid
Kiloliter	kl	1,000	1.31 cubic yards		
Hectoliter	hl	100	3.53 cubic feet	2.84 bushels	
Decaliter	dkl	10	0.35 cubic foot	1.14 pecks	2.64 gallons
Liter	l	1	61.02 cubic inches	0.908 quart	1.057 quarts
Deciliter	dl	0.10	6.1 cubic inches	0.18 pint	0.21 pint
Centiliter	cl	0.01	0.6 cubic inch		0.338 fluidounce
Milliliter	ml	0.001	0.06 cubic inch		0.27 fluidram
MASS AND WEIGHT					
Unit	Abbreviation	Number of Grams			
Metric Ton (Tonne)	MT or t	1,000,000	1.1 tons		
Quintal	q	100,000	220.46 pounds		
Kilogram	kg	1,000	2.2046 pounds		
Hectogram	hg	100	3.527 ounces		
Decagram	dkg	10	0.353 ounce		
Gram	g or gm	1	0.035 ounce		
Decigram	dg	0.10	1.543 grains		
Centigram	cg	0.01	0.154 grain		
Milligram	mg	0.001	0.015 grain		

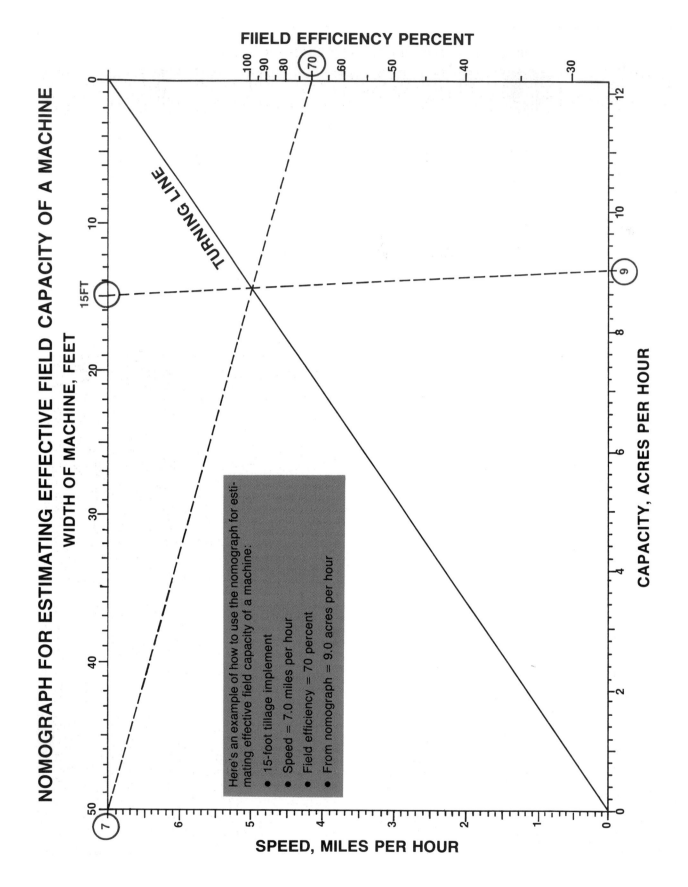

NOMOGRAPH FOR ESTIMATING EFFECTIVE FIELD CAPACITY OF A MACHINE

FIIELD EFFICIENCY PERCENT

TURNING LINE

WIDTH OF MACHINE, FEET

CAPACITY, ACRES PER HOUR

SPEED, MILES PER HOUR

Here's an example of how to use the nomograph for estimating effective field capacity of a machine:

- 15-foot tillage implement
- Speed = 7.0 miles per hour
- Field efficiency = 70 percent
- From nomograph = 9.0 acres per hour

NUMBER OF TRIPS THROUGH FIELD TO EQUAL ONE ACRE

Field Length (Feet)	4	6	8	10	12	14	16	20	25	30	35	40
300	36.3	24.2	18.2	14.5	12.1	10.4	9.1	7.3	5.8	4.8	4.1	3.6
400	27.2	18.2	13.6	10.9	9.1	7.8	6.8	5.4	4.4	3.6	3.1	2.7
500	21.8	14.5	10.9	8.7	7.3	6.2	5.4	4.4	3.5	2.9	2.5	2.2
600	18.2	12.1	9.1	7.3	6.0	5.2	4.5	3.6	2.9	2.4	2.1	1.8
700	15.6	10.4	7.8	6.2	5.2	4.4	3.9	3.1	2.5	2.1	1.8	1.6
800	13.6	9.1	6.8	5.5	4.5	3.9	3.4	2.7	2.2	1.8	1.6	1.4
900	12.1	8.1	6.1	4.8	4.0	3.5	3.0	2.4	1.9	1.6	1.4	
1000	10.9	7.3	5.4	4.4	3.6	3.1	2.7	2.2	1.7	1.5		
1100	9.9	6.6	4.9	4.0	3.3	2.8	2.5	2.0	1.6			
1200	9.1	6.1	4.5	3.6	3.0	2.6	2.3	1.8	1.5			
1300	8.4	5.6	4.2	3.4	2.8	2.4	2.1	1.7	1.3			
1400	7.8	5.2	3.9	3.1	2.6	2.2	1.9	1.6	1.2			
1500	7.2	4.8	3.6	2.9	2.4	2.1	1.8	1.5	1.2			
1600	6.8	4.5	3.4	2.7	2.2	1.9	1.7	1.4	1.1			

Implement Width, Feet

HOW TO DETERMINE FIELD SPEED

When tilling you should know how fast you are driving. You can determine your speed as follows:

1. Mark off a distance of 176 feet in the field.

2. Drive the measured distance at the speed you would like to till.

3. Check the number of seconds required to drive between the markers with a stop watch or watch with a sweep second hand.

4. Divide the time in seconds into 120 for speed in miles per hour (mph).

5. Adjust the tilling speed, if necessary, to the recommended speed.

The chart below lists the time in seconds for speeds up to 8 miles per hour.

Time To Drive 176 Feet or 33.3 meters	Speed mph or Km/h
120 seconds	1
60 seconds	2
40 seconds	3
30 seconds	4
24 seconds	5
20 seconds	6
17 seconds	7
15 seconds	8

ACREAGE PER MILE OF VARIOUS WIDTHS

Width	Acres
1 foot	0.121
5 feet	0.605
8 feet	0.968
10 feet	1.21
12 feet	1.452
14 feet	1.694
15 feet	1.815
16 feet	1.936
18 feet	2.178
20 feet	2.42
24 feet	2.904
25 feet	3.025

HECTARES PER KILOMETER OF VARIOUS WIDTHS

Width, Meters	Hectares
1	0.1
2	0.2
3	0.3
4	0.4
5	0.5
6	0.6
7	0.7
8	0.8
9	0.9
10	1.0

ACRES-PER-HOUR CHART

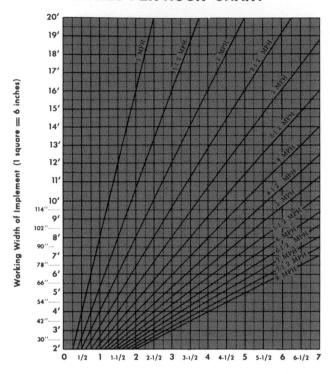

Working Width of Implement (1 square = 6 inches)

Acres per hour (1 square = 1/4 acre)

HECTARES-PER-HOUR CHART

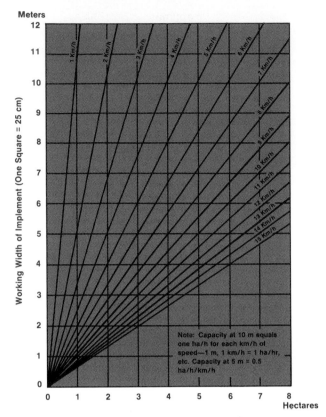

Working Width of Implement (One Square = 25 cm)

Note: Capacity at 10 m equals one ha/h for each km/h of speed—1 m, 1 km/h = 1 ha/hr, etc. Capacity at 5 m = 0.5 ha/h/km/h

Hectares Per Hour (One Square = 1/4 Hectare)

DIRECTIONS: In the left-hand column of the appropriate chart, find the line representing the working width of your equipment. Follow the line to the right until it reaches the diagonal line representing your speed of travel. From this point, follow the nearest vertical line directly to the bottom of the chart and estimate the acres per hour from the nearest figure. Note: If your equipment is wider than 20 feet (12 meters), or the horizontal and diagonal lines do not meet on the chart, make your estimate using half (or one-fourth) the working width and multiplying the result by two (or four, whichever is applicable).

The capacities indicated in these charts are theoretical. Actual capacity will be reduced by such factors as time spent turning, adjusting equipment, or operator personal time as well as overlapping the previous pass, etc. For information on field efficiency see Chapter 2.

EXAMPLE: (Using a 3-bottom, 14-inch plow cutting 42 inches, traveling at 3¼ mph.) Find 42 inches in the left-hand column. Follow the line at 42 to the right to a point midway between the diagonal lines marked 3 mph and 3½ mph. From this point, follow nearest vertical line to

bottom of chart. Note acreage per hour is slightly less than 1½.

Theoretical capacity in acres per hour equals working width in feet times speed in miles per hour, divided by 8.25. Theoretical capacity on hectares per hour equals width in meters times speed in kilometers per hour, divided by 10.

$$\text{Acres per hour} = \frac{\text{Width, feet} \times \text{speed, mph}}{8.25}$$

$$\frac{\text{Hectares}}{\text{per hour}} = \frac{\text{Width, meters} \times \text{speed, km/h}}{10}$$

EXAMPLE: (Using a disk harrow 3 m wide, traveling 8 km/h.) Find the 3 m size in the left-hand column. Follow the line at 3 to the right until it crosses the diagonal line marked 8 km/h. From this point, follow the nearest vertical line to the bottom of the chart. Capacity per hour is slightly less than 2.5 ha/h (actually 2.4).

354

GLOSSARY

AERATION—Providing optimum availability of air in soil for crop growth.

AUTOMATIC-RESET STANDARDS—Protect soil-working tools, such as plow bottoms, by swinging backward and up if solid soil obstruction is encountered, then immediately resuming plowing position when obstacle is passed. May be hydraulic or mechanical.

BALLAST—Weight added to tractor or implement to improve tractor traction and stability and/or improve penetration of soil-working tools.

BEDDER—See lister.

BED SHAPER—Makes wide beds on which up to four rows of crops are planted. Furrows between beds are used for irrigation and to guide cultivators.

BLADED CUTTERHEAD—Penetrates deeply, crushes clods, and firms soil when used with roller harrows and other field-finishing implements. Has rotating horizontal knives attached to supporting reels.

CABLE-DRAWN HITCHES—Permit pulling wide spans of rotary hoes and tooth-type harrows behind large tractors.

CENTER OF LOAD—Vertical and lateral center of resistance on tillage implements; affected by soil conditions, depth of tillage, type of soil-working tools, and operating speed.

CENTER OF PULL—Vertical and lateral center of pull of tractor; affected by tractor rear-wheel setting, convergence of 3-point hitch links with integral implements, and front pivot point of tractor drawbar with drawn implements.

CHILLED CAST-IRON BOTTOM—Plow bottom with maximum resistance to highly abrasive soils; subject to damage if large stones are present.

CHISEL PLOWS—Primary-tillage implements which leave rough, trash-covered surface, breaking and shattering soil with little trash mixdown. Basic frame has two or three rows of staggered, curved shanks (either semi-rigid or spring-cushion) for various types of soil-working tools, including points, sweeps, chisels, spikes, and shovels. Excellent vertical and lateral trash clearance.

CHISELS—Deep-working, narrow soil-engaging tools for chisel plows.

CLEARANCE—Vertical distance between implement frame bars and soil surface or lateral distance between standards for soil-working tools.

CLOD BUSTER—Tooth-type harrow attachment for rear of primary-tillage implements.

COMPACTION—Packing of soil from wheel traffic or repeated primary tillage to same depth.

CONE GUIDE—Used on cultivators for precision tillage of crops on bed-shaped land.

CONE-SHAPED BLADE—Concave disk blade which has equal distance between working surfaces of adjacent blades, top to bottom, for easier soil movement and less soil compaction.

CONSERVATION TILLAGE—Leaving some or much residue on soil surface when preparing soil, planting, and cultivating. Objectives are reduced soil erosion, moisture retention in soil, reduced compaction, and saving of fuel, time, and labor.

CONTROLLED TRAFFIC—All wheel traffic from tillage through harvest is confined to same tracks across field, with that area left untilled. Objectives are reduced compaction of crop area, improved water infiltration, reduced runoff and soil erosion, and reduced energy requirements.

CROP SHIELDS—Operate close to or over crop row (1) during cultivation to keep soil from covering small crop plants, or (2) during herbicide application to keep chemical from striking crop plants.

COULTERS—Sharp steel disks, usually flat, used to cut trash and define furrow slice ahead of moldboard plow bottoms, leaving clean furrow wall and reducing wear on share and shin; to provide lateral stability for rear-mounted row-crop cultivators and disk tillers; or assembled in rows and attached to the front of chisel plows (stubble mulch tillers) or field cultivators to cut through residue and reduce plugging. May have plain, notched or ripple-edge blades, and some are concave rather than flat. See disk coulter.

COULTER WHEEL—See tiller keel.

DEEP-TILLAGE BOTTOM—Plow bottom with extra-high moldboard for deep work in heavy soil.

DISK BEDDER—See lister.

DISK COULTER—Has same function as flat plow coulter, but is concave and turns ribbon of soil and trash from top of furrow slice into old furrow bottom and away from standard to improve trash control.

DISK HARROW—Tillage implement which moves soil both to right and left (disk plows and disk tillers move soil only one direction) by means of concave spherical or conical blades. Made in many types and sizes for uses ranging from seedbed preparation, weed control, summer fallowing, and chemical incorporation to deep primary tillage.

DISK HILLER—Cultivating tool with small concave blades instead of sweeps, shovels, or knives.

DISK PLOW—Primary-tillage implement which does not invert furrow slice but cuts, lifts, and rolls it. Differs from other disk implements in that each concave blade has its own bearing and standard, and revolves independently of the others.

DISKS—Concave, spherical or conical, rotating blades made in a variety of diameters for diversified purposes from comparatively shallow row-crop cultivation to secondary tillage and seedbed preparation (disk tillers, medium-weight disk harrows, and disk bedders) and deep primary tillage (disk plows and plowing disk harrows).

DISK TILLER—Falls between disk plow and disk harrow for primary and secondary tillage. Uses 18- to 26-inch (46 to 66 cm) concave spherical disks which turn together on common shaft. Often equipped with seeding or seeding-fertilizer attachments. Also known as one-way, seeding tiller, wheatland disk plow, and wide-level disk plow.

ECONOMY TILLAGE—See minimum tillage.

EFFECTIVE FIELD CAPACITY—Actual work accomplished (acres or hectares per hour) by an implement despite loss of time from field-end turns, inadequate tractor capacity, deficient tractor or implement preparation, adverse soil conditions, irregular field contours, lack of operator skill, or other factors. See theoretical field capacity.

FIELD CONDITIONER—Essentially heavy-duty wheeled spring-tooth harrow for secondary tillage; uses variety of soil-working tools.

FIELD CULTIVATOR—Similar to chisel plow but lighter in construction, designed for less-severe conditions, and with less vertical and lateral trash clearance. Has points, shovels, or sweeps on flexible shanks.

FIELD LAYOUT—Planning work to reduce number of row-end turns, point rows, and back and dead furrows.

FLAME CULTIVATION—Flares of burning LP gas kill young weeds between and sometimes (if crop plants are tall enough) within crop rows.

FLEXIBLE-TINE CULTIVATOR—Has sweeps or shovels on vibrant, S-shaped or double-curved spring-steel shanks; usually works deeper than spike- or tine-tooth harrows.

FLOTATION—Ability of tractor or implement tires to stay on top of soil surface; usually related to soil condition, tractor or implement weight, and contact area between tires and soil surface.

FROG—Part of plow bottom which connects moldboard, share, shin, and landside.

FURROW FILLERS—Smaller disk blades at outer ends of rear gangs of disk harrows.

FURROW WHEEL—Tractor or implement wheel which runs in furrow from previous implement pass.

GANG BOLT—Shaft on which blades of disk harrows are mounted.

GAUGE WHEELS—Control working depth and improve stability of many tillage implements; sometimes double as transport wheels.

GENERAL-PURPOSE BOTTOM—Plow bottom with long, fairly slow-turning (of furrow slice) moldboard.

GUIDE FINS—Used on some row-crop cultivators for lateral stability.

HIGH-SPEED BOTTOM—Plow bottom with longer moldboard than general-purpose bottom and less curve at upper end.

HYDRAULIC ACCUMULATOR—Charged with inert gas which maintains desired pressure on automatic-reset bottoms on plows and other implements when used behind tractors which lack sufficient hydraulic capacity or outlets.

IN-FURROW HITCH—One tractor rear wheel runs in previous furrow when plowing.

JOINTERS—Cut small ribbons of soil from surface just ahead of plow share points for improved trash coverage.

KNIVES—Shallow-working tools for killing weeds close to such crops as beets, beans, and vegetables with minimum surface disturbance.

LANDING—Adjusting moldboard plow so front bottom will cut desired width.

LANDSIDE—Part of plow bottom which bears against the furrow wall to offset side draft, improve plow stability, and sometimes carry part of weight of plow; may be flat or rolling.

LIFT-ASSIST WHEELS—For integral implements which tax or exceed hydraulic lift and/or front-end stability limits of tractors.

LINE OF DRAFT—Theoretical line between tractor center of pull and implement center of resistance.

LISTER—Primary-tillage implement which makes wide, V-shaped furrows in which row crops are planted (crops sometimes are planted on top of beds). May use lister bottoms which throw soil both right and left, or opposed disk gangs which do the same.

LISTER-PLANTER—Lister with planting attachment for once-over soil preparation and planting.

MIDDLEBREAKER—See lister; also name for attachment to remove ridge of soil between opposed gangs of disk harrows.

MINIMUM TILLAGE—Combines two or more of such tasks as primary tillage, seedbed preparation, residue mixdown, herbicide and/or fertilizer application, and planting. Also has such names as optimum, reduced, and economy tillage. Objectives are to save fuel, time, labor, and moisture, as well as to reduce soil compaction and erosion.

MOLE BALL—See subsoilers.

MOLDBOARD—Part of plow bottom which fractures, crumbles, and inverts furrow slice.

MOLDBOARD—EXTENSION—Used at rear of plow moldboard for more-positive control of furrow slice on hillsides or in heavy trash.

MOLDBOARD PLOW—Primary-tillage implement which severs bottom and one side of furrow slice, fractures and granulates the soil, inverts it into the previous furrow, and buries surface trash.

MULCH TILLAGE — Similar to conversation tillage.

MULCH TILLERS—"Hybrid" implements which combine disk-harrow and chisel-plow soil-working tools and functions. Some models have flat coulter blades instead of concave disk blades.

OFFSET DISK HARROW—Has two opposed disk gangs, working one behind the other.

ONE-WAY—See disk tiller.

ON-LAND HITCH—All tractor wheels run on untilled soil.

OPTIMUM TILLAGE—See minimum tillage.

PLOW BOTTOM—Three-sided wedge with landside and share as flat sides and shin and moldboard as curved side which inverts furrow slice severed by share and shin.

PLOWING DISK HARROW—Heavy-duty offset harrow with large blades and ample strength for deep primary tillage.

PLOW PAN—See plow sole.

PLOW SOLE—Compacted layer, restricting root and water movement, which may form in some soils just below the tilled area after several years of primary tillage to the same depth.

POWER HARROW—Secondary-tillage implement with reciprocating tines powered by tractor PTO. Tines move rapidly laterally from about 4 inches (100 mm) on front bar to 20 inches (500 mm) on fourth (rear) bar.

PRIMARY TILLAGE—First operation in preparing cropland, reaching full depth of intended root zone (unless deeper-working subsoilers are used).

PROFILE PLANTING—Variation of lister planting in which crop is planted on small ridge or raised bed in bottom of lister furrow.

REDUCED TILLAGE—See minimum tillage.

RESIDUE—Usually organic matter left on soil surface after harvest, such as stalks, straw, and stubble.

ROD WEEDER—Secondary-tillage implement used primarily for fallow-land weed control with minimum soil-surface disturbance and moisture loss. Employs round or square weeding rods, powered from gauge wheels, rotating under soil surface.

ROLLER HARROW—Field-finishing or pre-planting implement with one row of heavy rollers, two rows of spring teeth, then another row of rollers. Roller types include solid-rim, serrated, crowfoot, and sprocket-type; they operate independently on tubular gang axles.

ROLLER PACKER—Has rollers, similar to roller harrow, but no spring teeth.

ROOT CUTTER—Short horizontal blade attached to plow-bottom landside for more-positive cutting of heavy roots.

ROTARY CULTIVATOR—Has rigid, curved teeth mounted on "wheels" much like rotary hoe, but teeth are angled to slice and twist for more-aggressive weed control and tillage.

ROTARY HOE—Permits fast, shallow cultivation before or soon after crop plants emerge. Rigid, curved teeth, mounted on "wheels" which roll over ground, penetrate almost straight down but lift soil as they emerge.

ROTARY TILLERS—Use C-shaped or L-shaped overlapping blades, mounted on flanges and rotors, powered from tractor PTO or separate engine, for a variety of functions ranging from row-crop cultivation to full-depth soil conditioning for primary tillage.

ROW-CROP CULTIVATOR—Has gangs or rigs of sweeps, knives, shovels, disk weeders, disk hillers, or other soil-working tools which run between crop rows to break crust, kill weeds, aerate the soil, and improve moisture absorption.

SAFETY TRIPS—Mechanical devices on plows and other tillage implements which permit standards or the soil-working tools to release backward if they strike solid obstructions.

SCOTCH BOTTOM—Plow bottom with extra-long moldboard and narrow share which does not sever full width of furrow slice. Furrow slice is not inverted, but set on edge to catch rain and snow.

SCRAPERS—Attachments for tillage disks and steel or cast-iron implement wheels which help prevent dirt and trash buildup and plugging.

SECONDARY TILLAGE—Follows primary tillage to prepare soil for planting or to control weeds. Usually not as deep as primary tillage.

SEEDING TILLER—See disk tiller.

SHANKS—Standards extending down from implement frames to which soil-working points, shovels, and sweeps of row-crop cultivators and other tillage implements are attached; may be rigid, flexible, or spring-cushioned.

SHARE—Part of plow which severs bottom of furrow slice.

SHEAR BOLTS—Protect soil-engaging tools, such as plow bottoms, by shearing and releasing standards if solid soil obstruction is encountered.

SHIN—Part of plow at forward edge of moldboard which severs side of furrow slice.

SHOVELS—Deep-working (compared to sweeps and knives) soil-working tools for row-crop and field cultivators, chisel plows, and other implements.

SINGLE-ACTION DISK HARROW—Has two opposed disk gangs working side by side.

SLATTED BOTTOM—Plow bottom with slatted moldboard for better scouring in sticky soils.

SLED CARRIER—Toolbar with flat steel "sleds" for precision planting and cultivating of crops on bed-shaped land.

SLED GUIDE—Used on cultivators for precision tillage of crops on bed-shaped land.

SLIPPAGE—Reduced traction of tractor drive wheels caused by soil conditions. Excessive slippage wastes time and fuel and increases tire wear, but some slippage is desirable to cushion tractor engine and drive train from sudden overloads.

SMOOTHING HARROW—Tooth-type harrow attachment for rear of disk harrows and plows.

SMV EMBLEM—Used at rear of tractors and implements during road transport to warn overtaking car and truck drivers of "slow-moving vehicle."

SOFT-CENTER MOLDBOARD—Plow moldboard with low-carbon steel center layer, which helps absorb shocks, between two layers of high-carbon, hardened steel which resists erosion and scours well.

SOLID-STEEL MOLDBOARD—Plow moldboard made of one sheet of shock-resistant steel for use in soils where high abrasion and scouring are not problems.

SPIKES—Deep-working, narrow, soil-engaging tools for chisel plows; often used to rip hardpan or plow sole.

SPIKE-TOOTH HARROW—Secondary-tillage implement for seedbed preparation, crust breaking, killing small weeds, and similar purposes. Rigid square or diamond-shaped teeth extend down (adjustable angle) from several rows of frame bars.

SPRING-TOOTH HARROW—Aggressive secondary-tillage implement for such operations as deep seedbed preparation and destroying persistent weeds. Has vibrant spring-steel shanks for variety of soil-working tools.

SQUADRON HITCH—Wide hitch for using two or more implements working end to end; see tandem hitch.

STANDARDS—Supports for soil-working tools of primary-tillage implements, such as bottoms on plows. Have various devices to prevent damage from solid soil obstacles, including shear bolts, safety trips, and hydraulic or mechanical automatic resets.

STUB BARS—Short frame extensions for such multiple-framebar implements as field cultivators and chisel plows.

STUBBLE BOTTOM—Plow bottom with short, abruptly curved moldboard which turns furrow slice quickly; used where scouring is difficult; not suitable for fast speeds.

STUBBLE-MULCH PLOW—Has shallow-running sweeps or blades as much as 10 feet (3 m) wide (usually 5 or 6; 1.5 or 1.8 m) which cut weeds but leave residue anchored on soil surface with minimum soil disturbance and moisture loss.

SUBSOILERS—Chisel points or other soil-working tools operated below normal tillage depth to break up impervious soil layers and improve root and water penetration. Some models have "mole balls" behind points to make small water-drainage conduits.

SWEEPS—Soil-working tools for cultivators, chisel plows, stubble-mulch plows, and other tillage implements. Many types and sizes, from a few inches to several feet wide, but all have right and left fixed blades angled back from point.

TANDEM DISK HARROW—X-shaped disk harrow with two opposed front gangs and two opposed rear gangs.

TANDEM HITCH—Two or more implements hitched one behind another; see squadron hitch.

THEORETICAL FIELD CAPACITY—Work which could be done by an implement at optimum speed if no time were lost; see effective field capacity.

TILLAGE—Mechanical soil-stirring actions for nurturing crops by providing suitable soil environment for seed germination, root growth, and weed and moisture control.

TILLER KEEL—Flat, circular disk at rear of disk tiller to help keep the tiller running straight; also called coulter wheel.

TILL-PLANT—Shredding crop residue and planting with little or no preliminary tillage. Objectives are moisture conservation, reduced soil compaction and erosion, and saving of time, fuel, and labor.

TINE-TOOTH HARROW—Round, flexible teeth extending down from several rows of frame bars. Less aggressive than spike-tooth harrow, but used for same purposes, plus thinning small sugarbeet plants.

TOOLBARS AND TOOL CARRIERS—Basic frames for assembling "do-it-yourself" tillage and planting implements from scores of component parts, available from implement dealers, including gauge and transport wheels, standards and shanks, soil-engaging tools, hitches, braces, clamps, markers, hydraulic cylinders and hoses, unit planters, and others.

TRACTION—Effective force resulting from thrust of tractor tires against soil or other surface; depends on such factors as nature of surface, contact area between tires and surface, and tractor power and weight.

TRASH BARS—Flat bars which prevent buildup of soil and trash between disks of tillage implements.

TRASHBOARDS—Used above plow moldboards for more-positive trash deflection into furrow bottom.

TREADER-PACKER—Looks like heavy-duty rotary hoe pulled backward; has angled gangs of strong curved teeth which pack soil surface and seed zone.

TWISTED SHOVELS—Curved, somewhat like plow moldboard, to provide more soil inversion and trash coverage than flat shovels used on chisel plows.

TWO-WAY PLOW—Has two sets of plow bottoms or one set of disks used to throw soil opposite directions. Bottoms or disks are alternated at each end of field. Primarily used in irrigated land where back furrows and dead furrows would impede desired water flow.

WALKING BEAM—Improves surface conformity of tillage implements equipped with gauge wheels by having one wheel at each end of short support frame pivoted under main implement frame.

WEED HOOKS—Used on moldboard plows to hold tall weeds and bulky trash against furrow slice for improved covering.

WHEATLAND DISK PLOW—See disk tiller.

WIDE-LEVEL DISK PLOW—See disk tiller.

WIDE-SWEEP PLOW—See stubble-mulch plow.

ZERO-TILL PLANTING—Similar to till-planting.

INDEX

A

A-Frame Coupler 67, 68, 72
Accumulator System 308
Acreage Counter 163
Adjustable Depth Stop 308
Adjustable-Frame Plow 66
Adjustable Landside 58
Adjustments
    Cultivators 341
    Disk Coulters 80
    Gauge Wheels 81
    Integral-One Way Disk Plow 110, 111
    Integral, Reversible Disk Plow 112
    Jointers 81
    Moldboard Plow 76-81
    Rear Wheel 110, 112
    Rolling Coulters 80
    Rotary Tiller 193
    Semi-Integral Reversible Disk Plow 113
    Shield 190
    Top-Link 209, 211
    Turnbuckle 211
    Trash Boards 81
    Two-Way Plows 84, 85
    Vertical Hitch 76, 77
    Width of Cut 78, 79
Aeration of Soil 52
Angle
    Cutting 155
    Disk 181, 182
    Disk-gang 174, 178
    Gang 239
    Large 155
    Small 155
    Working 155, 205
Anhydrous Ammonia 142
Attachments
    Cultivator 330
    Field Cultivator 297
    Moldboard Plow 61-65
    Rolling Crumbler 281
    Spring-Tooth Planter 278
    Tine-Tooth Planter 274, 280
    Tine-Tooth Smoothing 297
    Smoothing 274, 276
Axial Shaft 150

B

Ballast
    Cast-Iron 29
    Liquid 29
Ball and Socket Coupling 164
Beams
    High Clearance 208
    One-piece 59
    Safety-trip 206
Beavertail Shovels 128
Bed Planting 204
Bed Shapers 202, 203, 208
Bedders 200-212
Bedder Toolbar 208
Beet-Bean-Vegetable Cultivator 326
Blackland Bottoms 205
Blackland Sweeps 333
Blades
    Breakage 244
    C-shaped 188
    Cone-shaped 232
    Coulter 62, 63, 64
    Flanges 186, 187
    Furrow-Filler 241
    Knife type 188
    L-shaped 188
    Rotary Tiller 188
    Sizes 232
    Spacing 234
    Spiral Cutter 282
Bolted Teeth 272
Bottoms
    Blackland 205
    Disk 205
    General purpose 205
    Hard-ground 205
    Lister 206
Bottom-link Draft Sensing 72
Box-beam Toolbars 36, 38

C

C-shaped Blades 188
Cable Hitch 164, 171, 321
Carriers, Sled 40
Caster Mounting 187
Cast-Iron
    Furrow Wheel 154, 156
    Gauge Wheel 43
Center-Link Adjustment 116, 117
Center-of-Load 73, 74, 155
Center-of-Pull 74
Chain Harrow 270
Chisels 4, 186
    Double-point 128
    Subsoiling 218
Chisel Plows 6, 121-139, 214
    Drawn 134
    Integral 133
    Maintenance 135
    Preparation 135
    Transport 136
    Safety 136
    V-shape Subsoiling 216
Chisel Points, Reversible 128
Chisel Sweeps 128
Clamps, Spacer 41
Closed-End Harrows 272
Combination-Frame Plows 66
Concave Disks 106-109
Conditioners, Field 5
Cone-Shaped Blades 232
Conservation
    Energy 6
    Moisture 6
    Soil 6
    Tillage 6
Contour Planting 4
Contour Plowing 90
Contour Tillage 4
Controlled-Traffic Farming 32
Conventional Tillage 5
Corrugated Rollers 256, 261
Coulter Blades
    Cutout or Notched 62, 63, 64
    Plain 62, 63, 64
    Ripple-Edge 62, 63, 64
Coulters 146, 176
    Disk 64
    Guide 340
    Rolling 62, 91, 144
Coupler
    A-Frame 67, 68, 72
    Ball and Socket 164
Crowfoot Rollers 257, 263

Cultimulchers 256
Cultipackers 256, 261
Cultivating 186
Cultivation
   Flame 320
   Row-Crop 318, 337
Cultivators 5
   Adjustments 341
   Attachments 330
   Beet-Bean-Vegetable 326
   Field 6
   Flame 337
   Flexible-tine 274
   Front-mounted 325
   Lister 329
   Rear-mounted 325
   Rotary 319, 329
   Row-Crop 5, 322
   S-tine 275
   Safety 342
   Shovel 319
   Sweep 319
   Transport 342
Cultivator Mounting 337
Cultivator Shovels 333
   Non-reversible, spearpoint 333
   Reversible, Double Point 333
   Reversible, Oval-Double Point 334
Cultivator Sweeps 332
   Blackland 332
   General-Purpose 333
   Mixed-Land 333
   Peanut 333
   Solid Triangular 333
   Universal 332
Curved Bed Knives 335
Curved Shank 132
Curved Standards 217
Cutterheads 257
Cutters, root 65
Cutting Angle 155
Cutting Width 108, 110
Cylinder Locks 114

D

Depth Stop, Adjustable 308
Disk Angle 109, 112, 181, 182
Disk Bedder 200, 212
Disk Bottoms 205
Disk Coulters 64, 80
Disk Cutting Depth 178
Disk Diameters 154
Diskers 150
Disk Gangs 206, 235
Disk Gang Angle 174, 178
Disk Gang Bearings 235
Disk Harrows 4, 5, 6, 200-253
   Drawn 225
   Double-Action 227
   Integral 225
   Leveling 240
   Double-Offset 227
   Offset 229
   Single-Action 226
   Tandem 227
   Wheeled 230
   Weight 237
Disk Harrow Frames 236
   Flexible 236
   Rigid 236
Disk Hiller 335
Disk Plows 105-119, 4
   Attachments 114
   Drawn Reversible 113

Integral One-Way 108-111
Reversible 111-114
Semi-Integral Reversible 113
Disks
   Concave 106-109
   Inside Beveled Edge 108
   Notched Edge 108
   Outside Beveled Edge 108
   Plowing 232
   Weeding 335
Disk Scrapers 111, 117, 237
   Hoe 114, 115
   Moldboard 114, 115
   Reversible 114, 115
Disk Tillers 4, 5, 6, 149-172
   Drawn 151
   Integral 151, 154
   One-Way 150
   Reversible 151, 154
Disk Tilt Angle 110, 112
Double-Action Disk Harrows 227
Double-Offset Disk Harrows 227
Double-Point Chisels 128
Double-Point Shovels 259, 295
   Reversible 297
Draft 82-84, 99, 123, 132, 155,
                       170, 184, 249, 263, 306
Drag Harrows 270
Drawbar 300
   Horsepower 185
   Integral Harrows 273
   Pull 85
Drawn Chisel Plows 134
Drawn Disk Harrows 225
Drawn Disk Tiller 151
Drawn Field Cultivators 292
Drawn Moldboard Plow 69, 70
Drawn Plow
   Leveling 78
   Line of Draft 74, 75
   Vertical Hitch Adjustment 76, 77
   Width of Cut 80
Drawn Reversible Disk Plow 113
Drawn Roller Harrows 261, 264
Drawn Stubble-mulch Tillers 179
Drawn Tillers
   Flexible Frame Disk 153
   Heavy-Duty Disk 153
   Rigid-Frame 151
Drive Wheels 306
Dual Wheels 30, 43
Duckfoot Sweep 334

E

Efficiency, Field 12, 17, 18, 20, 22
Elastomer Cushions 295
End-to-End Hitching 164
Extensions, Moldboard 65

F

Fall Chiseling 124
Fall Tillage 179, 194
Feed Gate, adjustable 162
Field Capacity 12, 175
Field Conditioners 5, 279
   Tine-teeth 274
Field Cultivators 6, 291-304
   Attachments 297
   Drawn 292
   Integral 292

Maintenance 300
Preparation 300
Safety 301
Transport 301
Field Efficiency 12, 17, 18, 20, 22
Field Layout 89, 90
Field Operation
  Bedders 209
  Chisel Plows 136
  Disk Harrows 246
  Disk Plows 114
  Disk Tillers 160
  Field Cultivators 301
  Listers 209
  One-Way Plows 89
  Packers 264
  Rod Weeders 312
  Roller Harrows 264
  Rotary Tillers 193
  Row Crop Equipment 339
  Stubble-Mulch Plows 145
  Stubble-Mulch Tillers 179
  Subsoilers 218
  Tooth-type Harrows 284
  Weed-Control Equipment 339
Field Pattern 246
Field Speed 17
Field Traffic 32, 33
Finger Harrows 274
Finishing Harrow 257
Fixed-Clamp Shanks 295
Fixed-Frame Plows 65
Flame Cultivation 320
Flame Cultivators 337
Flange, Rear-Wheel 116
Flexible-Disk Harrow Frames 236
Flexible-Disk Tiller Frames 153
Flexible Harrow 270, 273
Flexible-Tine Cultivators 274
Flexible-Tine Shanks 335
Flotation 30
Fluted Feed Rollers 162
Frames
  Disk Harrow 236
  Moldboard Adjustable 66
  Moldboard Combination 66
  Moldboard Fixed 65
  Spike-Tooth Harrow 272
  Weight 115
Friction-Trip Shanks 330
Frog 53, 57
Front-Mounted Cultivators 325
Full-Cut Shares 57
Furrow-Filler Blades 241
Furrow Openers 128
Furrow Wheels
  Cast-iron 156
  Front 169
  Rear 168, 170

G

Gang Angle 239
Gangs 151, 168-170, 227, 249
  Disk 206, 235
  Rotary Cultivator 335, 338
  Rotary Hoe 335
Gang Shaft 150
Gang Shoe 279
Gang Spools 236
Gauge Wheels 65, 113, 130, 210, 211, 310, 220
  Adjustments 81
  Cast Iron 43
  Dual Wheels 43
  Hydraulic Lift Assist 45

Single Wheels 43
Small Wheels 44
  Toolbar Transport 46
General Purpose Bottoms 205
General Purpose Sweeps 333
Guide Coulters 340
Guide Wheels 203, 204
Gumbo Shares 57

H

Hand Jacks 153
Hard Ground Bottoms 205
Hardpan 53
Hard-Surfaced Shares 57
Harrows
  Chain 270
  Closed-End 272
  Disk 4, 5, 6, 200-253
  Drag 270
  Finger 274
  Finishing 257
  Flexible 270, 273
  Open-End 272
  Heavy-Duty Spike Tooth 273
  Pegtooth 270
  Power 281
  Roller 5, 255-267, 281
  Section 270
  Smoothing 270
  Spike-Tooth 270, 319
  Spring-Tooth 276
  Tine-Tooth 274, 319
  Tooth-Type 269-288
  Wheeled Spring-Tooth 278
High Clearance Beams 208
High Crown Sweep 128
Hipping 201
History, Tillage 8
Hitch Cable 164, 171
Hitches
  Box-Beam Bar 38, 39
  Cable Drawn 321
  Integral 321
  Multiple Cable 310
  One Piece 36, 38
  On-Land 69
  Pull Behind 297
  Tandem 70
  Tiller 156
  Three-Point 66, 70, 109, 110, 112, 154, 187
  Toolbar Transport 46
  Wide Angle 169
Hitching
  End-to-End 164
  In-Furrow 73
  On-Land 73
  Tandem 153, 164
Hitch-level Screw 263
Hoes
  Rotary 5, 335
Hoe Scraper 114
Hood Adjustments 190
Hooded Shields 329, 332
Hoods, Weed 65
Horizontal-Axis Tillers 184
Horizontal Disk Angle 109, 110, 116
Horsepower 82
  Drawbar 185
  PTO 185
Hydraulic Accumulator System 60, 61
Hydraulic Automatic Reset Standard 60
Hydraulic Cylinders 48, 132, 153, 154
  One Double Acting 72
  Two Single Acting 72

Hydraulic Lift Assist Wheel 45
Hydraulic Lift Capacity 85, 109
Hydraulic Reset Shank 177
Hydraulic Reset Standards 101
Hydraulic Reset System 307

I

Indexing 111
In-Furrow Hitching 73
Insect Control 3, 52
Inside Beveled Edge Disk 108
Integral Disk Harrows 225
Integral Field Cultivators 292
Integral Harrow Drawbar 273
Integral Mounting 70
Integral Plows 65
    Chisel 133
    Leveling 77-78
    Line of Draft 74-75
    Load-and-Depth Control 84
    Moldboard 66, 67
    One-Way Disk 109-111
    Reversible Disk 112
    Two-Way 70-71
    Vertical Hitch Adjustment 76
    Width of Cut 79
Integral Disk Tiller 151, 154
Integral Stubble-Mulch Tillers 179
Irrigation Shovels 334

J

Jointers 64, 65, 81

K

Keel, Tiller 153, 166
Knife-Type Blades 188
Knives, Weeding 334

L

L-Shaped Blades 188
Land Rollers 261
Landsides 53, 57, 58
Land Wheel 156, 162, 170
Layout, Field 89, 90
Leveling Bar 261
Lift-Assist Wheels 299
Lift Links 112, 113
Listers
    Bedders 200-212
    Bottoms 206
    Cultivators 329
    Planters 200-212
    Toolbar 208
Lister Planting 201, 208
Low-Crown Sweep 128
Low-Point, Square-Turn Knives 334

M

Maintenance
    Bedders 208
    Chisel Plows 135
    Disk-Harrow 243
    Disk Tillers 158
    Field Cultivators 300
    Listers 208
    Moldboard Plows 87
    Packers 264
    Rod Weeders 312
    Roller Harrows 264
    Rotary Tillers 192
    Row-Crop Equipment 339
    Stubble-Mulch Plows 144
    Stubble-Mulch Tillers 178
    Subsoiler 218
    Weed-Control Equipment 339
Markers 46
Middlebreaker Attachment 227
Middlebuster 200-212
Minimum Tillage 5, 6, 184
Mixed-Land Sweeps 333
Moldboards 53, 56
Moldboard Extensions 65
Moldboard Plows
    Adjustments 76-81
    Attachments 61-65
    Bottoms 53-56
    Drawn 69, 70
    Frames 65, 66
    Integral 66, 67
    Maintenance 87-89
    Operating Sequence 91
    Preparation 87-89
    Reversible 70, 71
    Rolling 205
    Semi-Integral 67, 68, 69
    Two-Way 70, 71
Moldboard Scraper 114, 118
Mole Ball 215
Mole Drain 215
Mounting
    Caster 187
    Cultivator 337
    Front 204
    Integral 70
    Rigid 64
    Semi-Integral 187
    Single-Arm 64
    Spring Cushion 64
    Toolbar 204, 215
Mulch Tillage 6
Multiple Cable Hitches 310

N

Nitrogen Fertilizer 185
Narrow-Cut Shares 57
Narrow Double-Point Shovels 276
Narrow Reversible Alfalfa Point Shovels 297
Negative Draft 184
Non-Reversible, Spear Point Shovels 333
Notched Edge Disk 108

O

Offset Disk Harrows 229
Offset Tandem Wheels 294

One-Piece Beams 59
On-Land Hitch 69
On-Land Hitching 73
Open-End Harrows 272
Optimum Tillage 5
Orchard Tilling 186
Outside-Beveled Edge Disk 108

P

Packers 255-267
   Plow 257
   Treader 257
Packer Wheels 297, 263
Pattern, Working 18
Peanut Sweeps 333
Pegtooth Harrows 270
Pivot Arm 164
Plant Shields 186
Planters, Unit 186, 299
Planting
   Bed 204
   Contour 4
   Lister 201, 208
   Profile 201
   Ridge 204
   Till-Plant 6
   Zero-Till 6
Plates, Reversible Wear 57
Plow Bottoms
   Deep Tillage 56
   General Purpose 55
   High Speed 55
   Scotch 56
   Semi-Deep Tillage 56
   Slatted Moldboard 55
   Stubble 56
Plow Crowding 116
Plow Pattern 89, 90
Plow Shares 57
Plowing, Contour 90
Plowing Disks 232
Plows
   Adjustable 66
   Chisel 4, 6, 121-139, 214
   Disk 105-119
   Drawn Reversible Disk 113
   Frame, Adjustable 66
   Frame, Combination 66
   Frame, Fixed 65
   Integral 65
   Integral One-Way Disk 108
   Moldboard 4, 52
   Reversible Disk 111-114
   Semi-Integral 65
   Semi-Integral Reversible 113
   Stubble-Mulch 141-147
   Vertical Disk 150
   Wheatland Disk 150
   W Wide Sweep 4, 5, 142
Points
   Alfalfa, Narrow Reversible 297
   Straight Spike 188
Pointed Shoes 306
Power Harrow 281
Primary Tillage 4, 106, 187
Profile Planting 201
PTO Horsepower 185
PTO Master Shield 85
PTO Shaft 192, 195, 196
PTO Shields 195
Pull-Behind Implement Hitches 297
Pulverizers 261

Q

Quick Coupler 66, 67, 299
Quick-Adjustable Shanks 330

R

Rear-Mounted Cultivators 325
Rear Wheel Adjustment 110, 112, 116
Rear Wheel Flange 116
Rear Wheel Scraper 116
Rear Wheel Weights 115, 118
Reduced Tillage 5
Release-Type Standards 59
Remote Hydraulic Cylinders 130, 295
Reversible Chisel Points 128
Reversible Disk Plows 111-114
Reversible Disk Tillers 151, 154
Reversible Moldboard Plow 70, 71
Reversible Point Teeth 278
Reversible Scraper 115
Reversible Shovels 128, 333, 334
Reversible Wear Plates 57
Rice Fields 186
Ridge Planting 204
Ridging 4, 98, 117, 169
Rigid-Frame Harrows 236
Rigid-Frame Tiller 151
Rigid Mounting 64
Rippers 4
Rod Weeders 5, 136, 305-315
Roller Guide 164
Roller Harrows 5, 256-267
Roller Packers 5, 256-257
Rollers
   Crowfoot 263, 257
   Fluted Feed 162
   Land 261
Rolling Coulters 62, 80, 91, 144
Rolling Crumbler 281
Rolling Landside 58
Rolling Moldboards 205
Rolling Shields 332
Root Cutters 65
Rotary Cultivators 319, 329
Rotary Cultivator Gangs 335, 338
Rotary Hoes 5, 256, 320, 335
Rotary Hoeing 319
Rotary Hoe Gang 335
Rotary Tillage 185
Rotary Tillers 4, 183-197
Round-Turned Knives 334
Row-Crop Cultivators 5, 318, 322, 337
Row-Crop Equipment 317-345
Row Cropping 6
Row Shields 332

S

S-Tine Shank 275
S-Tine Cultivator 275
Safety
   Chisel Plows 136
   Cultivators 342
   Disk Harrows 248
   Disk Plows 114
   Disk Tillers 165
   Field Cultivators 301
   Moldboard Plows 102
   Packers 265

Rod Weeders 313
Roller Harrows 265
Rotary Tillers 195
Row Crop Equipment 342
Stubble-Mulch Plows 145
Stubble-Mulch Tillers 180
Subsoilers 220
Tooth-Type Harrows 286
Weed-Control Equipment 342
Safety-Trip Beams 206
Safety-Trip Standards 59, 100
Scrapers 163, 263
Disk 114-115
Hoe 114
Moldboard 114, 118
Rear-Wheel 116
Reversible 115
Scuffing Boards 281
Secondary Tillage 5, 187, 201
Section Harrows 270
Seedbed Preparation 184, 185, 187, 224
Seedbed Teeth 278
Seeding Tiller 150
Semi-Integral Moldboard Plows 67, 68, 69
Semi-Integral Mounting 187
Semi-Integral Plows 65
Leveling 78
Line of Draft 74, 75
Load-and-Depth Control 84
Vertical Hitch Adjustment 76, 77
Width of Cut 80
Semi-Integral Reversible Disk Plows 113
Semi-Integral Two-Way Plows 72
Semi-Rigid Shank 177
Serrated Wheels 263
Shaft
Axial 150
Gang 150
Shanks 45, 175, 176, 177, 295
Curved 132
Fixed-Clamp 295
Flexible-Tine 335
Friction-Trip 330
Hydraulic-Reset 177
S-Tine 275
Semi-Rigid 127, 177
Spear-Point 334
Spring-Cushion 295, 177
Spring-Loaded 321
Spring-Steel 123
Spring-Trip 330
Quick Adjustable 330
Shank Bars 293
Shank Clamps 127
Shank Mountings 127
Shares 91
Extra-Heavy Duty 57
Full-Cut 57
Gumbo 57
Hard-Surfaced 57
Heavy-Duty, Deep Suck 57
Narrow Cut 57
Plow 57
Shear Bolts 220, 216
Shear Bolt Standards 59, 100
Shear Pins 163
Shear-Pin Pendants 306
Shield Adjustment 190
Shields
Hooded 329, 332
Rolling 332
Row 332
Spray 332
Shims 134, 145
Shoes
Pointed 306
Stub-Pointed 306
Shovels
Beavertail 128

Cultivator 333
Double-Pointed 259, 295
Double-Point Reversible 297
Irrigation 334
Narrow Double-Point 276
Reversible Double-Point 128
Reversible Oval-Shaped, Diamond Point 334
Twisted 128
Side Draft 170
Single-Action Disk Harrows 226
Single-Arm Mounting 64
Spring-Cushion Mounting 64
Single Gang 261
Single Wheel 43
Sled Carriers 40
Sled Tool Carriers 202
Sloping Standards 217
Smoothing Attachments 274, 276
Smoothing Harrows 270
Soil-Engaging Tools 46, 128, 295
Soil Pulverizers 256
Solid-Rim Wheels 263
Solid Triangular Sweeps 333
Soil Shield 190
Spacer Clamps 41
Spear-Point Shanks 334
Spiders 321
Spider Gangs 329
Spikes 128
Spike Harrows 5, 270, 272, 319
Spiral Cutter Blades 282
Spray Tanks 299
Split Rockshaft Lift 325
Spray Shields 332
Sprayer Booms 186
Spring Automatic Reset Standards 61
Spring-Cushion Angling Linkage 271
Spring-Cushion Shanks 177, 295
Spring-Loaded Shanks 321
Spring-Trip Shanks 330
Spring Harrows 5
Spring Reset Standards 101
Spring-Steel Shanks 123
Spring Teeth 259, 277
Reversible Point 278
Seedbed 278
Utility 278
Weeding 278
Spring-Tooth Harrow 276
Spring-Tooth Planter Attachment 278
Spring Tillage 180, 194
Sprocket-Tooth Packer Wheels 263
Square Hollow Bars 36
Square-Turn Knives 334
Standards 45, 215, 220
Curved 217
Hydraulic-Automatic Reset 60
Hydraulic Reset 101
Moldboard Plow 58
Release-Type 59
Safety Trip 59, 100
Shear Bolt 59, 100
Sloping 217
Spring Automatic-Reset 61
Spring Reset 101
Steel Landsides 57
"Stop" Screw 78
Straight Spike Points 188
Strip Tillage 186
Stubble-Mulch Plows 141-147
Stubble-Mulch Tillage 123
Stubble-Mulch Tillers 173-182
Stub-Pointed Shoes 306
Subsoilers 4, 213-221
Subsoiling Chisels 218
Sweep Blades 145
Sweep Cultivators 319
Sweep Points 145

Sweeps 146, 174, 295, 297
  Blackland 333
  Chisel 128
  Cultivator 332
  Extra-Flat Duckfoot 334
  General Purpose 333
  High-Crown 128
  Low-Crown 128
  Mixed Land 333
  Peanut 333
  Regular Heavy Duty 128
  Solid Triangular 333
  Universal 332
  Wheatland 128
Symmetrical Shank Arrangement 134

T

Tandem Disk Harrows 227
Tandem Hitch 70
Tandem Hitching 153, 164
Tandem Wheels 131, 176
Teeth
  Bolted 272
  Spring 277
  Welded 272
Tillage
  Contour 4
  Conservation 6
  Conventional 5
  Fall 179, 194
  History 8
  Minimum 5, 6, 184
  Mulch 6
  Optimum 5
  Primary 4, 106, 187
  Reduced 5
  Rotary 185
  Secondary 5, 187, 201
  Spring 180, 194
  Strip 186
  Stubble-Mulch 123
Tillers
  Disk 4, 5, 149-172
  Drawn Disk 151
  Drawn Stubble-Mulch 179
  Flexible Frame Disk 153
  Heavy Duty Disk 153
  Horizontal-Axis 184
  Integral Disk 151, 154
  Integral Stubble-Mulch 179
  Reversible Disk 151, 154
  Rigid Frame 151
  Rotary 4, 183-197
  Seeding 150
  Stubble-Mulch 173-182
Tiller Keel 153, 166
Tiller Rotor 188
Till-Planting 6
Tilth 190, 191
Tine-Tooth Field Conditioners 274
Tine-Tooth Attachments 274, 280, 297
Tine-Tooth Harrow 274, 319
Tongue Jack 127
Toolbars
  Bedders 208
  Box-Beam 38
  Gauge Wheels 42, 43
  Hitches 36, 38
  Lister 208
  Markers 46, 47
  Mounting 204, 215
  Sled Carriers 40, 41
  Soil-Engaging Tools 46
  Spacer Clamps 41, 42

Transport 46
Tools, Soil Engaging 128
Tooth-Type Harrows 269-288
  Power 281
  Roller 281
  Safety 286
  Spring-Tooth 276
  Tine-Tooth 274
  Transport 286
Traction 26, 110
Traffic Farming 32
Transport
  Chisel Plow 136
  Disk Harrow 248
  Disk Plow 114
  Disk Tillers 165
  Field Cultivators 301
  Moldboard Plow 102
  Packers 265
  Rod Weeder 313
  Roller Harrows 265
  Rotary Tillers 195
  Row Crop Equipment 342
  Stubble-Mulch Plow 145
  Stubble-Mulch Tillers 180
  Subsoiler 220
  Toolbars 46
  Tooth-Type Harrows 286
  Weed Control Equipment 342
Transport Wheels 261, 293
Trash Bars 163
Trash Boards 64, 81, 91
Trash Plowing 91
Treader Packers 257
Triple Gang 261
Troubleshooting
  Bedders 211
  Chisel Plows 137
  Disk Harrows 249
  Disk Plows 116
  Disk Tillers 167
  Field Cultivators 302
  Listers 211
  Moldboard Plow ....
  Packers 265
  Rod Weeder 314
  Roller Harrows 265
  Rotary Tillers 195
  Row Crop Equipment 343
  Stubble-Mulch Plow 146
  Stubble-Mulch Tillers 181
  Subsoilers 220
  Toolbars 46
  Tooth-Type Harrows 286
  Weed Control Equipment 343
Turnbuckle Adjustment 211
Twisted Shovels 128
Two-Way Moldboard Plows
  Drawn Two-Way 73
  Integral Two Way 70, 71
  Semi-Integral Two-Way 72
Two-Way Plows 84, 85

U

Unit Planters 186, 299
Universal Sweeps 332
  Utility Teeth 278

V

V-Chisel 216
V-Ripper 216

V-Shape Subsoiling Chisel Plow 216
Vertical Disk Angle 108
Vertical Disk Plows 150
Vertical Hitch Adjustment 76, 77
Vertical Plow Clearance 65
Vibrating Chisels 126

W

Wedge 57
Weed Control Equipment 317-345
Weed Hooks 65
Weeding Disks 335
Weeding Knives 334
    Curved Bed 335
    Low-point, square-turn 334
    Round-turned knives 334
    Square-turn 334
Weeding Teeth 278
Weights 145, 155, 170, 191
    Disk Harrow 237
    Front-End 109, 114
    Rear-Wheel 115, 118
Welded Teeth 272
Wheatland Disk Plows 150
Wheatland Sweep 128
Wheeled Carts 273, 285
Wheels
    Drive 306
    Furrow 156, 169
    Gauge 65, 113, 130, 210, 211
    Guide 203, 204
    Land 156, 162, 170
    Lift-Assist 299
    Offset Tandem 294
    Packer 263, 297
    Rear Furrow 168, 170
    Roller 263
    Serrated 263
    Solid Rim 263
    Sprocket-Tooth Packer 263
    Tandem 131, 176
    Transport 261, 293
Wheeled Spring-Tooth Harrow 278
Wheel Slippage 27, 28, 85, 118, 181, 249
Wheeled Disk Harrows 230
Wick 53
Wide Angle Hitch 169
Wide-Sweep Plow 6, 142
Wings 228, 279, 294
Wing Flexing 242
Working Angle 155, 205
Working Pattern 18, 145

Y

Yield 7

Z

Zero-Till Planting 6